Communications
in Computer and Information Science 145

Ketan Shah V.R. Lakshmi Gorty
Ajay Phirke (Eds.)

Technology Systems and Management

First International Conference, ICTSM 2011
Mumbai, India, February 25-27, 2011
Selected Papers

Springer

Volume Editors

Ketan Shah
V.R. Lakshmi Gorty
Ajay Phirke

SVKM's NMIMS, MPSTME
Mukesh Patel School of Technology Management and Engineering
Mumbai – 400056, Maharashtra, India

E-mail:
{ketanshah, vr.lakshmigorty, ajay.phirke}@nmims.edu

ISSN 1865-0929
e-ISSN 1865-0937
ISBN 978-3-642-20208-7
ISBN 978-3-642-20209-4 (eBook)
DOI 10.1007/978-3-642-20209-4
Springer Heidelberg Dordrecht London New York

Library of Congress Control Number: 2011923678

CR Subject Classification (1998): C, D, H, J

Typesetting: Camera-ready by author, data conversion by Scientific Publishing Services, Chennai, India

Printed on acid-free paper

Springer is part of Springer Science+Business Media (www.springer.com)

Preface

The objective of the ICTSM conference is to provide a working professionals, researchers, faculty members, consultants and others to with common platform to share their views, knowledge and their achievements in the field of information and communication technology (ICT) and to develop aptitude and skills for the writing and presentation of technical papers.

After being held for five years, the National Conference on Information and Communication Technology (NCICT) went international in 2011 as the International Conference on Technology Systems and Management (ICTSM).

ICTSM 2011 attracted 276 submissions. Each paper was peer reviewed. The final program included 47 papers and 32 poster presentations covering research activities in the field of information technology and computer engineering electronics, telecommunication, and more. The poster details are available on the ICTSM website http://svkm-ictsm.org. ICTSM 2011 included invited talks from experts from industry and academia, papers presented by the delegates, exhibitions etc.

We appreciate the efforts made by the expert reviewers, Program Committee members and ICTSM 2011 student council members. We also thank the editorial board for their continued guidance and commitment. Thanks are also due to the reviewers who volunteered time amidst their busy schedules to help realize this volume. We appreciate the high-quality contribution by the authors to ICTSM 2011.

February 2011

Vijay Raisinghani
Ketan Shah
V.R. Lakshmi Gorty
Ajay Phirke

Organization

ICTSM 2011 was organized by SVKM's NMIMS Mukesh Patel School of Technology Management & Engineering and D.J. Sanghvi College of Engineering, two of India's premier engineering schools.

About the Organizers

SVKM

Shri Vile Parle Kelavani Mandal is a Public Charitable Trust registered under the Society's Registration Act and Bombay Public Trust Act. SVKM has always been committed to the cause of providing high-quality education at various levels. From its beginning with the Swadeshi Movement, the Mandal has now grown into an educational foundation promoting global thinking consistent with national interest and promoting values, professionalism, social sensitivity and dynamic entrepreneurship.

NMIMS

Located in the heart of India's financial capital, Mumbai, SVKM's Narsee Monjee Institute of Management Studies is among the nation's prime educational and research institutions. With more than 5,000 students spanning over 50 programs across 7 disciplines, NMIMS has transformed itself from Mumbai's pride to becoming India's most sought after academic community.

Today, 28 years after its inception, NMIMS continues its tradition of training young minds and finding solutions to address tomorrow's challenges. It continues to maintain its focus on developing students academically as well as allowing them to enjoy a socially and culturally rich life. Students of the university enthusiastically share the college's vision of transcending horizons.

MPSTME

Mukesh Patel School of Technology Management and Engineering (MPSTME) recognizes the changing face of industry and realizes the need for new education technology. The curricula at MPSTME are practically oriented. The college also encourages extracurricular and co-curricular activity, ensuring that students enjoy a healthy mix of academic and social life. The innovative MBA (Tech) course offered by the college combines engineering studies with management training to develop industry-ready students by the time they graduate. MPSTME also offers M Tech, B Tech, MCA and PhD courses in various fields.

DJSCOE

In a short span of time, Dwarkadas J. Sanghvi College of Engineering has come to be recognized as a premier institute of technical education. The favorable location of the institute in the heart of Mumbai along with state-of-the-art facilities and distinguished faculty has been a nurturing ground for students of high academic capabilities. Spacious classrooms, well-equipped laboratories and workshops, new-age computer facilities and a well-stocked library provide a stimulating educational environment within the college. The college has attracted qualified and experienced faculty members as well as top companies for student placements. The college is considered to be in the top echelons of quality engineering colleges with an 'A+' grade certificate from the Directorate of Technical Education, MS.

Patrons

Chief Patron

Amrish R. Patel	President, SVKM and Chancellor, NMIMS (Deemed to-be University)

Patrons

B.P. Sheth	Vice-President, SVKM
J.P. Gandhi	Hon. Jt. Secretary, SVKM
Bharat M. Sanghvi	Trustee, SVKM
Rajan Saxena	VC, NMIMS

Executive Committee

Convener

D.J. Shah	Dean, Mukesh Patel School of Technology Management and Engineering, Mumbai

Secretaries

Hari Vasudevan	Principal, D.J. Sanghvi College of Engineering, Mumbai
Vijay T. Raisinghani	Mukesh Patel School of Technology Management and Engineering, Mumbai

Chairman, Inaugural Function

Natrajan	Ex-Chairman, AICTE

Conference Chairs

H.B. Kekre	Mukesh Patel School of Technology Management and Engineering, Mumbai
Abhijit Joshi	D.J. Sanghvi College of Engineering, Mumbai

Finance Chair

R.C. Agarwal	Mukesh Patel School of Technology Management and Engineering, Mumbai

International Advisory Committee

K. Rao	UT Arlington, USA
Suyash Awate	Siemens Research, USA
V.C. Bhavsar	UNB, Canada
Chris Wert	TEM, France
Jeet Gupta	UA Huntsville, USA

Advisory Committee

Asoke Basak	CEO, SVKM
M.N. Welling	Pro-VC, SVKM's NMIMS
M.M. Shah	IEEE, Bombay Section
Zia Saquib	Executive Director, CDAC
Raju Hira	Chairman, IEEE, Bombay Section
Rajiv Gerela	Chairman, CSI, Mumbai Chapter
Varsha Parab	Registrar, SVKM's NMIMS

Program Co-chairs

R.R. Sedamkar	Mukesh Patel School of Technology Management and Engineering, Mumbai
A.C. Daptardar	D.J. Sanghvi College of Engineering, Mumbai
A.C. Mehta	Mukesh Patel School of Technology Management and Engineering, Mumbai
Ketan Shah	Mukesh Patel School of Technology Management and Engineering, Mumbai
Jayant Umale	Mukesh Patel School of Technology Management and Engineering, Mumbai

Technical Program Committee

U.B. Desai	Director, Indian Institute of Technology, Hyderabad
Durgesh Kumar Mishra	Chairman IEEE, Computer Society, Mumbai Chapter
Shubhalaxmi Kher	Arkansas State University, USA
S.C. Sahasrabudhe	DA-IICT
Ali El-Mousa	Chair IEEE, Amman, Jordan
A. Athputharajah	EE, Peradeniya, Sri Lanka
Sridhar Iyer	CSE, Indian Institute of Technology Bombay, Mumbai
Satish Devane	Ramarao Adik Institute of Technology, Mumbai
Sanjay Gandhe	Sardar Patel Institute of Technology, Mumbai
T.J. Mathew	Shreemati Nathibai Damodar Thackersey Women's University, Mumbai
B.K. Mohan	CSRE, Indian Institute of Technology Bombay, Mumbai
Sunita Sane	CSE, Veermata Jijabai Technological Institute, Mumbai
Geeta Kumta	School of Business Management, NMIMS, Mumbai
Sunita Mahajan	Mumbai Educational Trust, Mumbai
Nilay Yajnik	School of Business Management, NMIMS, Mumbai
Pravin Shrinath	Mukesh Patel School of Technology Management and Engineering, Mumbai
M.S. Panse	Veermata Jijabai Technological Institute, Mumbai
S.D. Bhagwat	Mukesh Patel School of Technology Management and Engineering, Mumbai
K.D. Desai	Mukesh Patel School of Technology Management and Engineering, Mumbai
Leena Wadia	Senior Fellow, Observer Research Foundation, Mumbai Chapter
Preetida. V. Jani	Mukesh Patel School of Technology Management and Engineering, Mumbai
Girish Kumar	EE, Indian Institute of Technology Bombay, Mumbai
N.S.T. Sai	Tech. Mahindra, Mumbai
Padmaja Joshi	Centre for Development of Advanced Computing, Mumbai
Kamal Shah	St. Francis Institute of Technology, Mumbai
Subhashree Dutta	Tata Consultancy Services, Mumbai
Uday P. Khot	Thadomal Shahani Engineering College, Mumbai
Sanjay Shitole	Shreemati Nathibai Damodar Thackersey Women's University, Mumbai

Asim Bannerjee	Dhirubhai Ambani Institute of Information and Communication Technology, Gandhinagar
Sasi Kumar	Centre for Development of Advanced Computing, Mumbai
Uttam Kolekar	Lokmanya Tilak College of Engineering, Navi Mumbai
Rekha Singhal	Centre for Development of Advanced Computing, Mumbai
Urjaswala Vora	Centre for Development of Advanced Computing, Mumbai
Vilas Kalamkar	Sardar Patel College of Engineering, Mumbai
Jinu Kurian	Mukesh Patel School of Technology Management and Engineering, Mumbai

PMI MC Track Co-chairs

V. Seshadri	Mukesh Patel School of Technology Management and Engineering, Mumbai
Rakesh Gupta	Project Management Institute, Mumbai
Saurabh Parikh	Project Management Institute, Mumbai
Priyesh Sheth	Project Management Institute, Mumbai

Marketing Chair

Nikhil Gala	Mukesh Patel School of Technology Management and Engineering, Mumbai

Proceedings Co-chairs

Pravin Shrinath	Mukesh Patel School of Technology Management and Engineering, Mumbai
Priybrat Dwivedi	Mukesh Patel School of Technology Management and Engineering, Mumbai
P.V. Srihari	D.J. Sanghvi College of Engineering, Mumbai
Vaishali Kulkarni	Mukesh Patel School of Technology Management and Engineering, Mumbai
Ajay Phirke	Mukesh Patel School of Technology Management and Engineering, Mumbai
V.P.N. Padmanaban	Mukesh Patel School of Technology Management and Engineering, Mumbai
Dhirendra Mishra	Mukesh Patel School of Technology Management and Engineering, Mumbai
S.S. Satalkar	D.J. Sanghvi College of Engineering, Mumbai

Tutorial Co-chairs

Meera Navrekar	D.J. Sanghvi College of Engineering, Mumbai
Sudeep Thepade	Mukesh Patel School of Technology Management and Engineering, Mumbai
Manoj Sankhe	Mukesh Patel School of Technology Management and Engineering, Mumbai

Collaboration Co-chairs

Vinod Jain	Mukesh Patel School of Technology Management and Engineering, Mumbai
Kishore Kinage	D.J. Sanghvi College of Engineering, Mumbai

Publicity Chair

Sanjeev Arora	Mukesh Patel School of Technology Management and Engineering, Mumbai
Harish Narula	D.J. Sanghvi College of Engineering, Mumbai
T.D. Biradar	D.J. Sanghvi College of Engineering, Mumbai
Trupti Malankar	D.J. Sanghvi College of Engineering, Mumbai

Awards Chair

Manali Godse	D.J. Sanghvi College of Engineering, Mumbai
Vishakha Kelkar	D.J. Sanghvi College of Engineering, Mumbai

Student Volunteer Co-chair

Narendra Shekokar	D. J. Sanghvi College of Engineering, Mumbai

Industry Track Co-chair

Mrinal Patwardhan	D.J. Sanghvi College of Engineering, Mumbai
Vivek Deodeshmukh	D.J. Sanghvi College of Engineering, Mumbai

Local Arrangements and Conference Director

S.M. Chaware	D.J. Sanghvi College of Engineering, Mumbai
Prasad Joshi	D.J. Sanghvi College of Engineering, Mumbai

Keynote Chair

Rekha Vig	Mukesh Patel School of Technology Management and Engineering, Mumbai

Program Webmaster

Neepa Shah	D.J. Sanghvi College of Engineering, Mumbai
Patidnya Hegde Patil	Mukesh Patel School of Technology Management and Engineering, Mumbai

Registration Chair

Pintu Shah	Mukesh Patel School of Technology Management and Engineering, Mumbai

Organizing Committee

Geeta Kumta	School of Business Management, NMIMS, Mumbai
J.M. Shah	School of Business Management, NMIMS, Mumbai
Anant Jhaveri	Mukesh Patel School of Technology Management and Engineering, Mumbai
Amit Deshmukh	D.J. Sanghvi College of Engineering, Mumbai
M.V. Deshpande	D.J. Sanghvi College of Engineering, Mumbai
A. Mahapatra	D.J. Sanghvi College of Engineering, Mumbai
Sonal Dhar	D.J. Sanghvi College of Engineering, Mumbai
Kedar Subramanium	Mukesh Patel School of Technology Management and Engineering, Mumbai
Rajesh Patil	Mukesh Patel School of Technology Management and Engineering, Mumbai
V.R. Lakshmi Gorty	Mukesh Patel School of Technology Management and Engineering, Mumbai
Sudarsana Sarkar	Mukesh Patel School of Technology Management and Engineering, Mumbai
R.S. Khavekar	D. J. Sanghvi College of Engineering, Mumbai
Nishita Parekh	Mukesh Patel School of Technology Management and Engineering, Mumbai
Sanjay Sange	Mukesh Patel School of Technology Management and Engineering, Mumbai
Abhay Kolhe	Mukesh Patel School of Technology Management and Engineering, Mumbai
V. Guite	Dy. Registrar, Mukesh Patel School of Technology Management and Engineering, Mumbai

Student Coordinators (Mukesh Patel School of Technology Management and Engineering, Mumbai)

Siddharth Mehta	Chief Student Coordinator and Publicity Head
Dishant Kapadia	Chief Student Coordinator and Marketing Head
Parth Jhaveri	Associate Event Organizer
Kautilya Sharma	Associate Event Organizer
Ratul Ramchandani	Associate Event Organizer
Abhash Meswani	Associate Event Organizer
Himank Shah	Associate Event Organizer
Viditi Parikh	Associate Event Organizer

Sponsoring Institutions

Central Bank of India
IEEE Bombay section
EMC^2

Table of Contents

Computer Engineering and Information Technology

Electronics and Telecommunication

Technology Management

Simple Message Passing Framework for Multicore Programming Development and Transformations

Prabin R. Sahoo

Tata Consultancy Services, Yantra Park, Thane,
Maharashtra, India
prabin.sahoo@tcs.com

Abstract. Multicore [1] is the latest processor technology in the commodity hardware market for achieving high performance. Though multicore processors are capable of producing high power computation but to exploit multicore requires a new way of program design. Therefore the existing applications require multicore transformation for effectively utilizing multicore computation capabilities. This research work demonstrates a framework for parallel computation on a multicore architecture which helps in migrating existing single threaded applications, development of new applications on multicore for achieving high performance. The contribution of this framework is to i) integrate single threaded applications to work in parallel ii) integrating threads and processes iii) simple interfaces for achieving inter thread and inter process communication together in one framework. This framework can transform single threaded C, C++, Java and Perl, Python, Shell scripts applications to run in parallel for achieving high performance in Linux environment with minimum effort.

Keywords: Multicore Technology, Shared Memory, Parallel Computation, Inter process communication, UML.

1 Introduction

Multicore processors are already into the commodity hardware market. Processor manufactures such as Intel®, AMD®, SUN® etc have started releasing multicore processors in the form of dual core, quad core, and processors with higher number of cores. The existing SMP [2] Symmetric multiprocessors, Single core processors have hit the bottleneck of scalability issues for achieving high performance. Increasing CPU speed for achieving high performance is no more a scalable solution at the cost of high heat and power consumptions, especially when the world is struggling to reduce global warming and reducing power consumptions. Multicore processors are going to play a major role in the IT computation. Though multicore processor compromises with reduced CPU cycles, but with introduction of multiple execution cores, it provides high computation ability. Currently parallel programming frameworks such as Cilk [3], MPI [4], openMP [5] etc are being revisited if these can be reused in as is form. However, the challenge is big as most existing applications are single threaded in nature, and requires code changes which consequently add huge man power and cost to redesign and implement.

K. Shah, V.R. Lakshmi Gorty, and A. Phirke (Eds.): ICTSM 2011, CCIS 145, pp. 1–9, 2011.

2 Literature Review

Message passing interface MPI appears close to simple message passing framework (**SMPF**) concept. However MPI is mainly used for massively parallel machines, SMP clusters, workstation clusters and heterogeneous networks. This framework is not easily adaptable for existing sequential applications as it requires extensive code changes. Perl based applications, single threaded C applications, Java applications are not fully compatible with it, though C, Java applications can be modified for compatibility. Even if changes are adapted for an existing application per se a C application, running it in a multicore server does not guarantee that performance would be improved on a multicore processor. Further the code changes involve the cost of development, testing would be high in such cases. openMP is another option for C/C++ based application where openMP pragma are used for splitting a for loop among several threads, but adapting this also not an easy choice. For example if C/C++ based application uses embedded SQL codes it is not easy to use openMP. Most legacy applications use embedded SQL. Similarly perl and Java based application cannot use openMP. This necessitates the need for building a new framework which can work not only applications with C and C++ but also with heterogeneous applications such as Java, Perl, Python, shell scripts etc in unix/linux environment. In addition to this, the framework should be simple for programmers to adapt and cost effective for business enterprises.

3 Model and Architecture

Multicore processors are equipped with multiple execution cores. This provides the backbone for executing tasks, processes in parallel. The experiments conducted for

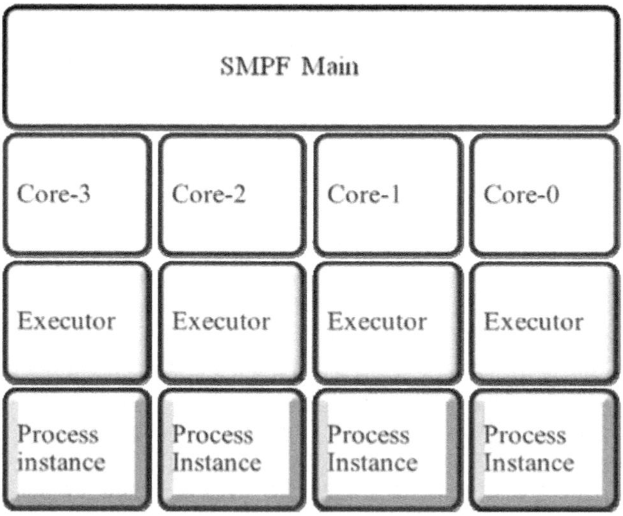

Fig. 1. SMPF Architecture (*Core-0,Core-1,Core2,Core3*) are the cores in multicore processor, *Executor()* is executed by the thread/process instance running on each core

this paper is based on Linux shown in Table 1 section 4. However, this can be extended to other UNIX/Linux OS. In addition, this uses POSIX thread model [6] [7] and C++ with UML [8] as underlying elements to build SMPF.

Figure 1 shows the architecture of **SMPF**. In this figure, **SMPF** main creates multiple thread objects which are mapped to the execution core in multicore server. Each thread object invokes its executor function with the binary and supplied arguments to create process instance. Its components are described as follows.

3.1 SMPF Executor

This method is a member of the **SMPF** class. It invokes the single threaded application binary with the necessary parameter and holds a pipe [9] onto the created process instance to read any message from it. Each process instance runs in a core of the multicore processor. Any message from the process instance is passed through the pipe to the executor. The executor stores the message which is finally retrieved by the main function.

3.2 SMPF Class

This class is a template class [10]. The *Execute* method in the following pseudo code is used for executing the command. The first argument to this method is the binary name along with any arguments. The arguments types can be any data type such as string, long, float, int etc. The Executor returns a pointer to a pipe which can be processed as a regular file processing. The *Execute_and_store* method executes the command and also stores output in a vector of strings. For example: two processes are trying to search and get the order information out of which one is processing a file and other one is searching a database to get related information. In this case the Execute_and_store method is useful. Each process can store the search information in the vector container. Once the process instances complete the search, the manger process instance processes the vector to retrieve the necessary information. The following pseudo code represents the **SMPF** definitions. Line number 4 represents the vector which is used by Execute_and_store method to store the output received from the application through pipe, and the **SMPF** main retrieves the messages from this vector.

```
1. template <class V, class X>
2. class SMPF{
3. public:
4. vector<string> vec;
5. fp  Execute(const char *cmd, V v, X x){ }
6. void Execute_and_store(const char *cmd, V v, X x){}
7. };
```

The variable V and X (line 1) can be of data types such as long, int, float, char *, string etc. The parameter cmd at line 5 represents the binary needs to be invoked and V, X are parameter which act as command line arguments. For arguments greater than 2 can be clubbed into V and X as strings with space in between. For example:

a C/C++ binary Execute("a.out V X)" can be invoked with more than 2 arguments as "a.out " "123 345 567" "890 xyz" where argument V="123 345 567" and X = "123 345 567". Similarly a python program can be invoked as Execute("a.py V X"). a shell script can be invoked as Execute("a.ksh V X"), and a perl script can be invoked as Execute("a.perl V X"). The basic idea here is to split the data among process instances to execute them in parallel. Inter process communication happens through the SMPF inter process communication channel described in following section 3.3.

3.3 SMPF Inter Process Communication

SMPF provides the integrated communication mechanism. It uses pipe and shared memory [11] for communicating with each other. SMPF synchronization is achieved through the combination of POSIX primitives [12] and token approach blended into object oriented paradigm for achieving powerful objected oriented interfaces for wide acceptance and reusability. The token approach uses two shared memory variables for synchronizations i.e. one shared memory acts as a token and the other as the data store. The token is required for achieving synchronizations among competing thread objects while updating data in the shared memory. The benefit out of this approach is that the processes invoked by **SMPF** frameworks do not directly act with shared memory. It is the thread object that reads the value from the process instance and sets the shared memory. Each process does its sequential computation without worrying about other process instances. The developer need not think much about shared memory, semaphores etc required for the parallel processing. Even a shell script program which has no features of shared memory, semaphore can be parallelized. For example: A file search application can be designed to search the file using the conventional program design. The only change if at all required is the application need to write the output to STDOUT using a delimiter **SMPF** understands. Currently **SMPF** can understand following separators such as comma, semicolon, colon, "->". The search string program can send a message like "orderid: 1234" for an order id "1234" to the SMPF Main through its pipe and the Main can update the shared memory.

```
this->Lock.lock();
 while(true){
            char *str = shmkey.get();
             if(atoi(str) == 0){
               // Update data in the shared memory
                  shm.set((char*)k.c_str());
               //Set the token for reader
                  shmkey.set("1");
             }
  }
```

4 Experimental Setup

The following table describes the experimental setup for both the case studies.

Table 1. Hardware Configuration

Parameter	Value
CPU	Genuine Intel$^{(R)}$ CPU 2.66GHz
No. Of CPUs	4
Cores per CPU	6
L3 Cache	16 MB
Main Memory	64 GB
Operating System	Redhat® Enterprise Linux 5

5 Case Study-I

To understand how **SMPF** works the following case study has been conducted. The investment companies collect various pieces of stock market data from data providers from various exchanges such as NYSE, Toronto stock exchange, nasdaq, London stock exchange, honk kong to name a few. The data providers for example could be Reuter, Bloomberg, UBS and so on. The problem is each data provider provides data in the format local to the vendor. So the subscribers to this data need to understand the layout for processing. The processing involves data extraction, validation, enrichment and loading into the database. In general each vendor file is processed by an application meant for it. Since file format may change time to time, rules for a vendor change time to time, so it makes sense to have separate application for each vendor. If everything is clubbed into one application it increases the maintenance cost. Over the year the volumes of transactions are increasing. The batch processing once completing in time is showing performance challenges. Following experiment demonstrates how performance has been improved using a multicore processor and **SMPF** framework.

5.1 Sequential Approach vs. Parallel Approach Using SMPF

Sequential approach is quite straight forward. Each file is processed by an application which is meant for a particular layout. Each application runs in batch for extraction, validation, enrichment and loading into the global data storage. The global data storage can be a database, but for this experiment I have taken it as a regular file system.

In parallel approach with **SMPF**, since the existing applications are already capable of extraction, validation, enrichment and loading into the database (for simplicity the experiment has been conducted with file system instead of database), the core logic is in place. Redesign into parallel program design involves almost writing from scratch which involves knowledge of parallel programming, coding, testing and finally deployment. **SMPF** becomes handy here as the changes are minimal. Fig. 2 represents the message flow. **SMPF** initializes, reads the binary information of the applications, create thread objects and calls the executor on each object to invoke the binary. Each thread object listens on the pipe to receive messages from each application and synchronize using shared memory key and lock mechanism as described in section 3.3.

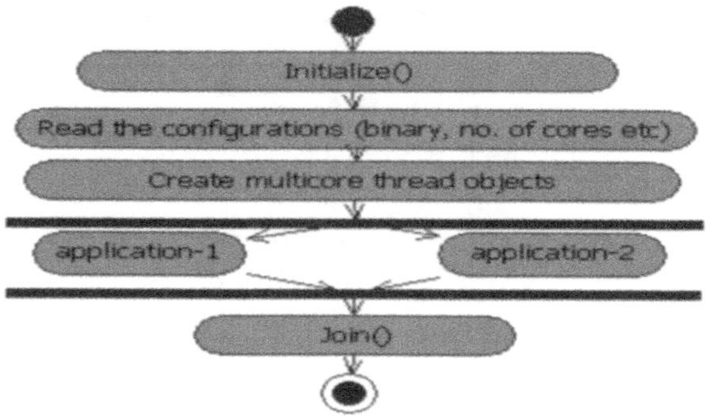

Fig. 2. SMPF Message Flow

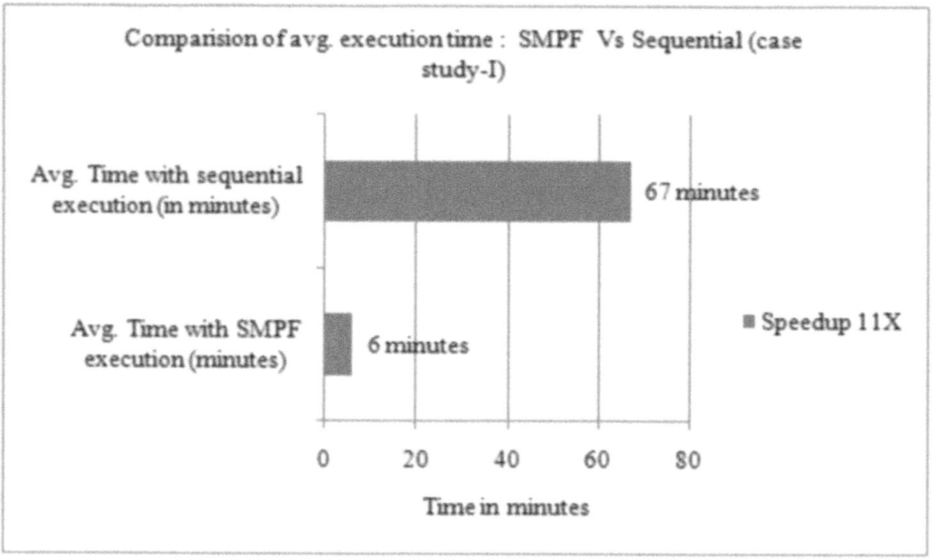

Fig. 3. Average Execution time of case study-I (*Parallel using SMPF vs. Sequential*)

In the case study there are 20 thread objects, out of which 19 thread objects invokes 19 perl based applications which do the extraction, validation, enrichment for 19 types of vendors with average file size 9.1 MB each. The remaining one is a java based application which reads from shared memory and writes to the file system. The results of execution are shown in fig. 3.

6 Case Study-II

In case study-II I have demonstrated π computation for proving the power of **SMPF** framework for exploiting multicore processor. This is a useful example as it involves purely computation compared to the case study-I in section 5.0. A typical π computation is represented by an equation as below.

$$\sum_{i=0}^{N} 4/(1+((i+0.5)/N)2) * (1/N) \tag{13}$$

The size of N determines the accuracy of π computation. Bigger the size of N, better is the precision. In the experiment the size of N is set to 1,000,000,000. The results of comparison between a sequential processing vs. parallel processing using **SMPF** is shown in fig. 4. The **SMPF** uses 24 cores for π computation.

There are 2 types of applications used in π computation using **SMPF** i.e **SMPF** with C based application to compute π, and **SMPF** with perl based application to compute the π.

6.1 Sequential Approach vs. Parallel Approach

Sequential logic is straight forward. With the value N=1,000,000,000 a simple for loop computes as per the equation in section 6 above.

In parallel approach the sequential program is invoked with **SMPF**. However, a minor change is required by the sequential program to accept N, start and end data set as command line argument. For example, the sequential program "computePi" can be invoked as "computePi 1,000,000,000 with data set 0 1000000000" to compute π sequentially. This program can be accessed by **SMPF** with different set of arguments. In this case the following data split has been conducted. With a 24 core server, the data spit can happen in the following way.

Range = 1000000000/24 = 41666667
First data set: 0 …. 41666667

…………

24th data set: 958333342….1000000000

Fig. 2 shows the flow of **SMPF**. The result is shown in section 6.2. Fig. 4 shows that the speedup of π computation using **SMPF** is 23X.

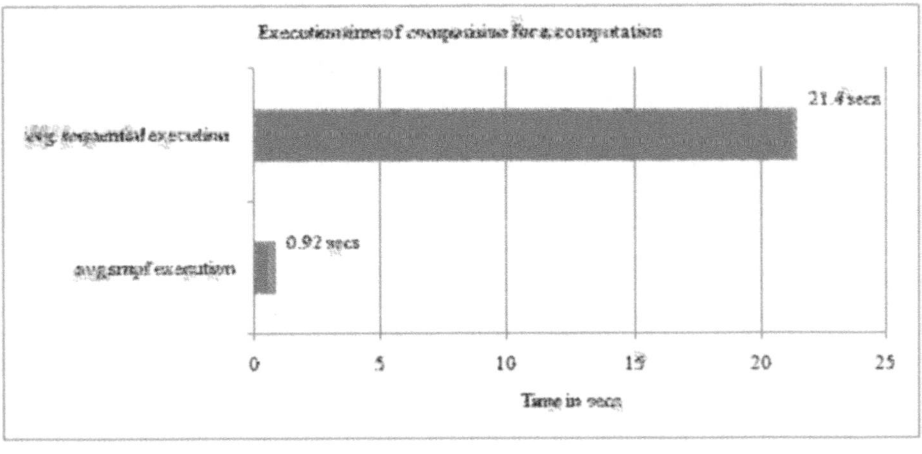

Fig. 4. Size (*N=1,000,000,000*) using SMPF vs. sequential processing (C based application) *speedup ~ 23X*

6.2 Results

The result of execution of case study-II is as follows.

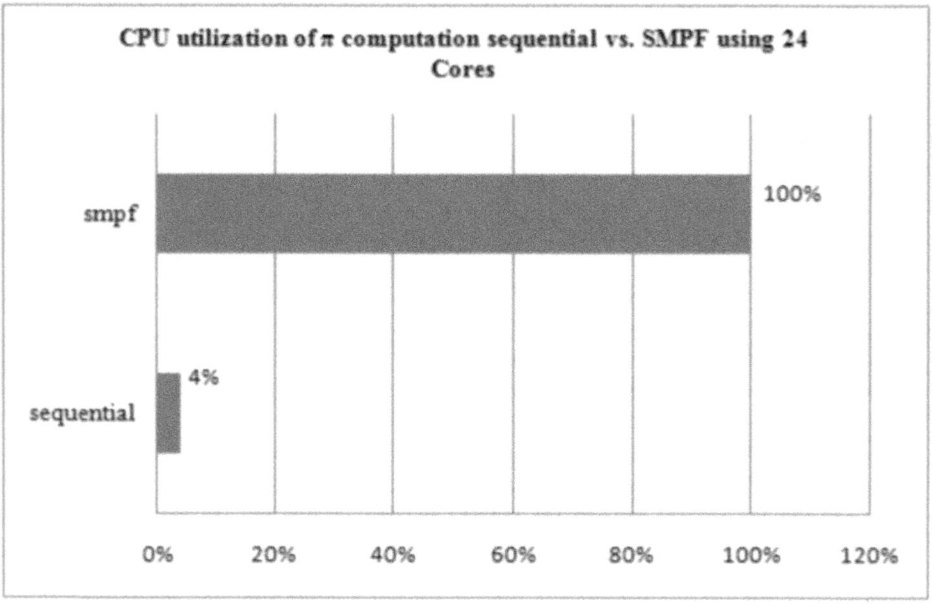

Fig. 5. Single threaded application Just uses 1 core with *(4% cpu)* and SMPF uses *(100% CPU)*

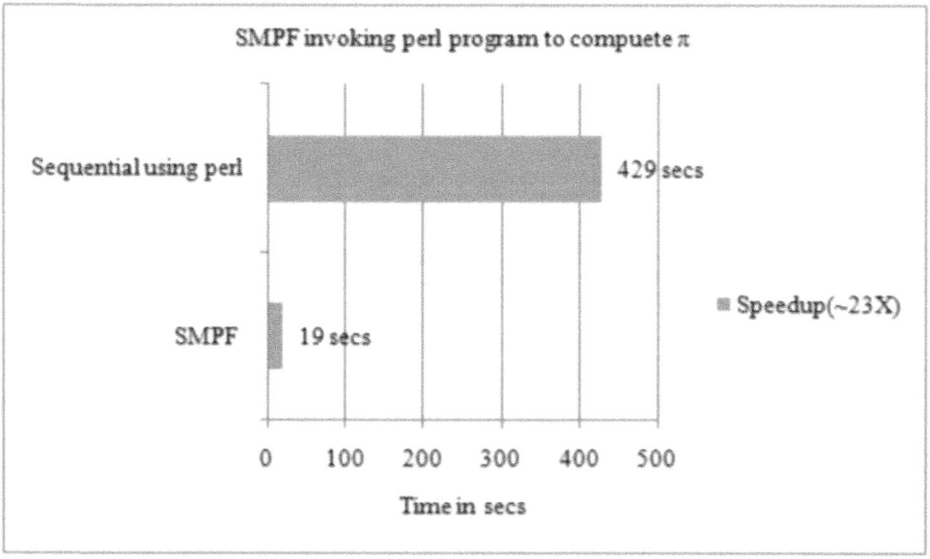

Fig. 6. SMPF invokes perl programs for *(π)* computation (*size N=1,000,000,000*)

7 Conclusions

From both the case studies it is evident that **SMPF** provides an easy way for multicore transformation. Applications involving computations have a greater speed up. Case study-I involves I/O and synchronization which shows 11X speed up (fig. 3.) where as in case study-II the speed up is 23X (fig. 4) as it involves pure computations. The CPU utilization is 100% as shown in fig. 5 for **SMPF** framework; whereas the single threaded application is able to use just 4% CPU. This implies a single threaded application when migrated to a higher multicore configuration it may underutilize the server which can be boost up by **SMPF** by simple transformation. **SMPF** transformation is simple as it involves minimum code changes, and also it supports heterogeneous languages(C, C++, Perl, Java, Shell scripts, Python) in one framework, and it offers opportunities for greater compatibility and widely acceptability. In addition, to speed up scripting languages, it works as a convenient framework. Further fig. 6 shows the result of π computation. In this case, **SMPF** invokes perl application. Since perl is script based application, the execution time is more than compared to the C based application. However, the speedup is still 23X. This implies **SMPF** speeds up the application performance relative to the speed of the serial execution of the application in a single core.

References

1. Thomas Evensen, CTO, Wind River, Multicore Challenges and Choices (June 2009), `http://whitepapers.opensystemsmedia.com/u/pdfs/WhitePaper.Wind_River_6648_WP_ESC-52.pdf`
2. Gropp, W.W., Lusk, E.L.: A Taxonomy of Programming Models for Symmetric Multiprocessors and SMP Clusters. IEEE Computer Society, Washington (1995)
3. Frigo, The Cilk Project (October 2007), `http://supertech.csail.mit.edu/cilk/`
4. Badrinath, R.: STSD Bangalore, Parallel Programming and MPI (September 2008), `http://hpce.iitm.ac.in/.../Parallel%20Programming%20and%20MPI.ppt`
5. OpenMP Architecture Review Board, The openmp api specification for parallel programming (January 2009)
6. Garcia, F., Fernandez, J.: POSIX thread libraries. Linux Journal (February 01, 2000)
7. Barney, B.: POSIX Thread Programming. Lawrence Livermore National Laboratory (2010)
8. Siegel, J.: OMG, Introduction to OMG's Unified Modeling Language™ (UML®) (June 2009), `http://www.omg.org/gettingstarted/what_is_uml.htm`
9. Mamrak, S., Rowland, S.: Unix Pipes (April 2004)
10. Moreno, C.: An Introduction to the Standard Template Library, STL (1999)
11. Marshall, D.: IPC: Shared Memory (1999), `http://www.cs.cf.ac.uk/Dave/C/node27.html`
12. Stallings, W.: Operating Systems, Internals and Design Principle (2009)
13. Hwang, W.K., Xu, Z.: Scalable Parallel Computing, Technology, Architecture, Programming-Mcgraw-Hill International Editions (2000)

Critical Variable Partitioning for Reconfiguration of Mobile Devices

Shweta Loonkar and Lakshmi Kurup

Faculty of Department of Computer Engineering,
D.J. Sanghvi College of Engineering, Mumbai, India
{shweta.loonkar,lakshmi.kurup}@djsce.ac.in

Abstract. Now a days multimedia application has increased in handheld devices and these applications needs a lot of power to operate. Power consumption is an important issue in these handheld devices, the problem starts when user feels that a device is running out-of-power. Dynamic reconfiguration of mobile devices to optimize power according to user is an active area for research. Further the remote reconfiguration of mobile devices is feasible only when Request Processing Time i.e. time required in sending the request, processing at remote server, transmission of new bit stream to mobile device and reconfiguration of mobile device as per the requested quality (RqoS) must be less. This is achieved through the static optimizations which greatly reduce the processing and response time of the server. The results available during static analysis and preprocessing are reduced further with the help of the Huffman compression. This will not only reduce the run time analysis of the application but also helps in optimizing power.

Keywords: Request Processing Time, Dynamic Reconfiguration, QIBO, QDBO Preprocessing, Critical Variables, Non-Critical Variables and Huffman Compression.

1 Introduction

Mobile devices work with dynamism i.e. they have to deal with the unpredicted network outage or should be able to switch to a different network, without changing the application. Reconfigurable systems have the potential to operate efficiently in these dynamic environments [1]. Dynamic Reconfiguration allows almost any customized design to be placed inside the hardware or an existing design to be updated or modified [2]. Remote reconfiguration is a strong solution for power optimization. Thus the proposed paper exhibits a satisfactory real time solution for operational performance of mobile devices [3].

Contributions: We therefore summarize the contributions of this paper as:

1) Bit-width optimization performed using QIBO and QDBO.
2) Preprocessing algorithm applied to the resulting variables and further compressed using Huffman compression algorithm.

K. Shah, V.R. Lakshmi Gorty, and A. Phirke (Eds.): ICTSM 2011, CCIS 145, pp. 10–17, 2011.

The remainder of the paper is structured as follows. We start by introducing client-server model of reconfigurable mobile framework described in Section 2. The Preprocessing Algorithm, which identifies the critical and non-critical variables, is described in Section 3. In section 4 experimental results are shown before and after inclusion of preprocessing algorithms using Huffman compression, conclusion is presented in Section 5 and finally in section 6 future scope has been included followed by references.

2 Client-Server Model of Reconfigurable Mobile Framework

The network view of framework explains that the client is the mobile device user while remote server provides the reconfiguration support to the user. Whenever user feels that the device is running out-of-power, he, initiates the process by selecting the required level of quality, which is then converted to technical specifications by client model of the framework and the parameters are sent to remote server through communication network. Server calculates the new optimized design and extracts the reconfigured bit-stream, which then sent to the client, reconfigures the device for reduced power consumption and compromised quality. The accuracy is reduced further till the limit of quality parameters is attained.

In this paper it will be seen that often the most efficient implementation of an algorithm in the reconfigurable hardware is one in which the bit width of each internal variable is fine-tuned to the best precision. Bit Width Optimization is done in two stages:

2.1 Quality Independent Bit Width Optimizations (QIBO)

QIBO is performed as a static process i.e. it is done only once for any source application and the results are stored for subsequent use by Quality Driven Optimization phase. QIBO phase is intended to remove the bits from Most Significant (MSB) and Least Significant (LSB) ends of a variable that do not introduce any error or inaccuracy. The QIBO is loss less approach. Algorithm for QIBO is as:

1. Assigning zero to *lsb_saving* and *msb_saving* vectors.
2. Removing one bit from LSB and MSB end of any variable (i.e. increasing the value of *lsb_saving* vector or *msb_saving* vector at the position corresponding to the variable under consideration, by one) and calculating the mean square error [6].
3. If *MSE* is found to be lower than the *Min_Err* value the algorithm proceeds towards the removal of further next bit. The process is repeated until the *MSE* remains lower than the *Min_Err*.
4. Similar steps are repeated for MSB end.
5. Repeat the steps 1-4 for all *n* variables in the application.

2.2 Quality Driven Bit Width Optimizations (QDBO)

The results of QIBO are made available for the QDBO at remote server after receiving the required quality level from the client. QDBO is a lossy approach and intends to enhance the statically determined bit widths (determined through QIBO) by allowing deterioration in quality up to an externally specified acceptable level.

1. In this algorithm *n* is the total number of variables in the application being optimized. The QDBO uses *lsb_saving* vector from QIBO.
2. Removing one bit from LSB end of any variable (i.e. increasing the value of *lsb_saving* vector at the position corresponding to the variable under consideration, by one) and calculating the mean square [6].
3. If *MSE* is found to be lower than the *Min_Err* value the algorithm proceeds towards the removal of further next bit. The process is repeated till the *MSE* remains lower than the *Min_Err*.
4. Repeat the steps 1-4 for all *n* variables in the application.

3 Preprocessing

To further reduce the preprocessing time of the variables obtained from QIBO and QDBO, pre-processing algorithm has been proposed in this paper The overall aim of this paper can be stated as "***To generate an optimum bit stream, in response to desired quality request from user, using preprocessing and Huffman compression algorithms, which can dynamically reconfigure the remote runtime of mobile devices for improved power efficiency against quality compromise***".

In this framework source code of algorithm to be optimized is preprocessed before applying QIBO and QDBO (Fig 1). Initially variables are identified and their bit width is documented (as per the language specification). E.g. source code to be analyzed is listed below

Example 1

```
# define W2 2 6 7 6
# define W6 1 1 0 8
x8 = x2 + x1;
if (x0)
{x1 = W6*(x3+x2);}
x2 = x1 – (W2+W6 ) * x2 ;
x3 = x1 + (W2–W6 ) * x3 ;
x1 = x4 + x6 ;
x4 –= x6 ;
for ( k = 0 ; k < x7 ; k++)
{x6 = x5 << x2 ;}
```

Identified variables and their bit widths from above source code are documented in the Table 1.

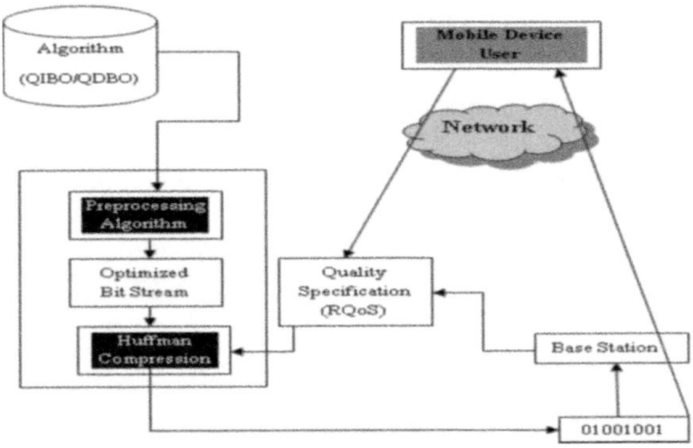

Fig. 1. Preprocessing at Remote Server

Table 1. Documentation of Variable and their bit widths

Variable	x0	x1	x2	x3	x4	x5	x6	x7	x8	W2	W6	k
Bitwidth	32	32	32	32	32	32	32	32	32	12	11	32

Table 1 data can be presented as Initial Variable Set (*Vi*).

$$Vi = \{x0, x1, x2, x3, x4, x5, x6, x7, x8, W6, W2, k\} \quad (1)$$

$$|Vi| = n = 12 \quad (2)$$

The optimization time requires by number of variables in algorithm as per eq. (3).

$$\text{Execution Time} = n * b \quad \text{where} \quad (3)$$

- **n** is the number of variables and
- **b** is the average bit width.

Therefore to reduce the execution time number of variables to be analyzed should be reduced. During preprocessing, variables are flagged as critical and non-critical. Critical variables are variables whose accuracy must not be alerted in order to retain functionality. These critical variables can be avoided in further analysis to reduce the execution time. Their usage patterns can identify critical variables. If variable

1. Participates in comparison statement.
2.. is used to calculate at least 80% of output variables.
3. are of Boolean type.
4. are a type on which number of loop iteration depends?

By analyzing variables based on usage pattern some of the variable in V_i can be identified as Critical Variable set (V_{cvup})

$$V_{cvup} = \{x0, x2, x7, k\} \quad (4)$$

$$|V_{cvup}| = m = 4 \quad (5)$$

The Approximate Effective Variable Set (V_{ae}) is

$$V_{ae} = Vi - V_{cvup} \quad (6)$$

$$V_{ae} = \{x1, x3, x4, x5, x6, x8, W6, W2\} \quad (7)$$

$$|V_{ae}| = n - m = 12 - 4 = 8 \quad (8)$$

Example 2: For a statement x8 = W3*(x6+x7) the injected code for virtual bit width reduction is

```
Mask(&x6 ) ;
Mask(&x7 ) ;
x8 = W3 * ( x6 + x7 )
Mask(&x8 )
f u n c t i o n Mask ( i n t  * x )
{ i f (* x > 0 )
*x = ( * x & Ma s k Pa t t e r n ) ;
e l s e
{ *x = -1 *  *x ;
*x = ( -1 ) *(* x & Ma s k Pa t t e r n ) ;}}
```

Now one variable from above set is chosen and usefulness of its bits is verified. Masking is used to verify usefulness of bit, we inject some code to reduce bit width virtually (during calculation a block of bits is masked to produce an effect of bit width reduction). It is done from LSB side. By masking we reduce the precision of variable. Advantages of masking process is that it provides a simulation of bit width specific calculation at system-level, thus a great potential for considering feedback and power saving without going to low-level. The modified program is executed and output is observed. If output is as accurate as with standard code, i.e. ***"Mean Square Error (MSE)"*** remains less than the visualized error threshold ***(MSE Minimum Threshold)***, the bits are flagged as *unused*. This process is called ***"Error Contribution Analysis"*** and repeated for every variable in V_{ae}. Output of this process is ***"Error Contribution Matrix"*** as shown in Table 2. Here MSE is calculated by adding difference i.e. error between all pixels of produced original and new image as given by-

$$MSE = 1/n^2 \sum_{i=0}^{n-1} \sum_{j=0}^{n-1} \{ f`(i,j) - f(i,j) \}^2 \quad (9)$$

Here masked column shows the number of bits masked from LSB Side and MSE introduced due to masking of bits. Thus we have an error contribution matrix for all variables V_{ae} set. If error contribution of LSB bit is greater than threshold (*TCriticalError*)

Table 2. Error Contribution Matrix for Variable *x*0

Masked	MSE	Masked	MSE	Masked	MSE	Masked	MSE
1	0	9	0.1362	17	191.463	25	1959.56
2	0	10	0.2352	18	393.053	26	1959.56
3	0	11	0.4259	19	859.466	27	1959.56
4	0	12	0.7683	20	1309.61	28	1959.56
5	0.0013	13	1.6258	21	1959.56	29	1959.56
6	0.0019	14	4.2247	22	1959.56	30	1959.56
7	0.0039	15	14.2902	23	1959.56	31	1959.56
8	0.0966	16	54.2229	24	1959.56	32	1959.56

Table 3. Error Contribution due to first bit for all variables of Approximate Effective Variable Set (V_{ae})

Variable	x0	x1	x3.	x4	x5	x6	x8	W6	W2
Error	0.00	0.00	0.00	0.002	14.9632	0.0011	0.00	0.0016	0.0287

then it can be flagged as critical variable. E.g. in Table 3 error introduced due to masking of first LSB bit is listed. Here value of *TCriticalError* is 10.0000 hence we can flag $x5$ as critical variable.

By including all identified critical variables we can reduce V_{ae} to Effective Variable set (V_{eff}). V_{eff} calculated for above example is -

$$V_{eff}=V_{ae}-\{x5\} \quad (10)$$

$$V_{eff}=\{x1, x3, x4, x6, x8, W6, W2\} \quad (11)$$

$$|V_{eff}|=7 \quad (12)$$

Thus preprocessing calculation takes significant time but as these are static tasks, these make the dynamic process faster.

4 Experimentation and Results

In this paper we have explored the preprocessing module in which processing of variables are done on various usage patterns mentioned above. In order to demonstrate the importance and benefits of the quality independent and quality driven bit-width optimizations in terms of performance and power consumption we have analyzed Huffman compression algorithm in MPEG-2 decoding.

4.1 Compression Using Huffman Algorithm

Compression is a technology for reducing the quantity of data used to represent any content without excessively reducing the quality of the image. It also reduces the

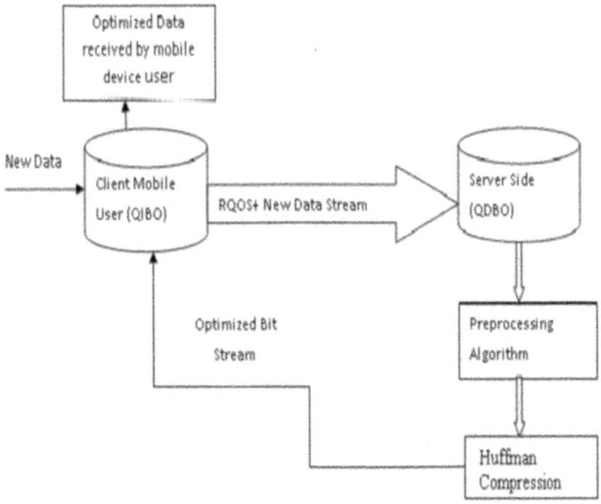

Fig. 2. Compression of Preprocessed Variables

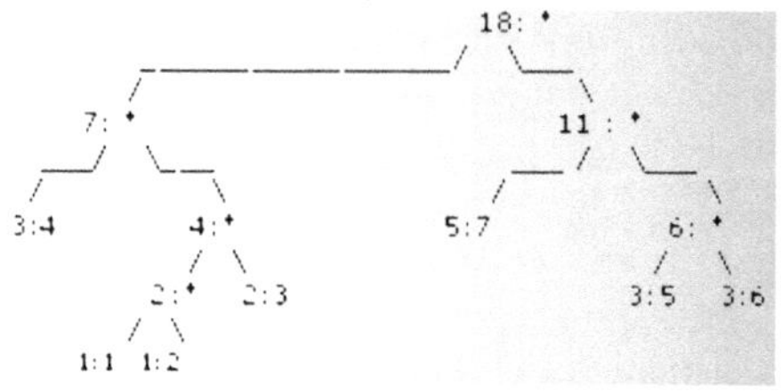

Fig. 3. Huffman Tree based on frequency of occurrence of preprocessed variables

number of bits required to store and/or transmit the digital data (Fig 2). Huffman coding [4] is an entropy encoding algorithm used for lossless data compression. Huffman coding uses a specific method for choosing the representation for each symbol, resulting in a prefix-free code [5] (that is, the bit string representing some particular symbol is never a prefix of the bit string representing any other symbol). Here our preprocessed sets of variables are $\{x1, x3, x4, x6, x8, W6, W2\}$. To generate a Huffman prefix code we traverse the tree to the value we want, outputting a 0 every time we take a left-hand branch and a 1 every time we take a right-hand branch shown in Fig 3.

Each preprocessed variable is an integer value i.e. 16*7=112 bits in length.

Table 4. Preprocessed variables and their Huffman Code

Preprocessed Variables	Frequency of Occurrence	Value	Huffman code
x8	1	1	0100
x4	1	2	0101
w2	2	3	011
x3	3	4	00
x6	3	5	110
w6	3	6	111
x1	5	7	10

Considering all the leaf nodes the partial tree for our preprocessed variables can be represented with 010001010110011011110, which is 21 bits long. So in our case, use of Huffman codes saved 91 (112-21) bits which is around 80% of the total preprocessed data Table 4).

5 Conclusion

Bit width analysis and compile time optimizations have emerged as a key technique for power and performance aware optimizations in future digital systems design. In this paper we have proposed a preprocessing algorithm for critical variable partitioning and then compressed using Huffman algorithm, which will minimize the total time taken in execution of a program, thereby minimizing the power consumption. Reduced power consumption will result into increased standby time of mobile devices.

6 Future Scope

This work can be implemented using hardware accelerators in real time application in order to reduce the Request Processing Time of Mobile Devices.

References

1. Martin, T.L.: Balancing Batteries, Power, Performance: System Issues in CPU Speed-Setting for Mobile Computing. Ph.D. dissertation, Carnegie Mellon University (1999)
2. Verma, S.S., Joshi, H., Sharma, G.K.: Quality Driven Dynamic low Power Reconfiguration of Handhelds. In: Proceeding of International Workshop on Applied Reconfigurable Computing, ARC 2006, Delft, Netherland (March 2006)
3. Luo, Jha, N.K.: Battery-Aware Static Scheduling for Distributed Real-Time Embedded Systems. In: Design Automation Conference, pp. 444–449 (2001)
4. Pham, H.-A., Bui, V.-H., Dinh-Duc, A.-V.: An Adaptive, Memory-Efficient and Fast Algorithm for Huffman Decoding and Its Implementation. In: ACM International Conference, Proceedings of the 2nd International Conference on Interaction Sciences: Information Technology, Culture and Human, Seoul, Korea, vol. 403, pp. 275–279 (2009)
5. Sharma, M.: Compression Using Huffman Coding. IJCSNS International Journal of Computer Science and Network Security 10(5) (May 2010)
6. Bins, J., Draper, B., Bohm, W., Najjar, W.: Precision vs. error in jpeg compression (1999)

A Markov Model Based Cache Replacement Policy for Mobile Environment

Hariram Chavan[1], Suneeta Sane[2], and H.B. Kekre[3]

[1] Information Technology, Terna Engineering College, Mumbai University, India
[2] Computer Technology, V.J.T.I., Mumbai, India
[3] Mukesh Patel School of Technology Management & Engineering, NMiMS University, India
chavan.hari@gmail.com, sssane@vjti.org.in, hb_kekre@nmims.edu

Abstract. The continued escalation of manageable, wireless enabled devices with immense storage capacity and powerful CPUs are making the wide spread use of mobile databases a reality. Mobile devices are increasingly used for database driven applications such as product inventory tracking, customer relationship management (CRM), sales order entry etc. In some of these applications Location Dependence Data (LDD) are required. The applications which use LDD are called Location Dependent Information Services (LDIS). These applications have changed the way mobile applications access and manage data. Instead of storing data in a central database, data is being moved closer to applications to increase effectiveness and independence. This trend leads to many interesting problems in mobile database research and cache replacement policies. In mobile database system caching is the effective way to improve the performance since new query can be partially executed locally. The desired caching can be achieved by convincingly accurate prediction of data items for the present and future query processing. It is important to take into consideration the location and movement direction of mobile clients while performing cache replacement. Due to cache size limitations, the choice of cache replacement techniques used to find a suitable subset of items for eviction from cache becomes important. In this paper we propose a Markov Model based Cache Replacement Policy (MMCRP) for mobile database environment. Our method predicts the new data item to be fetched by searching second and/or first order transaction probability matrix (TPM) of Markov Model for valid scope, access frequency, data distance and data size. The implementation of these policies has been done using java. Simulation results for query interval, client speed and cache size show that the MMCRP performs significantly better than Least Recent Used (LRU), Furthest Away Replacement (FAR) and Prioritized Predicted Region based Cache Replacement Policy (PPRRP).

Keywords: Location Dependent Information Services (LDIS), Markov model, MMCRP, TPM, PPRRP.

1 Introduction

The fast development of wireless communications systems and advancement in computer hardware technology has led to the seamlessly converged research area called

K. Shah, V.R. Lakshmi Gorty, and A. Phirke (Eds.): ICTSM 2011, CCIS 145, pp. 18–26, 2011.

mobile computing. The mobile computing research area includes the effect of mobility on system, users, data and computing. The seamless mobility has opened up new classes of applications and offers access to information anywhere and anytime.

The main reason for the growing popularity of the mobile computing is new kind of information services, called Location Dependent Information Services (LDIS). LDIS provides information to users depending upon his/her current location. As location plays a vital role in storing, representing and querying LDD for LDIS, it becomes an important piece of information. Users of LDIS face many challenges inherent to mobile environment such as client power, limited bandwidth and intermittent connectivity. Caching helps to address some of these challenges.

There are two common issues involved in client cache management: cache invalidation and cache replacement. A cache invalidation policy maintains data consistency between client and the server. However, a cache replacement policy finds suitable subset of data item(s) for eviction from cache when it does not have enough free space to store a new data item. Due to the limitations of cache size on mobile devices, an efficient cache replacement policy is vital to ensure good cache performance. Thus a cache replacement policy becomes unavoidable in order to utilize the available cache effectively.

The structure of the paper is as follows: section 2 describes Related Work, section 3 Motivation: Markov Model based caching, section 4 details proposed cache replacement policy. Section 5 details performance evaluation and comparisons. Section 6 concludes the paper.

2 Related Work

Prioritized Predicted Region based Cache Replacement Policy (PPRRP) [2] predict valid scope area for client's current position, assigns priority to the cached data items, calculates the cost on the basis of access probability, data size in cache and data distance for predicated region. The limitation of PPRRP policy is multiple replacements because of data size.

The Furthest Away Replacement (FAR) [5] depends on the current location and the direction of the client movement. FAR organizes data items in two sets: In-direction and out-direction. The selection of victim for replacement is based on the current location of user. The assumption for replacement is, locations which are not in the moving direction and furthest from the user will not be visited in near future. The limitation of FAR is it won't consider the access patterns of client and is not very useful for random movement of client.

Literature is full of cache replacement methods based on LRU. LRU-K [7] method is an approach to database disk buffering. The basic idea of LRU-K is to keep track of last K references to popular database pages. This historical information is used to statistically estimate and to better discriminate pages that should be kept in buffer. But the LRU-K takes into account only temporal characteristics of data access.

The dominating property for Location Dependent Data (LDD) [12] is the spatial nature of data item. None of these existing cache replacement policies are very useful if client changes its direction of movement quite often. It is necessary to anticipate the future location of the client, based on current location, in the valid scope using mathematically proven hypothesis. So, in this paper, we propose MMCRP based on Markov

Model. MMCRP predict next client location by first and/or second order Markov Models. The key idea of MMCRP is prediction of future user location and performance optimization by state pruning strategy. The basic cellular mobile network for a wireless communication used is similar as discussed by Vijay Kumar et al [8].

3 Motivation: Markov Model Based Caching

The main aim of prefetching is to reduce latency, optimal use of cache space and best possible utilization of available bandwidth. But when we enhance prefetching policy it may prefetch data items which may not be eventually requested by the users and result into server load and network traffic. To overcome these limitations we can use high accuracy prediction model such as Markov Model.

Specifically, in mobile database caching and prefetching can complement each other. The caching utilizes the temporal locality which refers to repeated users accesses to the same object within short time periods. The prefetching utilizes the spatial locality of the data items i.e clients' request, where accesses to same data items frequently stipulate accesses to certain other data items. So in this paper we present an integrated approach of effective caching and prefetching by convincingly accurate prediction of data items by Markov model for the present and future query processing.

4 Cache Replacement Policy

LDIS is spatial in nature. A data item can show different values when it is queried by clients at different locations. Note that data item value is different from data item, i.e., a data item value for a data item is an instance of the item valid for a certain geographical region. For example, "nearest restaurant" is a data item, and the data values for this data item vary when it is queried from different locations.

For cache replacement we should account the distance of data from clients' current position as well as its valid scope area. Larger the distance of data item from the clients' current position the probability is low that customer will enter into the valid scope area in near future. Thus it is better to replace the farthest data value when cache replacement takes place. Generally customer movement is random so it is not always necessary that customer will continue to move in the same direction. Therefore replacing data values which are in the opposite direction of customer movement but close to clients' current position may degrade the overall performance. In MMCRP we are considering previous few locations for the predication of next user location so there is very less probability that the data item with higher access probability related to previous (close) location will get replaced by new data item.

The system set up for this paper deals with following assumptions: Let the space under consideration be divided in physical subspaces (locations) as

$$S = \{L_1, L_2, \ldots, L_N\}$$

There are data items with

$$D = \{D_1, D_2, \ldots, D_M\}$$

Such that each data item is associated with valid scope which is either a set of one or more subspaces from S. Formally shown as

$D_k = \{L_{k1}, L_{k2}\}$ where L_{k1}, L_{k2} belong to S.

For example, consider the problem of predicting the next location client visits. The input data for building Markov models consists of the different locations on the path, where each location is on the path which client takes to travel. In this problem, the actions for the Markov model correspond to the different locations visited by the client, and the states correspond to all consecutive trips. In the case of first-order models, the states will correspond to single locations, in the case of second-order models, the states will correspond to all pairs of consecutive locations, and so on. To illustrate, we are assuming the predicted region routes between the five different locations of geographical area with locations L_1, L_2, L_3, L_4 and L_5.

Connections on the move for a trip are very large. However if only the cell address where the calls were made limits to the chances of intermittent contact during the travel. To counter this it is proposed that cells that contribute to the deflection from the previous path be noted and continued. This will cause significant reduction in state space search, thus reducing the complexity of the algorithm used as well as keeping the needed details in the data. In this paper we consider such pruned data for further analysis.

Once the states of the Markov model have been identified, the transition probability matrix (TPM) can be computed. There are many ways in which the TPM can be built. The most commonly used approach is to use a *training* set of action-sequences and estimate each t_{ji} entry based on the frequency of the event that action a_i follows the state s_j. For example consider the second trip of customer $TR_2(\{L_3, L_5, L_2, L_1, L_4\})$ shown in Fig-1.a. If we are using *first-order Markov model* then each state is made up of a single location, so the first location L_3 corresponds to the state s_3. Since state s_5 follows the state s_3 the entry t_{35} in the TPM will be updated (Fig. 1. b). Similarly, the next state will be s_5 and the entry t_{52} will be updated in the TPM. In the case of higher-order model each state will be made up of more than one actions, so for a second-order model the first state for the trip TR_2 consists of locations $\{L_3, L_5\}$ and since the location L_2 follows the state $\{L_3, L_5\}$ in the trip the TPM entry corresponding to the state $\{L_3, L_5\}$ and location L_2 will be updated (Fig. 1. c).

Once the transition probability matrix is built, making prediction for different trips is straight forward. For example, consider a client that has visited locations L_1, L_5, L_4. If we want to predict the location that will be visited by the client next, using a first-order model, we will first identify the state s_4 that is associated with location L_4 and look up the TPM to find the location L_i that has the highest probability and predict it. In the case of our example the prediction would be location L_5 (Fig. 1. b).

The fundamental assumption of predictions based on Markov models is that the next state is dependent on the previous *k* states. The longer the *k* is, the more accurate the predictions are. As client is mobile (random movement) its location at the time of next query will be different from its current position. Therefore predict the clients' presence in the near future, considering previous locations visited and current direction of movement for the replacement of data item from the cache. The MDS is developed with the following data attributes - valid scope, access probability, data size and distance related to respective location.

This centralized database is percolated to MSS in order to make sure that the data relevant in the valid scope is available in the nearest MSS thus making sure that the concept of Prioritized Region is taken into account while considering the cache replacement policy. We have valid scope as an attribute for the location to reduce computation and save the bandwidth. The search for the actual data is done in order as – first select the data as per region (location), get its valid scope and then data with higher access probability will be selected for access so that most probable data item is fetched to the cache. The access probability is updated in database with each access to the cached data item.

When cache has not enough space to store queried data item then, space is created by replacing existing data items from cache based on minimum access probability (Pa_i), maximum distance (ds_i) and scope invalidation (vs_i). If data items have same valid scope then replacement decision is based on minimum access probability. If valid scope and access probability is same then replacement decision is based on maximum data distance. In some cases, data size plays an important role in replacement. If fetched data item size is large enough and requires replacing more than three data items from cache then replacement is based on maximum equivalent size with minimum access probability, maximum distance and scope invalidation. Fig. 2 shows the representation of valid scope and Table 1 gives data instance for the same. The advantage is the complete knowledge of the valid scopes.

Trips	
Trip No	Locations
TR_1	$\{L_3, L_2, L_1\}$
TR_2	$\{L_3, L_5, L_2, L_1, L_4\}$
TR_3	$\{L_4, L_5, L_2, L_1, L_5, L_4\}$
TR_4	$\{L_3, L_4, L_5, L_2, L_1\}$
TR_5	$\{L_1, L_4, L_2, L_5, L_4\}$

a) Trips

1st Order	L_1	L_2	L_3	L_4	L_5
$s_1=\{L_1\}$	0	0	0	2	1
$s_2=\{L_2\}$	4	0	0	0	1
$s_3=\{L_3\}$	0	1	0	1	1
$s_4=\{L_4\}$	0	1	0	0	2
$s_5=\{L_5\}$	0	3	0	2	0

b) First Order TPM

2nd Order	L_1	L_2	L_3	L_4	L_5
$\{L_3, L_5\}$	0	1	0	0	0
$\{L_4, L_2\}$	0	0	0	0	1
$\{L_4, L_5\}$	0	2	0	0	0
$\{L_5, L_2\}$	3	0	0	0	0
$\{L_5, L_4\}$	0	0	0	0	1

2nd Order	L_1	L_2	L_3	L_4	L_5
$\{L_1, L_4\}$	0	1	0	0	0
$\{L_1, L_5\}$	0	0	0	1	0
$\{L_2, L_1\}$	0	0	0	1	1
$\{L_2, L_5\}$	0	0	0	1	0
$\{L_3, L_2\}$	1	0	0	0	0
$\{L_3, L_4\}$	0	0	0	0	1

c) Second Order TPM

3rd Order	L_1	L_2	L_3	L_4	L_5
$\{L_1, L_4, L_2\}$	0	0	0	0	1
$\{L_1, L_5, L_4\}$	0	0	0	0	0
$\{L_2, L_1, L_4\}$	0	0	0	0	0
$\{L_2, L_1, L_5\}$	0	0	0	1	0
$\{L_2, L_5, L_4\}$	0	0	0	0	0
$\{L_3, L_2, L_1\}$	0	0	0	0	0

3rd Order	L_1	L_2	L_3	L_4	L_5
$\{L_3, L_5, L_2\}$	1	0	0	0	0
$\{L_4, L_2, L_5\}$	0	0	0	1	0
$\{L_4, L_5, L_2\}$	1	0	0	0	0
$\{L_5, L_2, L_1\}$	0	0	0	0	1
$\{L_5, L_4, L_5\}$	0	0	0	0	0

d) Third Order TPM

Fig. 1. Sample location trips with the corresponding 1st, 2nd and 3rd order Transition Probability Matrix

MMCRP cache replacement policy works as follows:

- All customer data.
- Customer specific data.
- Pruned data for customer specific trips.
- Predicate where the customer will be in the near future by first and second order Markov model.
- The valid scope of the data item.
- Access probability of the data item.
- Distance of the data item.
- Data size in the cache.

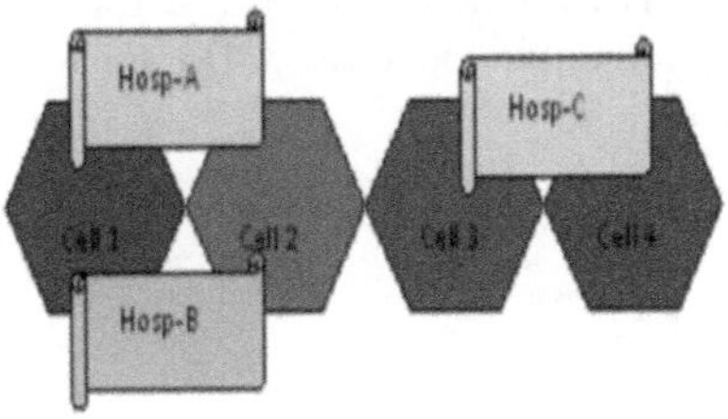

Fig. 2. Valid scope representation for Hospital

Table 1. Valid Scope for Hospital Data Instance

Data instance	Valid Scope
{ nearby-Hosp , {A,B} }	{ 1 , 2 }
{ nearby-Hosp , {C} }	{ 3 , 4 }

Client-side cache management algorithm:

```
Client requests for data item Di
 if Di is valid and in cache then
    validate and return the data item Di
else  if cache miss for data item Di then
          request for data item Di to DB server
          get data item Di from DB server
     if enough free space in cache then
          store data item Di  in cache
          update Pai
  else if not enough free space in cache then
          while ( not enough space in cache)
                        create enough space by replacing data
               item(s) from cache which has
                    ✓ Invalid scope (vsi)
                    ✓ Minimum Pai
                    ✓ Large distance (dsi)
    If (Multiple replacement )
                        Large size with scope invalidation.
            end
               Insert data item Di
                update Pai
        end
end
```

5 Performance Evaluation and Comparisons

For implementation of MMCRP the database is created with different regions with locations, location specific resources such as Hospital, Restaurant, ATM, Movies, Blood Bank, Police, Fire Station, Medical 24x7 and resources with different specialty such as Child and Maternity for hospital and as applicable for each resource. Data for user movement and query firing was collected. A data set consists of ten thousand records. Data for specific customer is obtained which is further pruned to form the data sets for evaluation. As Markov Model is computational intensive for large values of k, we have used only first and second order TPM with server side processing to reduce load of client processing. Fig. 3 shows a scenario of the request for information such as Food, ATM, Movies, Hospital etc. The client sends request for information. The server validates the client and for authenticated clients Database Server (DBS) responds with requested information.

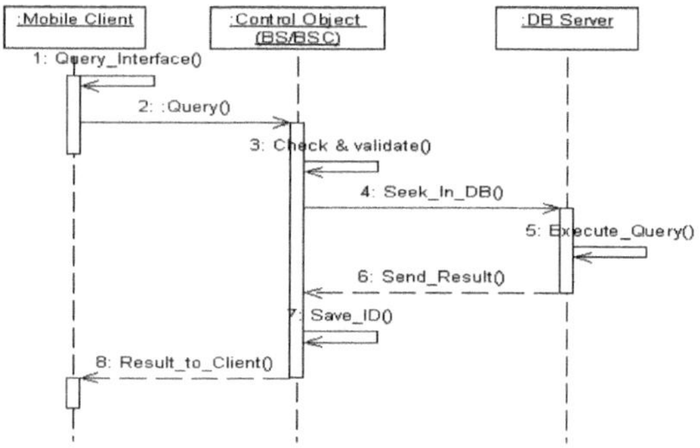

Fig. 3. Sequence diagram for Information Request

For our evaluation, the results are obtained when the system has reached the steady state, i.e., the client has issued at least 1000 queries, so that the warm-up effect of the client cache is eliminated. We have conducted experiments by varying the client speed, query interval and cache size. Query interval is the time interval between two consecutive client queries.

In order to predict and simulate the travel paths, close to realistic ones, some paths are chosen wherein no definite travel patterns are observed. One can say such mobile clients travel pattern show near random behavior. A typical output obtained from first and second order Markov model is shown in Fig. 4. a) and Fig. 4. b) with query interval from 10 to 200 seconds. Initially the result of PPRRP and MMCRP is same since client will be in the same predicated region if query interval is small. MMCRP outperforms when query interval is more because of accurate prediction of client movement based on historical data is better than PPRRP and far better than FAR and LRU.

The Fig. 5 depicts the effect of cache size on performance of LRU, FAR, PPRRP and MMCRP replacement policies. As shown and expected, the cache hit ratio of different policies increases with increase in cache since cache can hold more information which increases the probability of cache hit. The performance of MMCRP will be increased substantially with increase in cache size so this result could be used to decide the optimal cache size in the mobile client.

Clients' speed vs cache hit ratio is shown in Fig. 6. Four clients' speed ranges have been considered: 6~8 km/hr, 50~60km/hr, 100~150 km/hr and 400~500 km/hr for walking human, car, train and plain respectively. For higher speed range, the cache hit ration decreases since client spend less time at each location and first/second order model, valid scope becomes less effective.

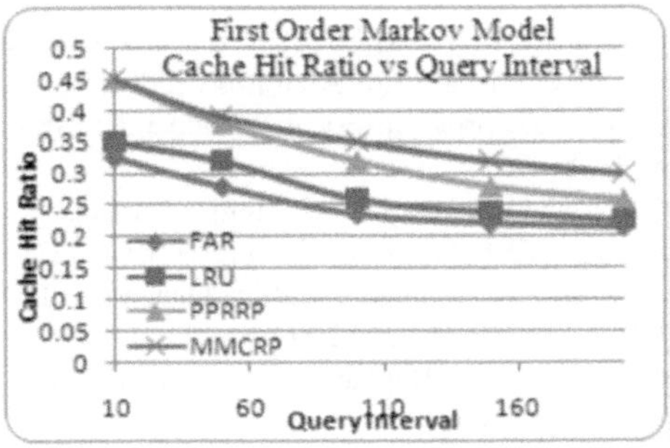

Fig. 4. a). Cache hit Ratio vs Query Interval Interval for first order

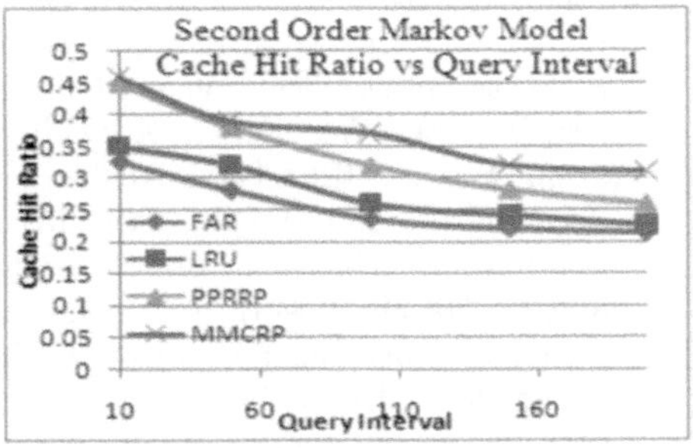

Fig. 4. b) Cache hit Ratio vs Query for second order

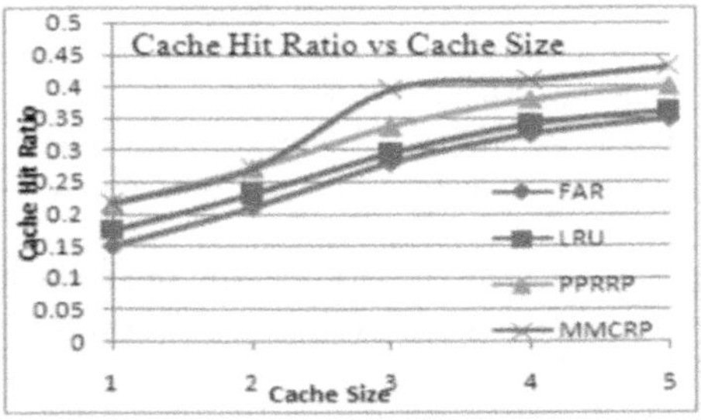

Fig. 5. Cache hit Ratio vs Cache Size

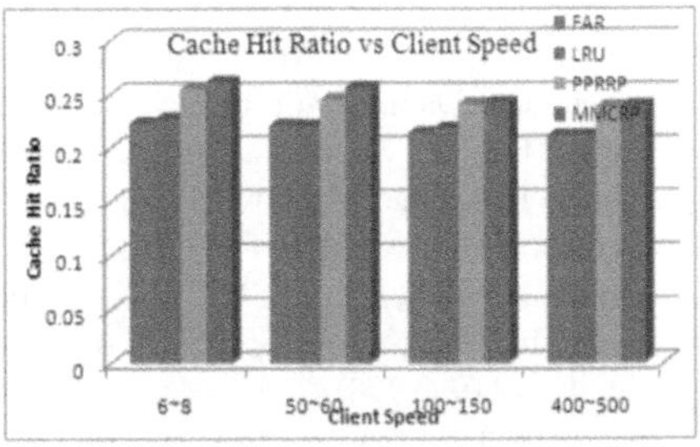

Fig. 6. Cache hit Ratio vs Client Speed

6 Conclusion

In this paper, we presented a cache replacement policy using first and second order Markov Model. The proposed MMCRP takes into account both the spatial and temporal properties of client movement and access patterns to improve caching performance. MMCRP takes into account valid scope, data size, data distance and access probability of data item for replacement. Whenever single (data item) storage results into multiple (three or more than three) replacements, we have considered as critical and treated differently for replacement. Consideration of already proven Markov model for prediction of client movement improves the performance. Simulation results for query interval, cache size and client speed show that the MMCRP has significant improvement in the performance than the LRU, FAR and PPRRP. In our future work we would like to incorporate a graph based Markov model for prediction and replacement.

References

1. Barbara, D.: Mobile Computing and Databases: A Survey. Proc. of IEEE Trans. on Knowledge and Data Engg. 1 1(1) (1999)
2. Kumar, A., Misra, M., Sarje, A.K.: A Predicated Region based Cache Replacement Policy for Location Dependent Data In Mobile Environment. IEEE, Los Alamitos (2006)
3. Lee, D.L., Lee, W.C., Xu, J., Zheng, B.: Data Management in Location-Dependent Information Services. IEEE Pervasive Computing 1(3) (2002)
4. Zheng, B., Xu, J., Lee, D.L.: Cache Invalidation and Replacement Strategies for Location-Dependent Data in Mobile Environments. Proc. of IEEE Trans. on Comp. 51(10) (2002)
5. Ren, Q., Dhunham, M.H.: Using Semantic Caching to Manage Location Dependent Data in Mobile Computing. Proc. of ACMIIEEE MobiCom, 210–221 (2000)
6. Balamash, A., Krunz, M.: An Overview of Web Caching Replacement Algorithms. Proc. of IEEE Communications Surveys & Tutorials 6(2) (2004)
7. O'Neil, E., O'Neil, P.: The LRU-k page replacement algorithm for database disk buffering. Proc. of the ACM SIGMOD, 296–306 (1993)
8. Kumar, V., Prabhu, N., Chrysanthis, P.K.: HDC- Hot Data Caching in Mobile Database System. IEEE, Los Alamitos (2005)
9. Getting, I.A.: The Global Positioning System. Proc. of IEEE Spectrum 12(30) (1993)
10. Kumar, A., Misra, M., Sarje, A.K.: A New Cache Replacement Policy for Location Dependent Data in Mobile Environment. IEEE, Los Alamitos (2006)
11. Li, K., Qu, W., Shen, H., Nanya, T.: Two Cache Replacement Algorithms Based on Association Rules and Markov Models. In: Proceedings of the First International Conference on Semantics, Knowledge, and Grid (SKG 2005)
12. Dunham, M.H., Kumar, V.: Location dependent data and its management in mobile databases. In: Proceedings of the 9th International Workshop on Database and Expert Systems, pp. 414–419 (1998)
13. Katsaros, D., Manolopoulos, Y.: Prediction in Wireless Networks by Markov Chains
14. Hiary, H., Mishael, Q., Al-Sharaeh, S.: Investigating Cache Technique for Location of Dependent Information Services in Mobile Environments. European Journal of Scientific Research 38(2), 172–179 (2009), ISSN 1450-216X
15. Mânica, H., de Camargo, M.S.: Alternatives for Cache Management in Mobile Computing. In: IADIS International Conference Applied Computing (2004)

Data Hiding in Gray-Scale Images Using Pixel Value Differencing

Shrikant S. Agrawal and Rahul M. Samant

MPSTME, NMIMS University, Mumbai, Maharashtra, India
Shrikantagrawal2003@gmail.com

Abstract. Steganographic method for embedding secret messages into a gray-valued cover image is proposed. It uses the fact that, Human visual system is having low sensitivity to small changes in digital data. It modifies pixel values of image for data hiding. Cover images is partitioned into non-Overlapping blocks of two consecutive pixels. Difference between the two consecutive pixel values is calculated. These difference values are classified into number of ranges. Range intervals are selected according to the characteristics of human vision's sensitivity to gray value variations from smoothness to contrast. A small difference value indicates that the block is in a smooth area and a large one indicates that the block is in an edged area. The pixels in edged area can tolerate larger changes of pixel values than those in the smooth area. So, in the proposed method we can embed more data in the edged areas than the smooth areas. The difference value then is replaced by a new value to embed the value of a sub-stream of the secret message. The number of bits which can be embedded in a pixel pair is decided by the width of the range that the **difference** value belongs to. The method is designed in such a way that the modification is never out of the range interval. This method not only provides a better way for embedding large amounts of data into cover images with imperceptions, but also offers an easy way to accomplish secrecy. This method provides an easy way to accomplish secrecy. This method provides an easy way to produce a more imperceptible result than those yielded by simple least-significant bit replacement methods. The embedded secret message can be extracted from the resulting stego-image without referencing the original cover image. Experimental results show the feasibility of the proposed method.

Keywords: Data, Experiment, Hiding, Image, Steganography.

1 Introduction

The explosion of Internet Applications leads people into the digital world, and transmission via digital data becomes frequent. Digital communication has become essential part of our life. Text, image, audio, and can be represented as digital data. Digital data is easily and rapidly available to everyone. This makes it ready for illegal use, misrepresentation of the original data. Security, privacy, and information protection are major concern in today's industrial informatics. Protection of during last few years such as data security in digital communication, copyright protection of digitized

K. Shah, V.R. Lakshmi Gorty, and A. Phirke (Eds.): ICTSM 2011, CCIS 145, pp. 27–33, 2011.

properties, invisible communication via digital media,etc. Information-hiding techniques have recently become important in a number of application areas. It is often thought that communication may be secured by encrypting the traffic, but this has rarely been in adequate practice. So the study of communication security includes not just encryption but also traffic security, whose essence lies in hiding information. This discipline includes technologies such as spread spectrum radio, which is widely used in military systems. Data hiding represents a class of processes used to embed data, such as copyright information, into various forms of media such as image, audio, or text with a minimum amount of perceivable degradation to the "host" signal; i.e., the embedded data should be invisible and inaudible to a human observer. One of the data hiding techniques is image watermarking. The main purposes of image watermarking include copyright protection and authentication. In such applications, it is required that the embedded information be unaltered, or altered only up to an acceptable degree of distortion, no matter how the watermarked image is attacked. Many other new applications of watermarking are also introduced such as broadcast monitoring, proof of ownerships, transactional watermarks, copy control, and covert communication. Data hiding in image can be done using spatial domain or transform domain techniques. Spatial domain techniques gives better data hiding capacity with less robustness and transform domain techniques gives less data hiding capacity with better robustness. LSB replacement, patchwork, masking and filtering are some of the examples of spatial domain techniques.

2 Proposed Data Embedding

Here for data embedding, a cover image is divided into a number of non-overlapping two-pixel blocks. Each block is categorized according to the difference of the gray values if the two pixels in the block. A small difference value indicates that the block is in a smooth area and a large one indicates that it is in edged area. The pixels in an edged areas mat tolerate larger changes of pixel values than those in the smooth areas. So, more data can be embedded in edged areas than in the smooth areas. And it is in this way the changes in the resulting stego-image are unnoticeable.

2.1 Quantization of Differences of Gray Values of Two-Pixel Blocks

The cover images are gray-valued ones. A difference value d is computed from every non-overlapping block of two consecutive pixels; say P_i and $P_{i+1,}$ of a given cover image. The way of partitioning the cover image into two-pixel blocks runs through all the rows of each image in zigzag manner, as shown in Fig. 1. Assume that the gray values of p_i and $p_{i+1}$are gi and g_{i+1} respectively, and then d is computed as g_{i+1}-g_i, which may be in the range from -255 to 255. A block with d close to 0 is considered to be an extremely smooth block, whereas a block with d close to -255 or 255 is considered as a sharply edged block. By symmetry, only absolute values of d(o through 255) are considered and classified into a number of contiguous ranges, say R_i where i =1,2,….,n. These ranges are assigned indices 1 through n. The lower and upper bound values of R_i are denoted by l_i and u_i, respectively, where l_i is 0 and u_n is 255. The width of R_i is u_i-l_{i+1}. The width of each range is taken to be a power of 2. This restriction of widths facilitates embedding binary data. The widths of the ranges which represent the difference values of smooth blocks are chosen to be smaller while those which represent the difference values of edged blocks are chosen to be larger. That is, ranges are created with smaller widths when d is close to 0 and with larger widths when d is far away from 0 for the purpose of yielding better imperceptible results. A difference value which falls in a range with index k is said to have index k.

All the values in a certain range (i.e., all the values with an identical index) are considered as close enough. That is, if a difference value in a range is replaced by another in the same range, the change presumably cannot be easily noticed by human eyes. Some bits of the secret message are embedded into a two-pixel block by replacing the difference value of the block with one with an identical index, i.e., a difference value in one range is changed into any of the difference values in the same range. In other words, the gray values in each two pixel pair are adjusted by two new ones whose difference value causes changes unnoticeable to an observer of the stego-image.

2.2 Data Embedding

Consider the secret message as a long bit stream. Every bit in the bit stream is embedded into the non overlapping two-pixel blocks of the cover image. The number of bits which can be embedded in each block is varied and is decided by the width of the range to which the difference value of the two pixels in the block belongs. Given a two pixel block B with index k and gray value difference d, the number of bits, say n, which can be embedded in this block, is calculated by

$$n=\log_2 (u_k-l_k+1) \tag{1}$$

Since the width of each range is selected to be a power of 2, the value of n is an integer. A sub-stream S with n bits is selected next from the secret message for embedding in B. A new difference d' then is computed by

$$d' = l_k + b \qquad \text{for } d >= 0 \tag{2}$$

$$d' = -(l_k + b) \qquad \text{for } d<0 \tag{3}$$

Where, b is the value of the sub-stream S. Because the value b is in the range from 0 to u_k - l_k, the value of d' is in the range from l_k to i_k. According to the previous discussions, if d is replaced with d' the resulting changes are presumably unnoticeable to the observer. Then b is embedded by performing an inverse calculation from d' described next to yield the new gray levels (g_i',g_{i+1}') for the pixels in the corresponding two pixel block (P_i',P_{i+1}') of the stego-image. The embedding process is finished when all the bits of the secret message are embedded. The inverse calculation for computing (g_i',g_{i+1}') from the original gray values (g_i,g_{i+1}) of the pixel pair is based on a function f((g_i,g_{i+1}),m) which is defined to be

$$f((g_i,g_{i+1}),m) = (g_i',g_{i+1}') \tag{4}$$

$$(g_i',g_{i+1}') = (g_i - \text{ceiling } m,\ g_{i+1} + \text{floor } m) \quad \text{If d is an odd number}$$

$$= (g_i - \text{floor } m,\ g_{i+1} + \text{ceiling } m) \quad \text{If d is an even number}$$

Where m=d'-d, ceiling m= [m/2], and floor m= [m/2] (5)

The above equation satisfies the requirement that the difference between g_i' and g_{i+1}' is d'. It is noted that a distortion reduction policy has been employed in designing equation (4) and (5) for producing g_i' and g_{i+1}' from gi and gi+1 so that the distortion over the so pixels in the block. The effect is that the resulting gray value change in the block is less perceptible.

2.3 Recovery of Embedded Data from Stego Image

The process of extracting the embedded message proceeds by using the same traversing order for visiting the two-pixel blocks as in the embedding process. Each time visit a two-pixel block in the stego-image and apply the same falling –off boundary checking as mentioned previously to the block to find out whether the block was used or not in the embedding process. Assume that the block in the stego-image has the gay values (gi*, gi+1*), and that the difference d* of the two gray values is with index k. Apply the falling-off boundary checking process to (gi*, g_{i+1}*) by using

$$f((g_i^*, g_{i+1}^*), u_k\text{-}d^*)$$

$$f((g_i^*, g_{i+1}^*), u_k\text{-}d^*)= (g^\wedge_i, g^\wedge_{i+1}) \qquad (6)$$

If either of the gray values of the computed values ($g^\wedge_i$*, $g^\wedge_{i+1}$*) falls off the range [0,255], then it means that the current block was not used for embedding data, or that the block was abandoned into the embedding process. On the contrary, if both of the values ($g^\wedge_i$*, $g^\wedge_{i+1}$*) do not fall of the range, it means that some data was embedded in the block. The value b, which was embedded in this two-pixel block, is then extracted out using the equation

$$b= d^*\text{-}lk \text{ for } d^*>=0 \qquad (7)$$

$$= \text{-}d^*\text{-}lk \text{ for } d^*<0 \qquad (8)$$

When we hide the data in the cover image means we are adding noise or distortion in that image. It is necessary to calculate Peak signal to noise ratio (PSNR) and root mean square error (RMSE).

$$MSE= 1/mxn \sum_{i=0}^{M-1} . \sum_{j=0}^{n-1} (\alpha_{ij} - \beta_{ij})^2$$

$$RMSE= \sqrt{MSE}$$

$$PSNR=10 \log_{10} [(255)^2 / MSE] \text{ dB}$$

Where

α_{ij} = Pixel of the cover image in which the coordinate is i, j.

$\beta_{i,j}$= Pixel of the stego image in which the coordinate is i, j.

(m x n)= Size of the cover and stego image.

3 Results and Discussion

3.1 Experimental Results

Here four cover images "Lena", "Peppers", "Cell" and "Mandrill" are used, each with size 512 X 512 as shown in Fig2. Four sets of widths of ranges of gray value differences are used in the experiments. The first experiment is based on selecting the range widths of 2, 2, 4, 4, 8, 8, 16, 16, 32, 32, 64, and 64, which partition the total range of [0 to 255], into [0,1],[2,3],[4,7], [8,11], [12,15], [16,23], [24,31], [32,47],[48,63],[64,95],[96,127], [128,191] and [192,255]. Let us say it as Range 1.

The second experiment is based on the use of the range widths of 4, 4, 8, 16, 32, 32, 64 and 64, which partition the total range of [0 to 255], into [0,3], [4,7], [8,15], [16,31],[32,63],[64,95],[96,127],[128,191], and [192,255]. Let us say it as Range2.

The third experiment is based on the use of range widths 8, 8, 16, 32, 64, and 138, which partition the total range of [0 to 255], into [0, 7], [8, 15], [16, 31], [32, 63], [64, 127], and [128, 255]. Let us say it as Range3.

The fourth experiment is based on the use of the range widths of 16, 16, 32, 64, and 128, which partition the total range of [0 to 255], into [0, 15], [16, 31], [32, 63], [64,127] and [128, 255]. Let us say it as Range 4.

The values of the capacities for embedding data by using the cover image and the four sets of range widths are given in Table 1. A word formatted file which consists of the Abstract of this paper is taken as the secret message in the experiments.

Table 1. Hiding capacity using pixel value differencing method

Cover image	Maximum hiding capacity in bytes			
	Embed using Range1	Embed using Range2	Embed using Range3	Embed using Range 4
Lena	28883	40497	51692.3	66074
Peppers	27496.9	38775.9	50684.9	65614
Cell	23159.6	36683.8	50282.3	59705.9
Mandrill	37189.6	49982	57116	67939

Resulting stego images enhanced differences images between the cover and stego images after data embedding are shown. One of the stego images resulting from the embedding the given secret data using the first set of range widths and corresponding enhanced difference images shown in figure 3(a) and 3(b). Range 2, Range 3 and Range 4 results are shown in figure 4, 5, 6 respectively.

The enhanced difference images are shown here to indicate the distortion resulting from data embedding process from them we see that most of the distortions are found on the edges in the images. This means that such distortion would be less noticeable because changes in edge parts of the images are generally less obvious to human eyes. For the purpose of comparison, one of the stego images resulting from embedding data into three LSBs of pixel values of cover images using conventional LSB replacement embedding steganographic technique and corresponding enhanced difference images are shown in figure 7(a) and 7(b). Here secret data is the abstract of this paper.

Fig. 1.

Fig. 2.

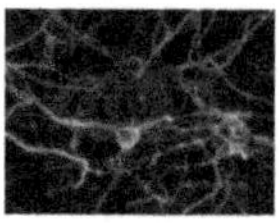

Fig. 3.

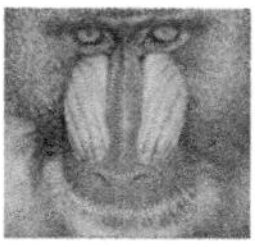

Fig. 4.

Fig. 3(a), 3(b), 4, 5(a) &5(b), 6(a) &6(b), 7(a) &7(b)

Table 2. Hiding capacity using LSB replacement method

Cover Image	Maximum hiding capacity in bytes			
	Embed using 1 LSB	Embed using 2 LSB	Embed using 3 LSB	Embed using 4 LSB
All cover Images	32768	65536	98304	131072

It is seen from these figures that distortions are spread equally in the image which are more obvious to observers than distortions resulting from pixel value differencing method which are limited essentially at edge areas, as shown in Figs 3, 4, 5 and 6.

3.2 Discussions

If we look at the stego images distortion are imperceptible to our eyesight. So simply looking at stego image you will not get any idea about secret communication via image.

Hiding capacity depends on the difference of the gray values of the two-pixels in the non-overlapping blocks of the cover image. Higher the difference values more will be the hiding capacity. Difference values are more in the edges of the cover image.

The enhanced difference images are shown here to indicate the distortions resulting from the data embedding process. Distortions are found on the edges of the images. These distortions are less noticeable because changes in the edge parts of the images are generally less obvious to human eyes. Variable numbers of bits are embedded into the blocks of two pixels. It does not replace the LSBs of pixel values directly; instead, it changes the differences of the two pixel values in a block. We cannot find obvious suspicious artifacts on the resulting stego images by simple visual inspection. Resulting stego images using Range1, Range2, and Range3 gives imperceptible results. Distortions are observed in stego images using Range4, but these distortions are less as compared to conventional LSB replacement method, the pixel value differencing method gives more imperceptible results.

4 Conclusions

An efficient computer-based steganographic method for embedding secret messages into images without producing noticeable changes is implemented. There is no need of referencing the original cover image while extracting the embedded data from a stego-image. This method utilizes the characteristic of the human vision's sensitivity to gray value variations from smoothness to contrast. Distortions in the cover images are less noticeable because changes in the edge part of the images are less obvious to human observer. The method not only provides a better way for embedding large amounts of data into cover images with imperceptions, but also offers an easy way to accomplish secrecy. This method provides better results as compared to LSB replacement method where the distortions are spread all over the image. The images are used for this algorithm are only 8-bit gray level images. It is suggested that this algorithm can be tried on color images, but the algorithm may get complicated.

This embedding method can be easily extended to efficiently carry content-related messages such as captions or annotations in audios and videos by embedding data in each adjacent data in each adjacent pair of signals of the data-streams.

References

1. Anderson, R.J., Petitcolas, F.A.P.: On the limits of Steganography. IEEE, 474–481 (1998)
2. Artz, D.: Digital Steganography:Hiding data within data. IEEE, 75–80
3. Delp, J.E, Lin, E.T.: A review of Data hiding in Digital images (n.d.)
4. Johnson, N.F., Jajodia, S.: Exploring Steganography: Seeing the unseen. IEEE Computer, 26–64 (February 1998)
5. Kharrazi, M., Sencar, H.T., Memon, N.: Image Steganography: Concepts and Practice
6. Petitcolas, F.A.P., Anderson, R.J., Kuhn, M.G.: Information hiding- a survey (1999)
7. Gonzalez, R.C.: Digital Image Processing, 3rd edn. Addision-Wesley Publishing Company, Reading
8. Gonzalez, R.C., Woods, R.E.: Digital Image Processing using MATLAB, 3rd edn. Addision-Wesley Publishing Company, Reading

Halftone Image Data Compression Using Kekre's Fast Code Book Generation (KFCG) Algorithm for Vector Quantization

H.B. Kekre[1], Tanuja K. Sarode[2], Sanjay R. Sange[1], Shachi Natu[3], and Prachi Natu[4]

[1,2] MPSTME, NMIMS University Mumbai-400056
[2,3] TSEC, Mumbai University
[4] G.V.A.I.E.T, Mumbai University
sanjay.sange@rediffmail.com

Abstract. Halftone technique is well known for printing where binary data is required. 8:1 compression ratio is achieved by half toning method. To get higher compression ratio the same technique can be used in image processing. In our earlier work different half toning operators are proposed. The half toning operator of size 3X3 which effectively take only one tap operation. Hence the computational complexity and memory space is reduced. For further compression of the image data Vector Quantization technique is used. Vector Quantization technique itself gives very high compression ratio. To avoid time and computational complexity Kekre's Fast Code Book Generation (KFCG) algorithm is used. In this paper we have used 8, 16, 32, 64, 128 and 256 codebook sizes are used. The pixel group of 2X2 size is used. The experimental results are obtained for various images. For image data compression and image quality measurement, we have used Compression Ratio and Mean Square Error as measuring parameters respectively. We have got result with good Compression Ratio and acceptable image quality. This proposed combination of two compression technique is suitable for video data streaming, where low bit rate for data transmission is the major constraint.

Keywords: Key words: Halftone, Vector Quantization, Code Book, and Code Vector.

1 Introduction

Now a day's extensive use of Multimedia application is continuously increasing. Multimedia application contents video, audio, graphics etc. objects. These applications require a lot of data over internet. Hence to reduce the bandwidth, it is essential to compress such huge data. So half toning technique is one of the lossy compression technique provide 87.5% compression ratio Various half toning operators are presented and each one has its own significance [1].Generally Half toning technique is used in printer, fax machines, advertising techniques and so on. Standard half toning techniques like Floyd-Stenberg and Jarvis operators are presented in [2]-[5]. Screened halftones and error diffused halftones have greatly differing artifacts. In this paper,

K. Shah, V.R. Lakshmi Gorty, and A. Phirke (Eds.): ICTSM 2011, CCIS 145, pp. 34–42, 2011.

the main focus is on error diffusion (ED). There are various techniques of error diffusion. In this paper we have used the low computation error diffusing operator [6].

Halftones and other binary images are difficult to process without causing severe degradation. Exceptions include cropping, rotation by multiples of 90, and logical operations. Halftones are difficult to compress lossless. Grayscale images, on the other hand, can be compressed efficiently [7], [8]. Many methods show good results, but several are iterative, which require large amounts of computation and memory. Most also make heavy use of floating-point arithmetic operations. Different types of half toning operators are presented in [6]. The paper concludes that these all half toning operator gives better results depend upon application. Jarvis half toning operator is used to process color image presented in [9].

This paper will discuss how to –

(1) Convert color image into halftone image.

(2) To get higher compression ratio using Kekre's Fast Code Book Generation (KFCG) algorithm an innovative Vector Quantization technique.

In this paper, we have introduced the combination of halftone and Vector Quantization technique compression techniques.

Vector Quantization algorithms are expressed in [10] and [14].Time complexity of Euclidian distance for codebook generation is presented in [15] and [16]. Kekre's Fast Code Book Generation (KFCG) algorithm is used to reduce time and computational complexity.

In this paper, section 2 and section 4 is the implementation of Halftone method and KFCG Algorithm. Section 3 gives the brief idea about vector quantization. In section 5 proposed algorithm is explained. In section 6 experimental results in the form of images and measuring parameters like Mean Square Error and Compression Ratio are given in Table-1. Section 7 gives parameter comparison in graphical format and brief discussion about the same. Conclusion of the paper is explained in section 8. Section 9 is the paper references.

2 Half Toning Metohd

The half toning operator shown in Fig.1 is operated on continuous image to obtain halftone image. The pixel value in each plane of image is represented by 8-bit. While the pixel value in halftone image is represented by 1-bit. In this way we get 8:1 compression ratio. In Fig.1 'X' represent the central pixel.

0	0	0
0	X	1
0	1	3

Fig. 1. Proposed Halftone operator

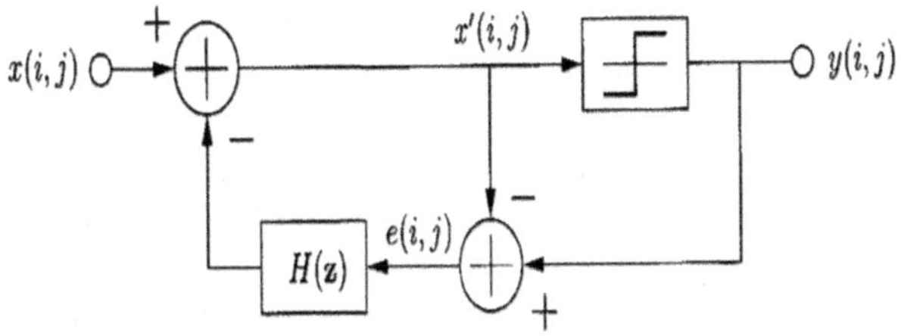

Fig. 2. Equivalent circuit of error diffusion

Floyd-Stenberg and Jarvis operator requires effectively 3-tap and 10-tap computations respectively [6]. Whereas the half toning operator is shown in Fig.1 is 3X3 in size requires the 3-tap, effectively 1-tap operation that it reduces the time and computational complexity. After operation of half toning operator on image, quantization is to be done as shown in Fig.2. Quantization process introduces error called blue noise that degrades the image quality.

3 Vector Quantization

A vector quantization function maps k-dimensional vectors in the vector space Rk of finite set CB = {C1, C2, C3,. . ., CN} Where CB is called as Code Book and it consist of N number of code vectors and k is the dimension of each code vector. Individual code vector is referred as Ci = {ci1, ci2, ci3, ..., cik} is of dimension k. Divide entire image into non-overlapping blocks and each block is treated as training vectors Xi = (xi1, xi2,..., xik). Table-1 shows the experimental results of block size2X2. Exhaustive search method is used which is optimal with high computational complexity in codebook to find nearest code vector. Code vector Cmin can be computed by using squared Euclidian distance-

$$d(Xi, Cmin)=\min 1\leq j\leq N \{d(Xi, Cj)\} \quad (1)$$

Where d(Xi, Cj) = ∑(xip, cjp)2

To obtain nearest code vector for training vector, it requires N number of Euclidian Xi distance computations where N is the size of the codebook. So for M image training vectors will require M x N number of Euclidian distance computations expressed by eq. (1). The size of the code book is represented by 2^n, where n=1, 2,., n. The code book size increases, the search time increases, on contrary to that Compression Ratio (CR), MSE get decreases with improvement in image quality.

The Encoder takes code vector as an input and search the codebook. Instead of sending that code vector on channel, search engine search the appropriate index from the codebook. This index can be send on channel, which is the compressed form of image. Decoding process in Vector Quantization is just selecting appropriate vector from the code book which is equivalent to that of input vector at Encoder. The code book size is represented by 2^n and ‘n’ is the index size (e.g. code book size (CB-8), 2^3 =8, therefore index size is of 3-bit only). Compression of image data is the main objective to improve upon some transmission parameters like bandwidth, low bit rate, low distortion rate as well memory storage.

4 Kekre’s Fast Codebook Genration (KFCG) Algorithm

Generation of code book with minimum computations that reduces time is the major task in this technique. Computation of Euclidian distance for codebook generation requires time that can be saved using algorithm presented in [15] and [16]. Instead of computing Euclidian distance, the whole image is divided into non-overlapping same blocks. Each block is converted to the vectors of size k. Consider the row wise pixel values, we will get a training vector. In this way, the entire image is represented as a cluster. Taking an average of first column a code vector C_1 will be the centroid. Generation of code book using sorting method is explained in paper [17]. Simple and efficient algorithm for codebook search, which reduces the computations for full code book search [18]. Kekre's fast codebook generation algorithm is used for segmenting the low-altitude aerial images. This technique is used as preprocessing step to form segmented homogeneous regions [19].

Bi-level codebook generation algorithm is used, which reduces mean squared error (MSE) for the same codebook size [20]. The novel technique for image retrieval using the color texture features extracted from images based on vector quantization with Kekre's fast codebook generation is proposed [21]. Segmenting the mammographic images into homogeneous texture regions representing disparate tissue types is often a useful preprocessing step in the computer-assisted detection of breast cancer. Such images get segmented using vector quantization. Kekre's Fast Codebook Generation algorithm (KFCG) is used for segmentation of mammographic images [22].

The entire cluster is split into two clusters by comparing first element of training vector with first element of code vector C1 [19]. The vector Xi is grouped into the cluster 1 if xi1< c11 otherwise vector Xi is grouped into cluster 2 as shown in Fig.3 where code vector dimension space is 2.

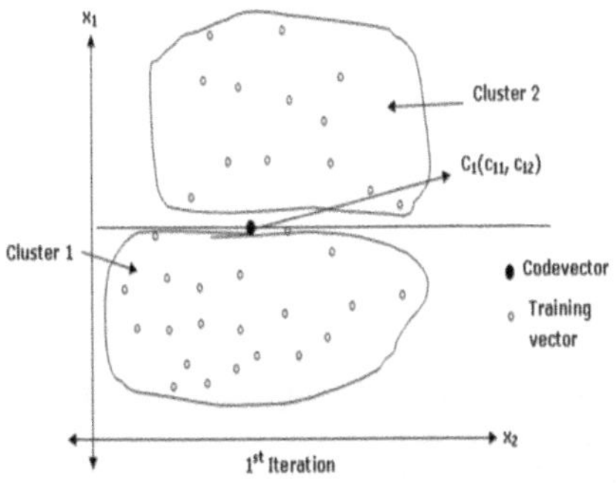

Fig. 3. Iteration1, Cluster 2

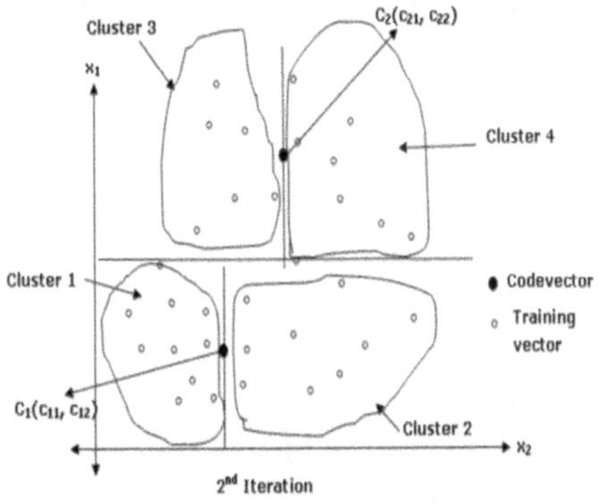

Fig. 4. Iteration2, Cluster 4

The two clusters are split into four clusters in the next iteration by comparing second element xi2 of vector Xi belonging to cluster 1 with that of the second element of the code vector which is centroid of cluster 1. Cluster 2 is split into two by comparing the second element xi2 of vector Xi belonging to cluster 2 with that of the second element of the code vector which is centroid of cluster 2, as shown in Fig.4.The same process is repeated till to get the desired codebook size. In this paper 8, 16, 32, 64, 128, and 256 codebook sizes are used.

5 Proposed Algorithm

To obtain the higher compression ratio the combination of Halftone and Vector Quantization is proposed. Detail explanation of the algorithm is given in [9].

Step1:- Split the color image into three primary color components. Apply the proposed operator as shown in Fig.1 on individual plane separately.

Step2:- Concatenate all the three plane so as to get color halftone image. This color halftone image is the input for further compression. Halftoning technique gives compression ratio 8:1. This ratio is further improved by using KFCG vector quantization.

Step3:- Apply Vector Quantization compression technique on individual plane to improve compression ratio.

Step4:- At the receiving end the decoder phase of Vector Quantization decodes image by selecting the appropriate code vector represented by index.

Step5:- To reproduce the color halftone image concatenate all the three planes.

6 Result

Mean Square Error (MSE) and Compression ratio (CR) are the performance measuring parameter for various Code Book (CB) sizes. Table-1 shows the MSE and average MSE (Avg. MSE) for different images and Compression ratio for various code book sizes. The sample original image of Rohit is shown in Fig.5a. Fig.2 is the halftone image. From Fig. 5c to Fig.5h are subsequent vector quantized different code book size images.Fig.6 to Fig.8 shows the results for five sample images comparing MSE, Avg. MSE, and Compression Ratio with code book sizes.

$$\text{Compression ratio (CR)} = \frac{512x512x3x8}{(CBx12x8) + Mx3} \quad \text{.......... (2)}$$

$$M = \frac{512x512}{2x2} \quad \text{....... (3)}$$

Where, CB stands for Code Book size and M stands for entire image (N x N size) is divided into number of blocks with group of pixels block size 2x2.

Equation number (2) gives the image data in bit is (512x512x3x8), where 512x512 is the number of rows and number of columns in image.

Fig. 5a. Original Image

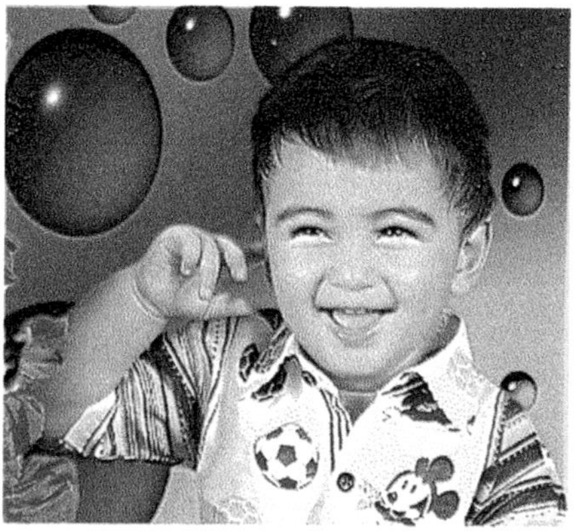

Fig. 5b. Halftone Image

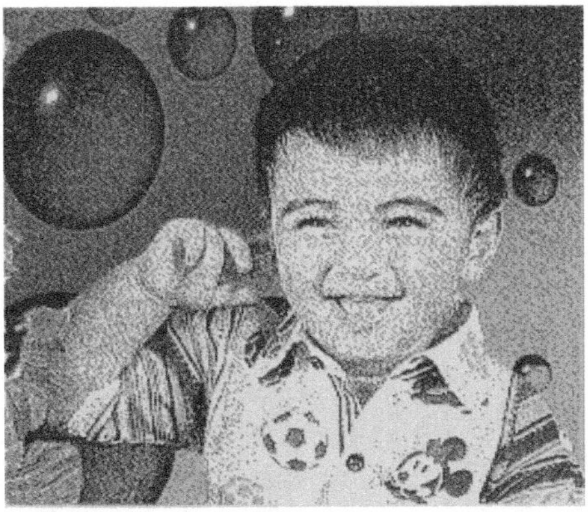

Fig. 5c. Vector Quantized Code Book size-8

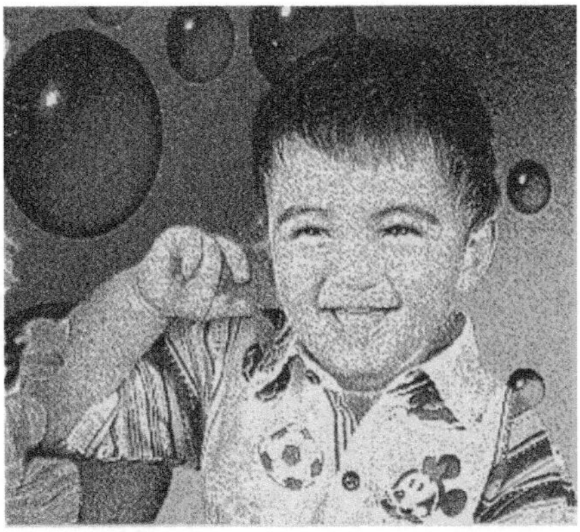

Fig. 5d. Vector Quantized Code Book size-16

Fig. 5e. Vector Quantized Code Book size-32

Fig. 5f. Vector Quantized Code Book size-64

Fig. 5g. Vector Quantized Code Book size-128

Fig. 5h. Vector Quantized Code Book size-256

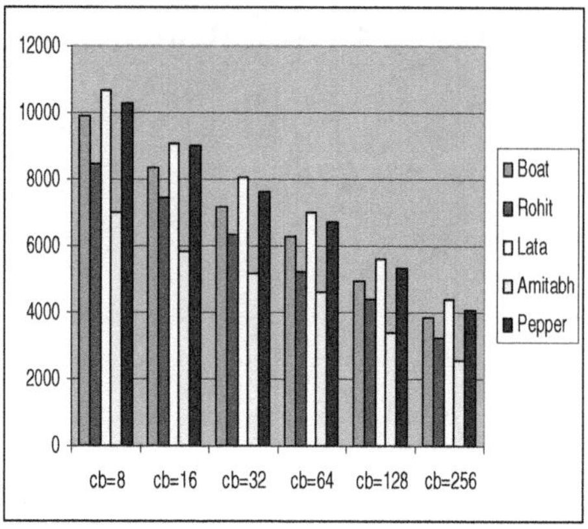

Fig. 6. Code Book size Vs. MSE

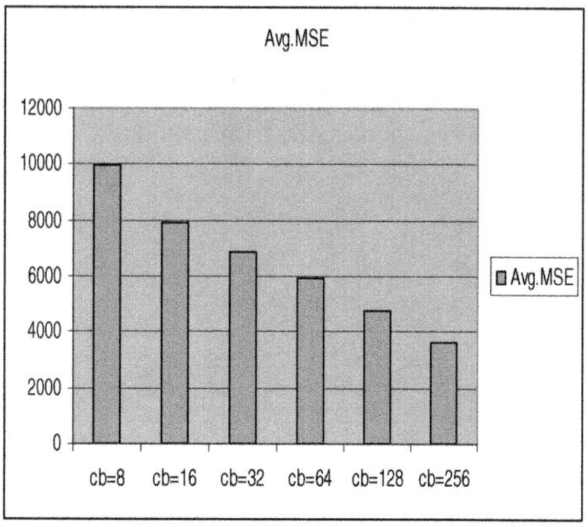

Fig. 7. Code Book size Vs. Avg. MSE

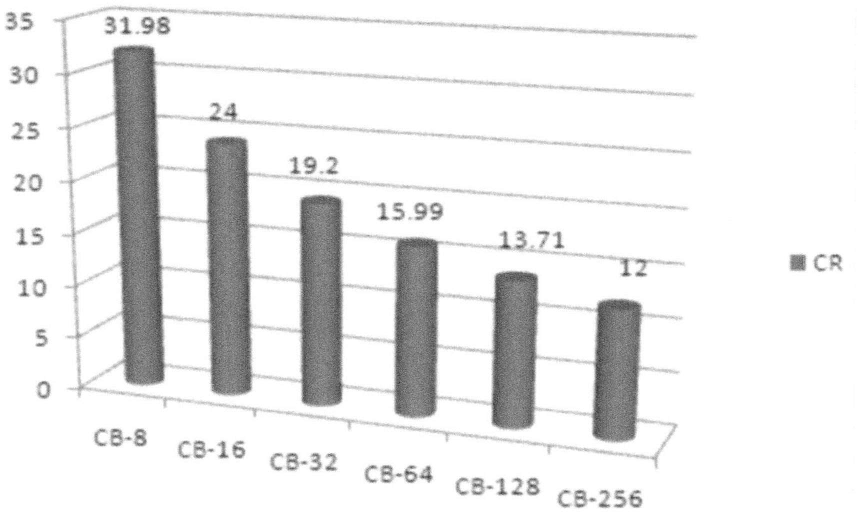

Fig. 8. Code Book size Vs. Compression Ratio

Table 1. Shows MSE, Avg. MSE, and CR for varing code book sizes 8,16, 32, 64, 128 and256

Image	CB - 8	CB-16	CB-32	CB-64	CB-128	CB-256
Boat	9873	8318	7177	6294.1	4968.7	3815
Rohit	8471	7432	6337	5213.4	4367.2	3249
Lata	10654	9067	8033	6984.5	5605.8	4371
Amitabh	6983	5854	5160	4602.1	3372.6	2567
Pepper	10296	8982	7626	6709.9	5319	4072
Avg:MSE	9256	7931	6867	5960.8	4726.7	3615
CR	**31.98**	**24**	**19.2**	**15.99**	**13.71**	**12**

Fig.5a shows Original Image Rohit, Fig.5b is the Halftone image and from Fig.5c to Fig.5h shows the Vector Quantized images for code book size varies from 8, 16, 32, 64, 128 and 256 respectively.

Fig.6 shows plot of MSE for images Boat, Rohit, Lata, Amitabh and Pepper w.r.t. variable code book sizes from 8, 16, 32, 64, 128 and 256. Fig.7 shows plot of Code Book size Vs Avg. MSE. Fig.8 shows plot of Code Book size Vs Compression Ratio.

7 Discussion

From images Fig.5a to Fig.5h, it is observed that the quality of image improves as the code book size increases. Instead of 2X2, if 3X3, 4X4 block of pixels is considered then the vector dimension 'k' will change to 27, 28 respectively. This higher vector dimension will give a patchy look. Fig.6 shows the variation of MSE w. r. t. code book size for sample images. For experiment sample images Rohit, Boat, Lata, Amitabh and Pepper are used. The Fig.7 and Fig.8 shows, as the code book size increases the MSE and Compression Ratio gradually decreases.

8 Conclusion

So far the proposed operator gives less computations and memory space. With the combination of proposed technique we got higher compression ratio and acceptable image quality. This technique is suitable for video data streaming where low bit rate of data transmission can be achieved. The study is being undertaken on video signal processing for video conferencing. Future work in the continuation to this experiment is to get Inverse of halftone image. As well as group size of pixels can be increased to 3X3 or 4X4 and so on to get higher compression ratio with degradation in image quality. Another scope is to use KFCG for higher size code book generation e.g. code book size 512 and 1024.Halftoning is extensively used in printing images, where large number of images are to be stored in halftone form. This technique will reduce the storage space requirement considerably.

References

1. Sange, S.: A Survey on: Black and White and Color Half toning Techniques, SVKM's NMIMS University, MPSTME. Journal of science, Engineering & Technology Management "Techno-Path" 1(2), 7–17 (2009), ISSN: 0975-525X
2. Floyd, R.W., Steinberg, L.: An adaptive algorithm for spatial grayscale. Proc. SID 17/2, 75–77 (1976)
3. Wong, P.: Inverse half toning and Kernel estimation for error diffusion. IEEE Trans. Image Processing 4, 486–498 (1995)
4. Hein, S., Zakhor, A.: Halftone to continuous–tone conversion of Error-diffusion coded images. IEEE Trans. Images Processing 4, 208–216 (1995)
5. kite, T.D., Evans, B.L., Bovik, A.C.: Modeling and Quality Assessment of Half toning by Error Diffusion. IEEE Transaction on Image Processing 9(5) (May 2000)

6. Sange, S.R.: Image data compression using new Halftoning operators and Run Length Encoding. In: 1st International Conference Thinkquest 2010, pp. 224–230. Springer Explorer and Springer CS Digital Library (February 2010)
7. Neuhoff, D., Pappas, T.: Perceptual coding of images for halftone display. IEEE Trans. Image Processing 3, 1–13 (1994)
8. Ting, M., Riskin, E.: Error-diffused image compression using a binary-to-grayscale decoder and predictive pruned tree-structured vector quantization. IEEE Trans. Image Processing 3, 854–858 (1994)
9. Sange, S.R.: Restoration of Color Halftone image by using Fast Inverse Half toning Algorithm. In: 2009 International Conference on Advances in Recent Technologies in Communication and Computing, pp. 650–655. IEEE, Los Alamitos (2009), doi:10.1109/ARTCom.2009, ISBN: 978-0-7695-3845-7/09 $25.00
10. Kekre, H.B., Sarode, T.K.: New Fast Improved Clustering Algorithm for Codebook Generation for Vector Quantization. In: Proc. of Int. Conf. ICETAETS, January 13-14. Saurashtra University, Gujarat (2008)
11. Gray, R.M.: Vector Quantization. IEEE ASSP Magazine, 4–29 (April 1984)
12. Begum, M., et al.: An Efficient Algorithm for Codebook Design in Transform Vector Quantization. In: WSCG 2003, February 3-7 (2003)
13. Vasuki, A., Vanathi, P.T.: Image Compression Using Lifting and Vector Quantization. ICGST International Journal on Graphics, Vision and Image Processing (GVIP), Special Issue on Image Compression 7, 73–81 (2009)
14. Hikal, N.A., Kountchev, R.: A Method for Digital Image Compression with IDP Decomposition Based on 2D-SOFM VQ. ICGST International Journal on Graphics, Vision and Image Processing (GVIP), Special Issue on Image Compression 7, 32–42
15. Kekre, H.B., Sarode, T.K.: New Fast Improved Codebook Generation Algorithm for Color Images using Vector Quantization. International Journal of Engg. & Tech. 1(1), 67–77 (2008)
16. Kekre, H.B., Sarode, T.K.: Fast Codebook Generation Algorithm for Color Images using Vector Quantization. Int. Journal of Computer Sci. and Information Technology 1(1), 7–12 (2009)
17. Kekre, H.B., Sarode, T.K.: An Efficient Fast Algorithm to Generate Codebook for Vector Quantization. In: First International Conference on Emerging Trends in Engineering and Technlogy, at G. H. Raisoni College of Engineering, Nagpur on July 16-18 (2008), Cited online at IEEE Xplore , ACM Portal
18. Kekre, H.B., Sarode, T.K.: Fast Codebook Search Algorithm for Vector Quantization Using Sorting Technique. In: ACM International Conference on Advances in Computing, Communication and Control (ICAC3), Fr. CRCE Mumbai, January 23-24 (2009); Available on ACM Portal
19. Kekre, H.B., Sarode, T.K.: Color Image Segmentation Using Kekre's Fast Codebook Generation Algorithm Based on Energy Ordering Concept. In: ACM International Conference on Advances in Computing, Communication and Control (ICAC3), Fr. CRCE Mumbai, January 23-24 (2009); Available on ACM Portal

Designing a Dynamic Job Scheduling Strategy for Computational Grid

Varsha Wangikar[1], Kavita Jain[2], and Seema Shah[3]

[1] Lecturer K.C. College of Engineering, Thane
varshawangikar@gmail.com
[2] Lecturer Xavier Institute of Engineering, Mumbai
kavita.rjain@gmail.com
[3] Asst. Prof. Vidyalankar Institute of Technology, Mumbai
seema.shah@vit.edu.in

Abstract. Grid computing is an emerging computing infrastructure which is a collection of computing resources connected by a network, to form a distributed system used for solving complex problems. This system tries to solve these problems or applications by allocating the idle computing resources over a network or the internet commonly known as the computational grid. In Computational Grid main emphasis is given on performance in terms of Execution time. Resources in Computational Grid are heterogeneous and are owned and managed by other organizations with different access policies. This can be achieved by Scheduling. Scheduling algorithms increases performance by reducing the execution time so Scheduling is an important issue. Scheduling is the decision process by which application components are assigned to available resources to optimize various performance metrics. Hence in this paper we have specifically focused on improving Computational grid performance in terms of Execution time. We have first presented a concept of Scheduling in Computational grid. Followed by detailed analysis of two scheduling strategies simulated by GridSim. Next we have proposed a dynamic Scheduling strategy based on this analysis. We are sure that this will improve the performance by reducing execution time.

Keywords: Grid Computing, **Computational** Grid, Scheduling, GridSim etc.

1 Introduction

The Grid is emerging as a new paradigm for next-generation computing. It support resource sharing and coordinated problem solving in dynamic, multi-institutional Virtual Organization. In majority of organizations, there are large amounts of underutilized computing resources existing. Most desktop machines are busy less than 5 percent of the time[1],[2]. Grid computing provides a framework for exploiting these underutilized resources and thus has the possibility of substantially increasing the efficiency of resource usage. One of the important categories of grid is Computational Grid. It provides secure access to a huge pool of shared processing power required

K. Shah, V.R. Lakshmi Gorty, and A. Phirke (Eds.): ICTSM 2011, CCIS 145, pp. 43–48, 2011.

for high throughput applications in homogenous and heterogeneous environment. Computational Grid appears to be a promising trend for three reasons: (1) its ability to make more cost effective use of a given amount of computer resources, (2) as a way to solve problems that can't be approached without an enormous amount of computing power, and (3) because it suggests that the resources of many computers can be cooperatively and perhaps synergistically harnessed and managed as a collaboration toward a common objective. To achieve the promising potentials of Computational Grids, an effective and efficient scheduling system is fundamentally important.

The resources in grid are heterogeneous in terms of their architecture, power, configuration, and availability. They are owned and managed by different organizations with different access policies that vary with time, users, and priorities. Different applications have different computational models that vary with the nature of the problem. It also offers an incentive to resource owners for sharing resources on the Grid and end-users trade-off between the timeframe for result delivery and computational expenses. This can be achieved with the help of scheduling only. So Scheduling is the most important issue in Computational Grid.

In this paper we focus on Scheduling strategy to give a better performance in Computational Grid. This paper is ordered as follows. Section 2 and 3 will be discussed about detail concept of Scheduling in Computational Grid followed by a brief explanation of simulation with GridSim. Section 4 focuses on result analysis of Scheduling algorithms using GridSim. Based on this discussion the design of the dynamic Scheduling strategy is explained in sections 5. Finally we end the paper with concluding remarks.

2 Scheduling in Computational Grid

Scheduling acts as a “resource selector”. It is a gateway to the grid by selecting resources from a global directory and then allocating the job to one of the available grid nodes. Grid Scheduling is the decision process by which application components are assigned to available resources to optimize various performance metrics.

Grid scheduling is defined as the process of making scheduling decisions involving allocating jobs to resources over multiple administrative domains [3]. This can include searching multiple administrative domains to use a single machine or scheduling a single job to use multiple resources at a single site or multiple sites. A logical grid scheduling architecture is described as shown in figure 1. Grid Scheduler (GS) or Grid Resource Broker (GRB) receives applications from grid users, selects feasible resources for those applications according to acquired information from the Grid Information Service (GIS) and finally generates application-to-resource mappings.

Grid Schedulers usually cannot control grid resources directly and are not necessarily located in the same domain with the resources, which are visible to them. GIS specify CPU capacities, memory size, network bandwidth, software availabilities and load of a site in a particular period. On the basis of application properties and performance of a job with resource, cost estimation computes a cost. A Local Resource Manager (LRM) responsible for local scheduling as well as reporting resource

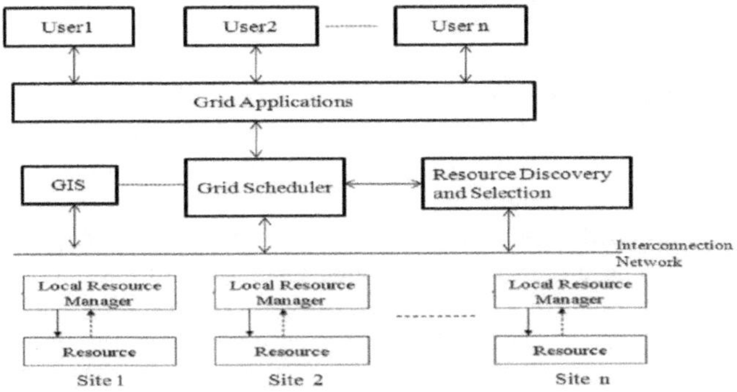

Fig. 1. Scheduling Process

information to GIS [5]. Job Scheduler is a service responsible for queuing jobs, allocating resources, dispatching the jobs to the compute nodes, and monitoring the status of the job, and nodes.

A proper scheduling algorithm can lead to an improved overall system performance and a lower turnaround time. Scheduling algorithms in Grid fall into a hierarchy as Static scheduling and dynamic scheduling. In Static scheduling, resources and jobs are fixed and each job is assigned to a resource. Thus the placement of an application is static, and a firm estimate of the cost of the computation can be made in advance of the actual execution. Dynamic scheduling is applied when it is difficult to estimate the cost of applications, or jobs are coming online dynamically. Dynamic scheduling is advantageous when the system need not be aware of the run-time behavior of the application before execution. It is particularly useful in a system where the primary performance goal is maximizing resource utilization, rather than minimizing runtime for individual jobs, but difficult to implement. Contrary, one of the major benefits of the static model is that it is easier to program from a scheduler's point of view as details of resources and job are known. Some dynamic and static scheduling algorithms are Minimum Completion Time (MCT), Min-Min, Max-Min [4], BLBD [5], OD [6] etc.

3 Simulating Computational Grid Using GridSim

Grid environment is a complex and challenging process, these systems offer speedup in computational performance. Therefore, researchers need to ensure that their newly designed systems are feasible and can perform as expected before proceeding on with the actual development. In Computational Grid environment, it is hard and even impossible to perform scheduler performance evaluation in a repeatable and controllable manner as resources and users are distributed across multiple organizations with their own policies. To overcome this limitation, a Grid simulator is used.

GridSim is a Java-based simulation toolkit based on the SimJava library [7]. It was designed specifically to analyze and compare the performance of resource scheduling algorithms for the grid. It supports static and dynamic scheduler. GridSim simulates different Scheduling strategies.

4 Result Analysis of Scheduling Algorithms Using GridSim

GridSim Version 4.0 simulator is installed on Windows XP platform. Various Scheduling strategies are simulated by GridSim simulator. Two strategies Space based and Time based are analyzed in terms of number of jobs which users want to execute, resources available and job length.

In Computational Grid important performance metric is time required to complete execution of a job. The effect of performance has studied by simulating the system consisting of 10 jobs each of size 500 MI with 1 resource. For these variables we simulate both the strategies. The result is tabulated in following graph.

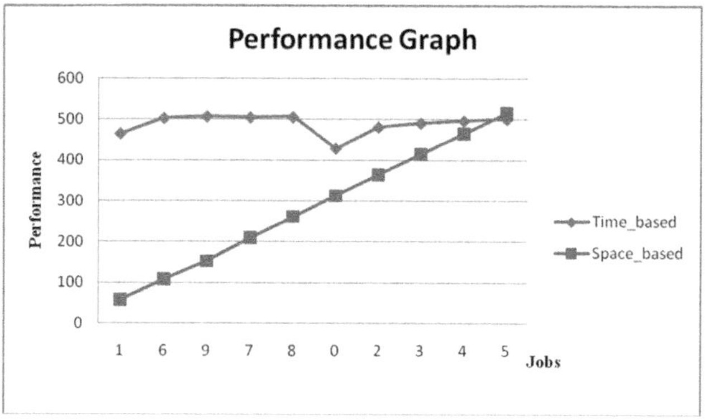

Fig. 2. Performance based on Space based and Time based algorithms

Analysis of Graphs

Performance metric such as execution time for all jobs on Grid is observed for. Time based and Space based methods. In above graph Y-axis shows the execution time with unit MIPs(millions of instructions per second) and position of job on X-axis.

To study the effect of performance we simulate these job in a simulated environment. In Space based when job arrives, it starts execution immediately if resource is available. Whereas in Time based when jobs arrive, it starts their execution immediately and share resources among all jobs. As we can observe that Space based method gives better performance for initial jobs compare to Time based method.

5 Proposed Algorithm

Based on the analysis of two strategies we argue that when resources are less than number of jobs then initial jobs in a queue gives better result for Space based strategy. For rest of the jobs in queue Time based will be suitable. we have designed an algorithm which will switch between these two algorithms refering previous performance results. A threshold value will be evaluate based on these results. Therefore Space based is always better for jobs before threshold as waiting time is less. The remaining jobs will give better performance for Time based as waiting time is reduced.

Algorithm:

Step 1: Create a list of all individual requests by validating the user specification(s). i.e Jobs= i
Step 2: Create list to obtain a list of available resources. i.e. resource=j
Step3: Specify information of fixed parameters like 5000 bandwidth etc.
Step 4: For each job user[i] does the following steps.
Step 5 Specify length of Gridlet in MI.
Step 6: Calculate threshold (Value not yet finalized)
Step 7: For each combination of job i and resource j compare position of each job for different job_position. If it is less than Threshold then the job will execute with Space_based.
Step 8: Else execute with Time_based.
Step 9: If the list is empty, then the algorithm terminates, otherwise go to Step7.

The layered structure will be as given below.

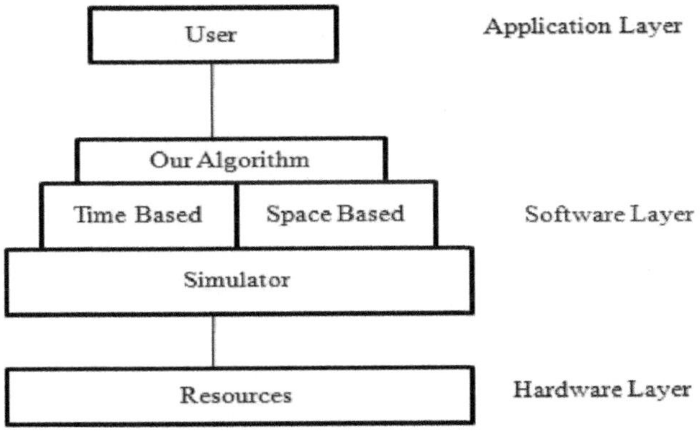

Fig. 3. Layered structure

The proposed dynamic scheduling algorithm is a software layer on GridSim toolkit, not as a separate layer but running with GridSim toolkit. Based on the inputs given by the user at runtime our proposed strategy switches between Space based and Time based, thus each job will execute with minimum execution time increasing performance in Computational Grid.

6 Conclusion

The motivation of Grid computing is to aggregate the power of widely distributed resources, and provide non-trivial services to users. To achieve this goal, an efficient Grid scheduling system is an essential part of the Grid. Scheduling is the decision process by which application components are assigned to available resources to optimize various performance metrics. Performance metric such as execution time for all jobs on Grid is observed for various Scheduling strategies. In this paper we have first presented a detailed analysis of various Scheduling strategies like Time based and Space based methods simulated by GridSim by varying different parameters. We propose a dynamic Scheduling strategy which switches between Time based and Space based. We are sure that, this will certainly improve the performance of a job by minimizing execution time.

Future plan may focus on finding the threshold value to implement this proposed algorithm and proposed algorithm will be compared with existing algorithm.

References

1. Foster, I., Kesselman, C., Tuecke, S.: The Anatomy of the Grid Enabling Scalable Virtual Organizations. Intl J. Supercomputer Applications (2001)
2. Foster, I., Kesselman, C. (eds.): The Grid: Blueprint for a New Computing Infrastructure. Morgan Kaufmann, San Francisco (1999)
3. Technical Report No. 2006-504. Scheduling Algorithms for Grid Computing: State of the Art and Open Problems. Fangpeng Dong and Selim G. Akl School of Computing, Queen's University Kingston, Ontario (January 2006)
4. Yu, J., Buyya, R., Ramamohanarao, K.: Workflow Scheduling Algorithms for Grid Computing
5. Wang, T., Zhou, X.-s, Liu, Q.-r., Wang, Y.-l.: An Adaptive Resource Scheduling Algorithm for Computational Grid. In: Proceeding of the, IEEE Asia-Specific Conferenceon Services Computing (2006)
6. Buyya, R., Murshed, M.: GridSim: A Toolkit for the Modeling and Simulation of Distributed Management and Scheduling for Grid Computing. Concurrency and Computation: Practice and Experience (CCPE) 14, 13–15 (2002)

Network Clustering for Managing Large Scale Networks

Sandhya Krishnan, Richa Maheshwari, Prachi Birla, and Maitreya Natu

Tata Research Development and Design Centre, Pune, India

Abstract. In this paper we present algorithms to intelligently divide the network into smaller sub-networks. We systematically define the problem of finding minimal number of small-size clusters, and present a reduction of this problem to the Minimum Vertex Cover problem. We then use this reduction to present two algorithms that vary in their optimality and execution time. We present extensive experimental evaluation to evaluate the proposed algorithms on different network topologies.

1 Introduction

With the increasing size and complexity of today's networks, solutions to many network analysis problems, such as probe station placement, placement of verification servers, fault localization, network monitoring, etc., observe a problem of scalability to such sizes. In order to demonstrate these problems and the application of network clusters to solve them, we consider the problem of probe station selection [7]. Probe stations are the nodes that send probes in the network to test network health. The probe station selection problem is to select a smallest set of nodes as probe stations such that each node and link in the network can be probed by at least one probe station even in the presence k node or link failures in the network. This problem has been proved to be NP-Complete. Furthermore, the approximation algorithms have been proved to be compute-intensive [5]. Probe station selection problem can be addressed using network clustering as follows. The network can be divided into smaller network clusters (sub-networks) and probe stations can then be selected within each cluster. A hierarchical solution for network probing such as [8] can then be built to address intra-cluster and inter-cluster probing. This paper addresses the problem of intelligently dividing a large network into smaller manageable clusters. We name this problem as the *Minimum Cluster Set* problem and define it as follows: Given a network, divide the network into smaller clusters such that - (a) number of clusters is minimal, (b) size of clusters is less than a threshold, and (c) cluster sizes are uniform.

Challenges: The problem becomes challenging due to various reasons: (a) The above defined problem can be proved to be NP-Complete. (b) The proposed algorithm should provide effective solution to varying types of topologies.(c) Clustering criteria can be based on various network parameters such as network delay, traffic, affinity/exclusion of network components, etc.

K. Shah, V.R. Lakshmi Gorty, and A. Phirke (Eds.): ICTSM 2011, CCIS 145, pp. 49–54, 2011.

Related work: Clustering of networks has been addressed in the past primarily using domain knowledge or manual selection of cluster centers [8]. This might not be feasible in case of large networks. Clustering done based on IP addresses [3] does not consider other metrics such as physical distance, communication patterns, affinity/exclusion of nodes, etc. Authors in the area of network monitoring [4] have proposed heuristics to incrementally select network monitors. Authors in [2] perform clustering of requests by computing request similarity. Multi-dimension clustering is performed in the area of text mining by computing distance measures across strings.

Contributions: The main contributions of this paper addressing the above mentioned challenges are as follows: (a) We demonstrate the reduction of the Minimum Cluster Set problem to the Mini-mum Vertex Cover problem. (b) We present two algorithms (to address trade-off between optimality and execution time) to solve the Minimum Cluster Set problem (c) We also discuss how the proposed algorithms can be used for various clustering criteria such as communication patterns, affinity/exclusion of components, etc. (d) We present an extensive experimental evaluation of the proposed algorithms.

2 Problem Description and Design Rationale

We represent the network as a graph *G(V,E)* where the set *V* represents the network nodes and the set *E* represents the network links. A *network cluster* is a sub-network represented by a graph G'(V ', E')such that $V' \in V$ and $E' \in E$. In this section, we reduce the Minimum Cluster Set problem to the Minimum Vertex Cover problem. Using this reduction, we show that a solution to the Minimum Vertex Cover problem can provide a solution to the Minimum Cluster Set problem. For simplicity, we assume a non-weighted graph and construct clusters using hop distance or spatial closeness as the clustering criteria.

2.1 Reduction to the Minimum Vertex Cover Problem

To facilitate the reduction, we define both the Minimum Cluster Set problem and the Minimum Vertex Cover problem precisely.

Minimum Cluster Set problem
Instance: Graph *G(V,E)* and a cluster size threshold *T*.

Problem: Find the smallest set *S* of subgraphs of *G*, such that $\forall G_i \in S : |V_i| \leq T$, and $\bigcup_{\forall G_i \in S} (G_i) = G$

Minimum Vertex Cover problem
Instance: Graph G '(V' , E').

Problem: Find the smallest set $V_c^{'} \subseteq V^{'}$ such that every edge in $E^{'}$ is incident on at least one vertex in $V_c^{'}$.

We now provide a reduction of the Minimum Cluster Set problem to the Minimum Vertex Cover problem such that a good solution for the Minimum Vertex Cover problem (small number of vertices in vertex cover) can provide a good solution for the Minimum Cluster Set problem (small number of clusters).

Construction: Given an instance of the Minimum Cluster Set problem (Graph *G(V,E)*, Threshold *T*), the instance of the Minimum Vertex Cover can be formed as follows.

Step 1: Construct a graph G' from graph *G* by adding additional edges between each node and its k hop neighbors, where k ranges from 1 through k_{max}.

Step 2: Threshold T is mapped to k_{max} as follows. Consider that average node degree of the network is represented by AD. The maximum number of nodes that can be present in a cluster with radius k_{max} and average node degree AD is $AD^{k_{max}}$. To ensure that average number of nodes in a cluster is less than T, $(AD^{k_{max}}) \leq T$ .Hence, $k_{max} \leq \log_{AD}(T)$.

It can be seen that the above construction can be obtained in polynomial time. Considering a graph G with n nodes, the graph G' can be constructed in $O(n^2)$ operations. The vertex cover V_c computed on the graph G' can then be converted to the Minimum Cluster Set S of graph G as follows: For each node $V_i' \in V_c'$:

1. Identify the corresponding node $V_i \in G$.
2. Create a subgraph $S_i \in S$ consisting of node V_i, its k-hop neighbors, and their inter-connecting edges, where $k = 1 \ldots \log_{AD}(T)$.

The set S thus formed is the Minimum Cluster Set. Based on the above reduction, we present two algorithms to compute the Minimum Cluster Set in the next section.

3 Proposed Approach

We first present a less computation-intensive approach, called the Randomized approach which is fast but leads to less optimal solution (computes more number of clusters than optimal). This approach is suitable when the clustering needs to be performed periodically or when the size of network is very large. We then present another approach, called the Heuristic-based approach which provides closer to optimal solutions by generating fewer number of clusters but is more computation-intensive. This approach is suitable when the clustering needs to be performed less frequently or when the size of the network is relatively small.

3.1 The *Randomized* Approach

In order to keep the Randomized algorithm computationally light-weight, the algorithm does not perform the construction step explained in the reduction in the previous section. Instead, it reads the graph representing the original network topology and randomly selects any node as part of a Vertex Cover and hence a cluster center. It then populates the cluster members by iteratively adding 1-hop, 2-hop...k-hop neighbors until the cluster size becomes greater than the desired threshold. The process is repeated until all nodes are covered. For a graph consisting of n nodes, the complexity of this algorithm is $O(n)$. The Randomized approach does not intelligently select cluster centers and hence leads to sub-optimal solution.

3.2 The *Heuristic-Based* Approach

Given a graph representing the network topology, this algorithm uses a greedy approach to select a cluster center. The algorithm iteratively selects the best node that can be a cluster center and populates its 1-hop, 2-hop... k-hop neighbors as its cluster members. The algorithm first computes a desired hop distance to which the cluster of any node should be expanded. This hop distance is computed based on the desired maximum cluster size and the aver-age node degree. The algorithm first performs a construction step explained in the reduction in the Section 2. In the graph of the network topology additional links are added connecting every node to its 1-hop, 2-hop...k-hop neighbors, where k is chosen using the criteria explained in Section 2. Unlike the Randomized approach, the Heuristic-based approach makes the best choice for the cluster center by identifying the node that can cover maximum number of uncovered nodes in its cluster. For a graph consisting of n nodes, the complexity of this algorithm is $O(n^2)$.

3.3 Incorporating Other Clustering Criteria

The above approaches can also be used for other criteria such as loss rate, network traffic, network delay, etc. The network topology is then represented as a weighted graph where the weights on the nodes represent the average amount of traffic observed by a node. The Minimum Cluster Set problem can then be modified as follows:

Minimum Cluster Set problem

Instance: Graph $G(V,E)$, $\forall$ $Vi \in V$ Weight(Vi), and a cluster size threshold T.

Problem: Find the smallest set S of subgraphs of G, such that

$$\forall G_i \in S : \sum (\text{Weight}(V_i)) \leq T \text{, and } \bigcup_{\forall G_i in S} (G_i) = G$$

where Weight (V_i) refers to the individual weight of each node based on the criteria such as traffic observed. The proposed Randomized and Heuristic-based approaches can be then be used for the modified definition.

4 Experimental Evaluation

4.1 Experimental Setup

We generated network topologies using BRITE and evaluated the proposed algorithms on different network sizes (100 to 1000 nodes), and different average node degrees (10 to 50) We ran the proposed Randomized and Heuristic-based approaches on these networks and evaluated the two approaches on number of clusters, the distribution of cluster sizes, and the execution time. Each point plotted on the graphs is an average of 10 runs. We have also plotted the 95% confidence intervals.

Analysis on different network sizes. We fix an average node degree to 50 and vary the number of nodes from 100 to 1000. The results of this experiment are presented in Figure 1a. It can be seen that the number of clusters computed by the Heuristic-based approach is smaller than that by the Randomized approach. However, the execution time of the Randomized approach is less than that of the Heuristic-based approach. The difference between the cluster sizes computed by the two algorithms for the same network, increases with increasing number of nodes in the network. The confidence intervals for the Heuristic-based approach are much smaller than those of the Randomized approach. The random selection in Randomized approach results in non-determinism in the results and hence larger confidence intervals.

Analysis on different average node degrees. In Figure 2a, we fix the number of nodes to 1000 and vary the average node degree of the network from 10 to 100. The graph shows a significant difference between the numbers of clusters computed by the two algorithms in case of sparse networks. The Heuristic-based approach outperforms the Randomized approach significantly. However, in denser graphs, there are larger number of good candidates that can be chosen as cluster centers. This increases the probability of selection of a good candidate as a cluster center using even the random node selection technique of the Randomized approach. We now evaluate the properties of clusters, such as size of clusters and uniformity of cluster sizes, constructed by the two algorithms. Figures 2b and 2c show the minimum, average, and maximum size of clusters computed by the Randomized and Heuristic-based approaches respectively for networks varying in their average node degree. Figures show the evaluation of 1000-node networks with their average node degree varying from 10 to 50. The cluster sizes computed by the Randomized approach are smaller than the cluster sizes computed by the Heuristic-based approach. Randomized approach hence results into computing a large number of small size clusters. Also, there exists a higher variation between minimum, average and maximum cluster sizes computed by the Randomized algorithm. This variation in cluster sizes computed by the Heuristic-based algorithm is much less.

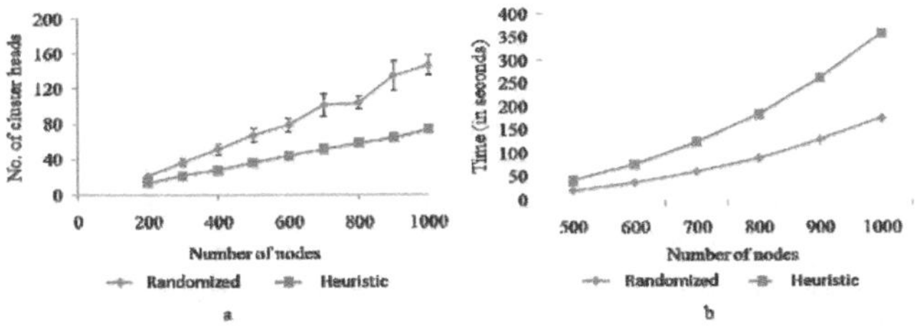

Fig. 1. (a)Number of clusters and (b) Execution time for Randomized and Heuristic-based approaches for different network sizes

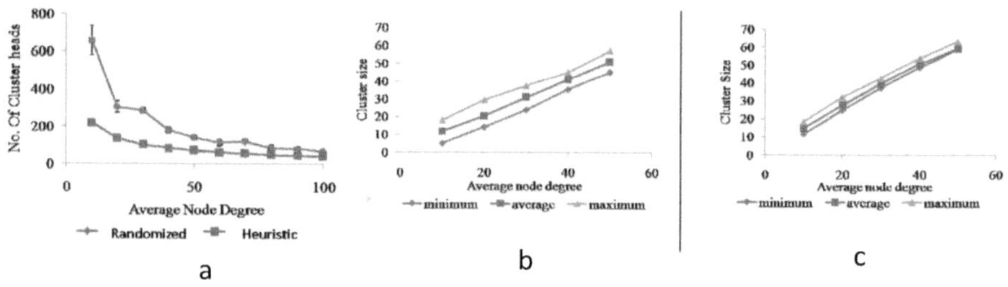

Fig. 2. (a) Number of clusters computed by Randomized and Heuristic-based approaches for networks with different average node degree; Minimum, average, and maximum sizes of the clusters computed by the proposed approaches for networks with different average node degree. (b) Randomized approach, and (c) Heuristic-based approach.

5 Conclusion and Future Work

In this paper, we addressed the problem of clustering a large network into smaller sub-networks. Given the large scale of the networks, we argue that only simple and computationally light-weight solutions can provide feasible solutions to this otherwise a very complex and computation-intensive problem. We systematically defined the Minimum Cluster Set problem and demonstrated its reduction to the Minimum Vertex Cover problem. We presented two generic algorithms for clustering a network that address the computation-optimality trade-off in two different ways. As part of ongoing and future work, we are extending the experimental evaluation by incorporating various network properties such as traffic, loss rate, bandwidth, affinity/exclusion of components.

References

1. Cooperative association for Internet data analysis (CAIDA). the skitter project. Technical report (2001), `http://www.caida.org/tools/measurement/skitter/index.html`
2. Barham, P., Isaacs, R., Mortier, R., Narayanan, D.: Magpie: Online modeling and performance-aware systems. In: 9th conference on Hot Topics in Operating Systems, HOTOS 2003, Berkeley, CA, USA (2003)
3. Bouloutas, A., Gopal, P.: Clustering schemes for network management. In: INFOCOM, pp. 111–120 (1991)
4. Jamin, S., Jin, C., Kurc, A.R., Raz, D., Shavitt, Y.: Constrained mirror placement on the Internet. In: INFOCOM, pp. 31–40 (2001)
5. Jeswani, D., Korde, N., Patil, D., Natu, M., Augustine, J.: Probe station selection for robust network monitoring. In: Student Research Symposium, International Conference on High-Performance Computing, Kochi, India (Best Paper Award) (December 2009)
6. Kumar, R., Kaur, J.: Efficient beacon placement for network tomography. In: Internet Measurement Conference, IMC (2004)
7. Natu, M., Sethi, A.S.: Probe station placement for robust monitoring of networks. Journal of Network and Systems Management (2007)
8. Steinder, M., Sethi, A.S.: Multidomain diagnosis of end-to-end service failures in hierarchically routed networks. IEEE Transactions on Parallel and Distributed Systems 8(3) (March 2007)

Sectorization of DCT-DST Plane for Column Wise Transformed Color Images in CBIR

H.B. Kekre[1] and Dhirendra Mishra[2]

Sr. Professor[1], Associate Professor & PhD Research Scholar[2]
MPSTME, SVKM's NMIMS (Deemed-to be-University)
Vile Parle West, Mumbai -56, India
hbkekre@yahoo.com, dhirendra.mishra@gmail.com

Abstract. We have introduced a novel idea of sectorization of DCT-DST plane of column wise transformed color images and feature vector generation with and without augmentation of extra row components. We have proposed augmentation of average value of zeroeth row component of DCT plane and absolute values of highest row components of DST plane column transformed color images to the feature vector. Two similarity measures such as sum of absolute difference and Euclidean distance are used and results are compared. The cross over point performance of overall average of precision and recall for both approaches on different sector sizes are compared. The DCT-DST plane sectorization is experimented on DCT-DST planes of transformed image The proposed algorithm is worked over database of 1055 images spread over 12 different classes. Overall Average precision and recall is calculated for the performance evaluation and comparison of 4, 8, 12 & 16 DCT-DST sectors done. We have also proposed two new performance measuring parameters namely LIRS (Length of initial relevant strings) and LSRR (Length of string to recover all relevant images). The use of sum of Absolute difference as similarity measure always gives lesser computational complexity and better relevant image retrieval rate compared to Euclidian distance.

Keywords: Keywords-CBIR, DST, DCT, Euclidian Distance, Sum of Absolute Difference, Precision and Recall, LIRS, LSRR.

1 Introduction

Content-based image retrieval (CBIR), [1][2] is the technique that helps to access, organize digital image databases by their content i.e. feature, shape, color etc. CBIR has various applications like pattern matching, security, large image database archival etc. CBIR holds enough ideas to be useful for many more applications [3][4]. For example, many commercial organizations are working on image retrieval despite the fact that robust text understanding is still an open problem. Of late, there is renewed interest in the media about potential real-world applications of CBIR and image analysis technologies, There are various approaches which have been experimented to generate the efficient algorithm for CBIR like FFT sectors [5-8], Transforms , Vector quantization, bit truncation coding ,Walsh. In this paper we have introduced a novel concept sectorization of combined plane of DCT-DST for feature extraction (FE).

K. Shah, V.R. Lakshmi Gorty, and A. Phirke (Eds.): ICTSM 2011, CCIS 145, pp. 55–60, 2011.

Two different similarity measures namely sum of absolute difference and Euclidean distance are considered. Two new algorithm performance measuring parameters namely LIRS and LSRR are used. The performances of these approaches are compared.

2 Discrete Transforms

DCT is made up of cosine functions taken over half the interval and dividing this half interval into N equal parts and sampling each function at the center of these parts. The discrete cosine transform matrix is formed by arranging these sequences row wise. The most common DCT definition of 1D Sequence of Length N is

$$C(u) = \alpha(u)\sum_{x=0}^{N-1} f(x)\cos\left[\frac{\pi(2x+1)u}{2N}\right], \quad \alpha(u) = \begin{cases} \sqrt{\frac{1}{N}} & for \quad u = 0 \\ \sqrt{\frac{2}{N}} & for \quad u \neq 0. \end{cases} \tag{1}$$

For u= 0,1, 2….N-1.

The discrete sine transform matrix is formed by arranging these sequences row wise. The NxN Sine transform matrix y(u,v) is defined as

$$y(u,v) = \sqrt{2/(N+1)} * \sin[\pi(u+1)(v+1)/(N+1)]$$

$$\text{for } 0 <= u, v <= N\text{-}1 \tag{2}$$

3 Extraction of Image Features

The generation of feature vector is described below:

Step1: The DCT-DST plane formed [17].
Step2: The DCT-DST Plane formed in step 1 is sectored[7-13] in various sector sizes.
Step3: The segment average value of a particular sector was taken as feature vector components.

To form the DCT-DST plane we extracted the DCT co-efficient from row 2 to row N of column wise DCT transformed version of the image (where N is number of columns in the image) and DST co-efficient from the row 1 to row N-1 of column wise DST transformed version of the image. The combination of these extracted co-efficient values are considered to form the DCT-DST plane where DCT coefficient values are on X axis and DST co-efficient on Y axis thus taking these components as coordinates we get a point in X-Y plane. The work has been experimented with sum of absolute difference [7-13] and Euclidean distance [7-9] [11-14] as similarity measures. In addition to these the feature vectors are augmented by adding two components which are the average value of zaroeth and the last row components of DCT and DST plane respectively. Thus for 4, 8, 12 & 16 sectors 8, 16, 24 and 32 feature components along with augmentation of two extra components for each color planes i.e. R, G and B are generated. Thus all feature vectors are of dimension 30, 54, 72 and 102 components.

4 Results and Discussion

The sample Images of the database of 1055 images of 12 different classes such as Flower, Sunset, Barbie, Tribal, Puppy, Cartoon, Elephant, Dinosaur, Bus, Parrots, Scenery, Beach is shown in the Fig. 1.

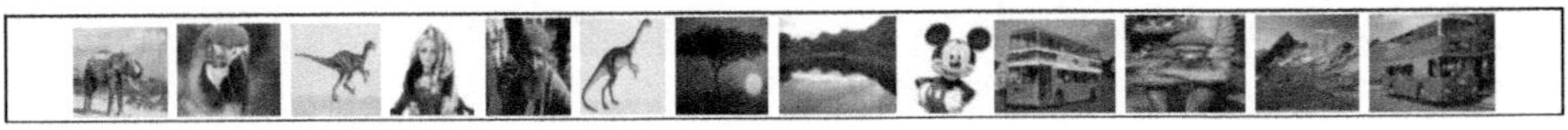

Fig. 1. Sample Image Database

The Barbie class image is taken as sample query image .The first 21 images retrieved for 16 sectors with augmentation of average value of zeroeth and highest row components of DCT and DST transformed image used for feature vectors and sum of Absolute difference as similarity measure is shown in the Fig. 2. It is seen that only 3 images of irrelevant class.

Fig. 2. First 21 Retrieved Images of 16 Sectors (column wise) with augmentation of average value of zeroeth and highest Row components

Once the feature vector is generated for all images in the database a feature database is created. A query image of each class is produced to search the database. The image with exact match gives minimum absolute difference and Euclidian distance. To check the effectiveness of the work and its performance with respect to retrieval of the images we have calculated the precision and recall The precision is defined as ratio of number of relevant images retrieved to total number of images retrieved whereas recall is ratio of number of relevant image retrieved to total number of relevant images in the database .Two new parameters LIRS and LSRR has been calculated. LIRS is given by ratio of length of initial relevant string of images to total relevant images retrieved; LSRR is given by ratio of length of string to recover all relevant images to total images in the database. All these parameters lie between 0-1 hence they can be expressed in terms of percentages.

The Fig. 3 shows the class wise comparison of average precision and recall cross over points for all sectors with respect to both similarity measures. It can be very clearly seen that as far as individual class performance is concerned Flower, Barbie, Diana sour, Sunset and horses class has retrieval of more that 50% in all sectors whilst the sum of absolute difference most of the time outperforms the ED.

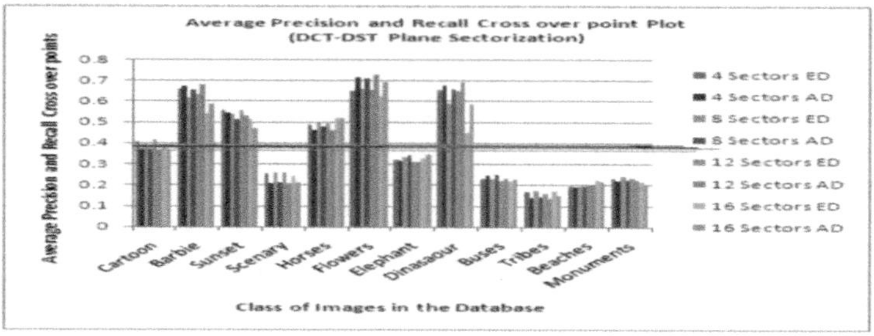

Fig. 3. Class wise Average Precision and Recall cross over point plot. The average value of each sector is shown with horizontal lines [0.399 (4 Sector ED), 0.395 (4 Sector AD), 0.392 (8 Sectors ED), 0.394 (8 Sectors AD), 0.398(12 Sectors ED) ,0.396(12 Sectors AD),0.368 (16 Sectors ED), 0.3800 (16 Sectors AD)].

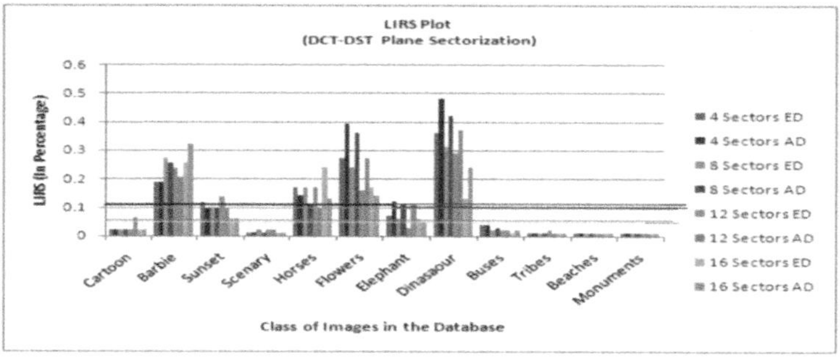

Fig. 4. LIRS Plot. The average value of each sector is shown with horizontal lines [0.10 (4 Sector ED), 0.12 (4 Sector AD), 0.10 (8 Sectors ED), 0.12 (8 Sectors AD), 0.09(12 Sectors ED) ,0.10 (12 Sectors AD),0.082 (16 Sectors ED) ,0.085(16 Sectors AD)].

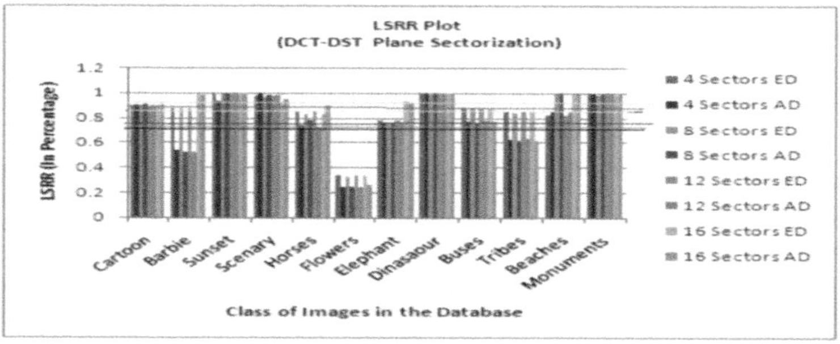

Fig. 5. LSRR Plot The average value of each sector is shown with horizontal lines [0.85 (4 Sector ED), 0.77 (4 Sector AD), 0.86 (8 Sectors ED), 0.79 (8 Sectors AD), 0.85(12 Sectors ED), 0.78 (12 Sectors AD), 0.88 (16 Sectors ED) ,0.85 (16 Sectors AD)].

The overall performance shown with horizontal line which is not lesser than 40% for all approaches. Two new parameters LIRS and LSRR are used for performance measurement of the said algorithm shown in Fig. 4 and Fig.5 For the best retrieval performance LIRS must be more and LSRR must be least as it is very clearly seen for the Flower class. These parameters are different for combination of all sector sizes and similarity measures.

5 Conclusion

The novel idea of sectorization of combined plane of DCT-DST column wise transformed colour images proposed in the paper. The work has been experimented with the approach of with and without augmenting the feature vector with extra components. It was found that the segmentation gives the better performance of retrieval almost more than 30% .The proposed method sectorizes the DCT-DST plane in various sizes i.e. 4,8,12,16 and each sector sizes are used with the combination of two different similarity measuring parameters i.e. ED and AD. It has been observed that the performance of the algorithm varies for each sectors and similarity measures it can be seen in the precision and recall cross over point comparison in Fig. 3. The sum of absolute difference (AD) is preferred to be used as similarity measure than Euclidian distance (ED) since it has lesser computational complexity and better support in the final result of retrieval. We have also demonstrated the algorithm performance with respect to LIRS and LSRR .These parameters can make us to easily recognise the performance. LIRS talks about how many relevant image retrieved at one go before the retrieval of the first irrelevant image in the result whereas LSRR shows how quickly all relevant images are retrieved in the result. The Fig.4 and Fig. 5 shows the clear understanding of retrieval with respect to each class of images. Thus these two parameters are highly recommended. Sector wise performance of the proposed method Where 4,8 and 12 sectors give overall average performance of precision and recall cross over point of all classes close to 45% .

References

1. Kato, T.: Database architecture for content based image retrieval in Image Storage and Retrieval Systems. In: Jambardino, A., Niblack, W. (eds.) Proc. SPIE, vol. 2185, pp. 112–123 (1992)
2. Datta, R., Joshi, D., Li, J., Wang, J.Z.: Image retrieval:Idea,influences and trends of the new age. ACM Computing Survey 40(2), Article 5 (April 2008)
3. Berry, J., Stoney, D.A.: The history and development of fingerprinting. In: Lee, H.C., Gaensslen, R.E. (eds.) Advances in Fingerprint Technology, 2nd edn., pp. 1–40. CRC Press, Florida (2001)
4. Newham, E.: The biometric report. SJB Services (1995)
5. Kekre, H.B., Mishra, D.: Digital Image Search & Retrieval using FFT Sectors. In: Published in Proceedings of National/Asia Pacific Conference on Information Communication and Technology (NCICT 2010), March 5-6, SVKM'S NMIMS MUMBAI (2010)

6. Kekre, H.B., Mishra, D.: Content Based Image Retrieval using Weighted Hamming Distance Image hash Value. Published in the Proceedings of International Conference on Contours of Computing Technology (Thinkquest 2010), March 13-14 (2010)
7. Kekre, H.B., Mishra, D.: Digital Image Search & Retrieval using FFT Sectors of Color Images. Published in International Journal of Computer Science and Engineering (IJCSE) 2(02), 368–372 (2010), http://www.enggjournals.com/ijcse/doc/IJCSE10-02-02-46.pdf, ISSN 0975-3397
8. Kekre, H.B., Mishra, D.: CBIR using upper six FFT Sectors of Color Images for feature vector generation. Published in International Journal of Engineering and Technology (IJET) 2(2), 49–54 (2010), http://www.enggjournals.com/ijet/doc/IJET10-02-02-06.pdf, ISSN 0975-4024
9. Kekre, H.B., Mishra, D.: Four Walsh transform sectors feature vectors for image retrieval from image databases. Published in international journal of computer science and information technologies (IJCSIT) 1(2), 33–37 (2010), http://www.ijcsit.com/docs/vol1issue2/ijcsit2010010201.pdf, ISSN 0975-9646
10. Kekre, H.B., Mishra, D.: Performance comparison of four, eight and twelve Walsh transform sectors feature vectors for image retrieval from image databases. Published in International Journal of Engineering, Science and Technology(IJEST) 2(5), 1370–1374 (2010), http://www.ijest.info/docs/IJEST10-02-05-62.pdf
11. Kekre, H.B., Mishra, D.: Density distribution in Walsh transforms sectors as feature vectors for image retrieval. Published in International Journal of Computer Applications (IJCA) 4(6), 30–36 (2010), http://www.ijcaonline.org/archives/volume4/number6/829-1072, ISSN 0975-8887
12. Kekre, H.B., Mishra, D.: Performance comparison of density distribution and sector mean in Walsh transform sectors as feature vectors for image retrieval. Published in International Journal of Image Processing (IJIP) 4(3) (2010), ISSN 1985- 2304
13. Kekre, H.B., Mishra, D.: Density distribution and sector mean with zero- sal and highest-cal components in Walsh transform sectors as feature vectors for image retrieval. Published in International Journal of Computer Science and Information Security (IJCSIS) 8(4) (2010), http://sites.google.com/site/ijcsis/vol-8-no-4-jul-2010, ISSN 1947-5500
14. Ross, A., Jain, A., Reisman, J.: A hybrid fingerprint matcher. In: Int'l conference on Pattern Recognition (ICPR) (August 2002)

Construction of Test Cases from UML Models

Vinaya Sawant[1] and Ketan Shah[2]

[1] Lecturer, D.J. Sanghvi College of Engineering, Mumbai
vinaya.sawant23@yahoo.com
[2] Associate Professor, Mukesh Patel School of Technology Management & Engineering, NMIMS University, Mumbai
ketanatnmims@gmail.com

Abstract. The aim of system testing is to ensure that a fully developed and integrated system is error free. System testing is often considered to be the most complex and intricate among all types of testing. Therefore, automatic design of system test cases is assuming crucial importance. The paper presents a technique to create the test cases from UML models. In this technique, the UML diagrams such as Use Case Diagram, Class Diagram & Interaction Diagram of any application are considered for creating the test cases. A graph is created to store the necessary information that can be extracted from these diagrams & data dictionary expressed in OCL for the same application. The graph is then scanned to generate the test cases that are suitable for system testing.

Keywords: Testcases, UML Diagrams, OCL expressions.

1 Introduction

The systematic production of high-quality software, which meets its specification, is still a major problem. Although formal specification methods have been around for a long time, only a few safety-critical domains justify the enormous effort of their application. The state of the practice, which relies on testing to force the quality into the product at the end of the development process, is also unsatisfactory. The need for effective test automation adds to this problem, because the creation and maintenance of the testware is a source of inconsistency itself and is becoming a task of comparable complexity as the construction of the code.

This strong traceability can already be achieved for certain parts of the requirements. For example, steps in a use case relate to messages in dynamic UML models, which map (in some contexts) to methods of a programming language and to steps in a test case. For other parts, the systematic exploitation of the information throughout all phases in the software process is not yet defined.

Model Based Testing (MBT) is gaining its popularity in both academia and in industry. As systems are increasing in complexity, more systems perform mission-critical functions, and dependability requirements such as safety, reliability, availability, and security are vital to the users of these systems.

In recent years, Unified Modeling Language (UML) has emerged as the de facto standard for modeling software systems and has received significant attention from

K. Shah, V.R. Lakshmi Gorty, and A. Phirke (Eds.): ICTSM 2011, CCIS 145, pp. 61–68, 2011.

researchers as well as practitioners. The importance of UML models in designing test cases has been well recognized.

2 Problem Definition

The aim is to develop an automatic test case creation tool using UML model. The sequence diagram is considered as a source of test case generation. The generated test suit aims to cover optional and use case dependency faults, various iterations as well as scenario faults. For generating the different components of a test case, i.e. input, expected output and pre- and post-condition, use case diagram, class diagram, and dictionary in the form of OCL expressions along with sequence diagram is considered. The generated test cases can be stored separately in a different file for future use.

3 The Proposed Approach

Given a use case diagram (UD), class diagram (CD), sequence diagram (*SD*), transform it into a representation called sequence diagram graph (*SDG*) [1]. Each node in the *SDG* stores necessary information for test case generation. This information are collected from the use case template (also called extended use case), class diagrams, and data dictionary expressed in the form of object constrained language (OCL) [3], which are associated with the use case for which the sequence diagram is considered. Then traverse *SDG* and generate test cases based on a coverage criteria and a fault model. A schematic diagram of the approach is as shown in the fig .1[1].

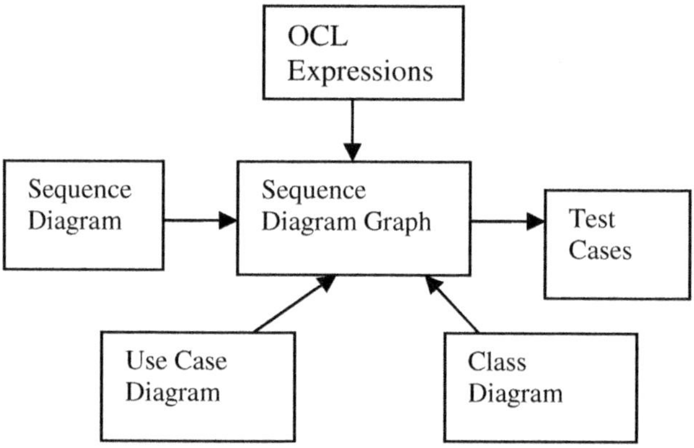

Fig. 1. Schematic Block Diagram for the proposed approach

4 Implementation Work

The different UML tools that such as MagicDraw and Rational Rose that has support for drawing UML diagrams as well as OCL expressions can be used to draw the UML diagrams according to the specification of any application that has to be considered.

These UML diagrams are then exported to XML format. The process of exporting generates XML file from UML diagrams. This file contains all the XML tags describing all the UML diagrams. This XML file needs to be parsed to generate the graph. The parser is written that reads XML file and generate the different nodes for the graph by considering the sequence diagram of an application. These nodes are then mapped into the different scenarios according to the flow of messages in the sequence diagram. The nodes of the graph stores the information such as attributes of the corresponding objects at that state, arguments in the method, and predicate of the guard if any, involved in the interaction. The Use case template is also considered while generating the test cases. The Use Case template provides the information such as precondition & postcondition for a particular scenario that has to be considered. The OCL syntax will be followed to represent the data dictionary [3]. For the specification of a test case, the test specification language according to the IEEE Standard 829 is followed. Test cases generated will be recorded in a temporary file for future references. The case study for Bank ATM System is considered to generate the test cases. The Use Case Diagram, Class Diagram and Sequence Diagram is drawn which clearly gives the detailed description of the application. These diagrams are drawn using Magic Draw tool. Using the same tool, the diagrams are exported to XML format. The parser has been written in java that reads XML file and generates the different nodes of the graph. The sequence diagram is scanned to identify the set of the scenarios from the start node to end nodes. Now these set of scenarios along with Use Case Template and OCL data dictionary need to be traversed to generate the test cases. The generated test cases are recorded into a separate temporary file.

5 Case Study: Bank ATM System

The case study of Bank ATM system is considered to show the implementation work of the project. The Bank ATM system allows user to perform login with authenticated PIN number. If the user logs in with authorized PIN no, then the user is allowed to do deposit the money in his bank account, withdraw money from his bank account. He is able to check the balance and also can request for the mini statement. The following are the steps that need to be executed for generating test cases for the system from the UML Design Diagrams [6].

The implementation for this is done using JAVA programming language. The User Interface for the tool is provided so that on clicking on the desired buttons, the different steps needed to generate test cases can be easily done.

5.1 Step 1: Drawing UML Diagrams from the Problem Statement

The Use Case Diagram, Class Diagram and Sequence Diagram for the system are drawn to understand the system as a whole. The UML diagrams are drawn using the MagicDraw tool since this tool has the good support for OCL expressions as well as compared to any other tool. The fig. 2 represents the Sequence Diagram for the Bank ATM System. The UseCase Diagram, Class Diagram with OCL expressions are also used for this tool.

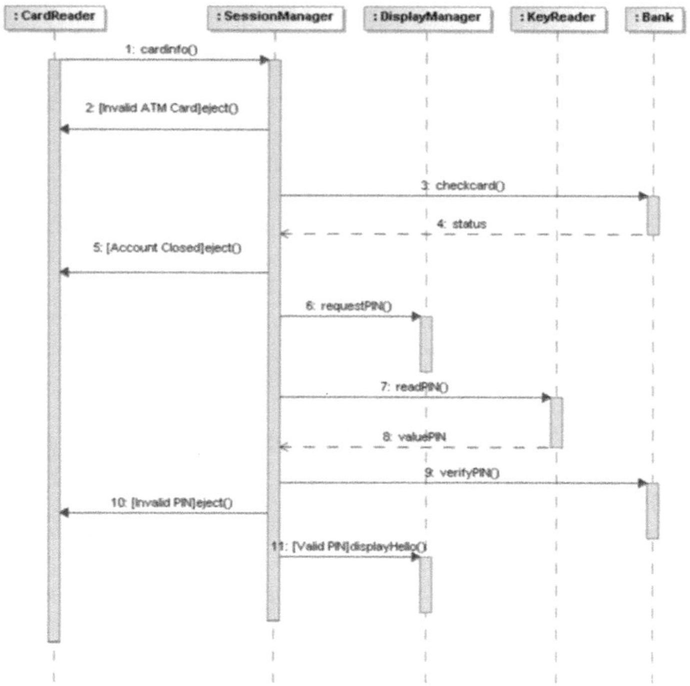

Fig. 2. Sequence Diagram of Bank ATM System

5.2 Step 2: Generating XML File

The UML design tool has also the support for generating XML file from the UML diagrams which are easily transportable without installing the UML design tool. The MagicDraw Software has the option of exporting the UML diagram to XML file. The XML file thus generated contains the tags for all the properties of the diagrams including the color properties that are not required for generating the test cases. Thus, this XML file needs to be edited according to the requirements for further processing. This is one of the disadvantages of this method. Once the necessary changes in the XML file is made, this XML file can be used as input for the ATCUM (Automatic Test Case generation from UML Models) tool for generating test cases. Using the ATCUM tool, the option of opening the XML file is provided.

5.3 Step 3: Parsing the XML File

The parser is written that reads XML file that is selected from the previous step and gives the description about all the tags as well as the attributes of every tags from the XML file. This information will be useful for generating the description for the Sequence Diagram Graph that contains the nodes which stores the necessary information to generate scenarios.

The project makes use of a tree-based API (such as Document Object Model, DOM) builds an in-memory tree representation of the XML document. It provides classes and methods for an application to navigate and process the tree [8].

In general, the DOM interface is most useful for structural manipulations of the XML tree, such as reordering elements, adding or deleting elements and attributes, renaming elements, and so on. For example, for the XML document above, the DOM creates an in-memory tree structure.

5.4 Step 4: Generating Scenarios

The input to this step requires the parsed XML file of the previous step. The nodes for the graph are generated at this step. Each node stores the information such as message passed between the two objects, object that sends the message, the object that receives the message, guard condition if any and the OCL expression of that message [3]. Depending upon the message sent from one object to another object, the number of nodes is determined. The sequence diagram is then scanned again using the XML file to generate the scenario which is nothing but the combination of different nodes. Initially scenario begins with StateX. Then the message detail that is in the form of node is added to the scenario. If the receiver object is same as the sender object of the first node, then that scenario is completed and finally ends with StateY or StateZ. In this manner all the remaining scenarios from the sequence diagram is determined. These scenarios are used as test cases for the test case generator module. From the above sequence diagram, four different scenarios can be generated. Out of these, the following figure Fig. 3 represent two scenarios from Bank ATM System.

```
contents of scenario:1:
<stateX>
S1: null cardinfo :SessionManager :CardReader uml:Message
S2: Invalid ATM Card eject :CardReader :SessionManager
uml:Message context CardReader::eject();pre:
SessionManager.cardtype="not ATM";post: result="eject card
& Displays Welcome Message" OCL2.0
uml:OpaqueExpression
<StateY>
contents of scenario:2:
<stateX>
S1: null cardinfo :SessionManager :CardReader uml:Message
S2: null checkcard :Bank :SessionManager uml:Message
S3: null reply status :SessionManager :Bank uml:Message
S4: Account Closed eject :CardReader :SessionManager
uml:Message context CardReader::eject();pre:
SessionManager.cardtype="ATM" and
SessionManager.status="OK" and
SessionManager.account="closed" ;post: result="eject card &
Displays Welcome Message" OCL2.0 uml:OpaqueExpression
<StateY>
```

Fig. 3. Output of Scenario Generation

5.5 Step 5: Display Graph

The graphical representation of the nodes starting from StateX to StateY or StateZ is displayed in this frame. Initially all the nodes of the first scenario is displayed, then after clicking on the button one by one all the nodes of the remaining scenarios can be

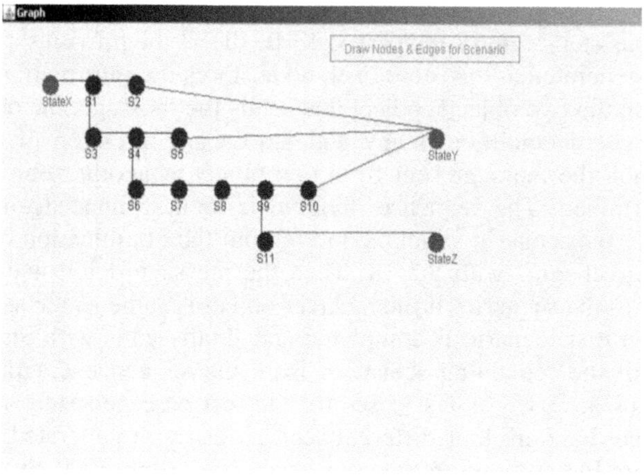

Fig. 4. Displaying Sequence Diagram Graph

Test Case Generation

PreCondition : ATM is idle and displaying welcome message.

The user enter ATM card

Test Scenario:Invalid ATM Card

Input: cardtype="not ATM"

Output: "eject card & Displays Welcome Message"

Test Scenario:Account Closed

Input: cardtype="ATM" status="OK" account="closed"

Output: "eject card & Displays Welcome Message"

Test Scenario:Invalid PIN

Input: cardtype="ATM" status="OK" account="open" PIN_type="invalid"

Output: "Invalid PIN, Try again"

Test Scenario:Valid PIN

Input: cardtype="ATM" status="OK" account="open" PIN_type="Valid"

Output: "Displays Hello & Menu for Transaction "

PostCondition : Displays Menu for transaction

Fig. 5. Test Cases Generation

displayed. Thus, the visual representation of the graph as well as scenarios can be obtained on clicking the required button of the ATCUM tool.

The fig. 4 represents the required graph for the above sequence diagram.

5.6 Step 6: Test Set Generation

This step reads the different scenarios that are generated from the previous step. Each scenario corresponds to test case. All the paths from the start node to final node need to be scanned to generate the test cases. The information that is stored in the nodes is used to determine the input and expected output of the scenario. The OCL expression plays a major role in determining the test cases. Form the OCL expression, the pre-condition and the post condition for a message can be determined. A message having the guard condition and OCL expression is selected for generating the test case. The following figure represents the output of test case set generation.

5.7 Step 7: Creating and Saving Temporary File

The last step is to save the test cases generated from the previous step in a temporary file for future reference. Also this file can be used software testers to create the test plans for the desired software. The temporary file can be saved with .txt extension so that it can be easily available for other users. The following figure fig. 5 represents the text file.

6 Conclusion

The focus was on automatic construction of test cases from UML diagrams. But while generating test cases the Use Case Diagram, Class Diagram, Use Case template, sequence diagram & data dictionary using OCL is also considered. A methodology has been proposed to convert the UML sequence diagram into a graph called sequence diagram graph. The information those are required for the specification of input, output, pre- and post- conditions etc. of a test case are retrieved from the extended use cases, data dictionary expressed in OCL 2.0, class diagrams (composed of application domain classes and their contracts) etc. and are stored in the *SDG*. The approach does not require any modification in the UML models or manual intervention to set input/output etc. to compute test cases. Hence, the approach provides a tool that straightway can be used to automate testing process. A graph based methodology is followed and run-time complexity is governed by the breadth-first search algorithm to enumerate all paths. In fact deciding test data, which are embedded in design artifacts, is computationally intensive task and the approach significantly is able to score in this issue.

The ATCUM tool takes the input in the form of the .XML file and this serves as a very important feature since the UML diagrams can be drawn using any UML tool. The Rational Tool allows you to draw the diagram using the Rational Tool and then generate test cases [6]. The Rational Tool takes into consideration only the State Chart Diagram while ATCUM Tool considers Use Case Diagram, Class Diagram & Sequence Diagram of a particular application.

References

1. Sarma, M., Kundu, D., Mall, R.: Automatic Test Case Generation from UML Sequence Diagrams. Department of Computer Science & Engineering, IIT Kharagpur, IEEE (2007)
2. Fröhlich, P., Link, J.: Automated Test Case Generation from Dynamic Models
3. Object Constraint Language 2.0 is available from Object Mangement Group's web site, http://www.omg.org
4. McGregor, J.D., Sykes, D.A.: A Practical Guide to Testing Object-Oriented Software. Addison-Wesley, Reading (2001)
5. Abdurazik, A., Offutt, J.: Using UML Collaboration Diagrams for static Checking and Test Generation. In: Proceedings of the Third International Conference on the UML
6. A Rational Approach to Software Development using Rational Rose 4.0, IBM's Rational Rose, http://www.ibm.com/
7. Sarma, M., Mall, R.: System Testing using UML models. Department of Computer Science & Engineering, IIT Kharagpur, 16th IEEE Asian Test Symposium
8. http://www.w3schools.com/dom/dom_node.asp

Genetic Algorithmic Approach for Personnel Timetabling

Amol Adamuthe[1] and Rajankumar Bichkar[2]

[1] Departemnt of IT, College of Engineering Pandharpur, Pandharpur, MS, India
amol.admuthe@gmail.com
[2] Department of Computer and IT, G. H. Raisony College of Engineering and Management, Pune, MS, India
bichkar@yahoo.com

Abstract. This paper presents a genetic algorithmic approach to the solution of the problem of personnel timetabling in which the objective is to assign tasks to employees. The problem is multi-constrained and having huge search space which makes it NP hard. The problem considered is that of the timetabling of laboratory personnel. Genetic algorithm is applied to a problem instance with 14 employees and 9 tasks. Canonical genetic algorithm demonstrates very slow convergence to optimal solution. Hence, a knowledge augmented operator is introduced in genetic algorithm framework. This helps to get the near-optimal solution quickly.

Keywords: Genetic algorithm, Personnel timetabling, Scheduling.

1 Introduction

Personnel scheduling problem is mostly encountered in service organizations such as call centres, airport ground personnel, security agencies, hospitals, railway and bus personnel. Causmaecker [1] has classified personnel scheduling problem in four planning categories namely, permanence centred, mobility centred, fluctuation centred and project centred planning. The main difficulties in solving these problems are their highly constrained nature and the environmental conditions that are different for each organization such as working hours, planning periods, existence of breaks for employees, existence of part-time employees in addition to full-time ones etc.

Research in personnel scheduling focuses on three main objectives.

1) Allocation of personnel to shifts, for example nurse scheduling [2, 3, 4, 5, 6], hospital personnel scheduling [7].
2) Assignment of tasks to personnel, for example laboratory personnel scheduling [8].
3) Minimization of personnel cost, for example personnel scheduling on ship [9].

First objective has received more importance as compared to other two.

In literature, the terms '*scheduling*' and '*timetabling*' are used interchangeably. According to Wren [20]: "*Scheduling is the allocation, subject to constraints, of resources to objects being placed in space-time, in such a way as to minimize the total cost of some set of the resources used*". Common examples of scheduling are driver

K. Shah, V.R. Lakshmi Gorty, and A. Phirke (Eds.): ICTSM 2011, CCIS 145, pp. 69–76, 2011.

scheduling [10, 11] which seek to minimize the total cost and job shop scheduling [12, 13, 14] which may seek to minimize the number of time periods used or some physical resources. *"Timetabling is the allocation, subject to constraints, of given resources to objects being placed in space-time, in such a way as to satisfy as nearly as possible a set of desirable objectives"*. Examples of timetabling are class timetabling [15, 16] and examination timetabling [17] and some forms of personnel allocation. In view of these definitions the problem described by Philip and Post [8] is a timetabling problem.

NP hard and NP complete problems can not be solved in reasonable time with traditional search and optimization methods. Genetic algorithms are proven to be one of the effective techniques to solve different scheduling problems [6, 13, 15, 16]. In this paper, we are presenting a genetic algorithmic solution to problem of assignment of tasks to personnel in particular, the laboratory scheduling [8].

The rest of the paper is organized as follows: In section 2 genetic algorithms is briefly described. Section 3 is about personnel scheduling problem formulation. Section 4 gives the implementation details that include genetic algorithm, chromosome representation and operators. In Section 5 we give a detailed description of the problem instances and results. Finally, in Section 6 we outline the conclusions of our study.

2 Genetic Algorithm

Genetic Algorithms (GAs) are randomized yet structured search and optimization algorithms based on the evolutionary ideas of natural selection and genetics [18]. Genetic algorithms exploit historical information to direct the search into the region of better solutions within the search space. They do not require problem specific information for their working. Hence genetic algorithms are used to solve highly constrained, combinatorial optimization problems having huge search space.

GAs simulates the survival of the fittest among individuals over consecutive generations for solving a problem. Each generation consists of a population of individuals. Each individual represents a point in a search space and a possible solution. The individuals in the population are then made to go through a process of evaluation.

GAs are based on the following foundations:

1. Individuals in a population compete for resources.
2. Individuals with above average population fitness produce more offspring than those individuals that are below average population fitness. This indicates exponential speed up in the search process.
3. Genes from "good" individuals propagate throughout the population so that two good parents will sometimes produce offspring that are better than either parent.
4. Thus, each successive generation will become more suited to their environment.

After an initial population is randomly generated, the algorithm evolves through three operators. Selection operator equates the survival of the fittest. Crossover operator represents mating between individuals. Mutation introduces random modification.

Strengths of Genetic Algorithm. Genetic algorithms are intrinsically parallel. These are well-suited to solve problems having huge and complex search space. Also they are good at solving problems with multiple objectives.

3 Personnel Scheduling

The scheduling problem we are presenting here is described by Philip and Post [8]. It is an assignment problem in which the required numbers of tasks are known in advance and the employees are to be assigned to tasks by satisfying constraints.

Hard Constraints. Hard constraints are those that must be satisfied. Violation of these constraints (also called as conflicts) will cause the solution to be infeasible.

1. Coverage constraint: Every task must be allotted the required number of personnel.
2. Constraints by work regulations: Number of work hours assigned to personnel must satisfy his/her work regulations.
3. Skill set constraint: Task should not be assigned to an employee who is not skilled for it.
4. Constraints defined by task types: Tasks are categorized into three types namely, half-day task, day task and week task. If the task is week task then it must be assigned to same person for the entire week. Whereas, day task is given to same person in both slots (morning and evening) and half day task is to be assigned to different personnel during morning and afternoon slots of a day.

Soft Constraints. Soft constraints are those that are desirable in order to produce a good quality timetable but violations are allowed to satisfy hard constraints. In this problem, the task assignment should be according to the skill set of employees. Here history cost (that is number of hours worked by an employee on a task) is considered as skill set.

4 Genetic Algorithm for Personnel Timetabling

One of the popular implementation of genetic algorithm is GAlib [21], a C++ library of Genetic Algorithm Components developed by Matthew Wall at Massachusetts Institute of Technology. The GAlib source code is available at no cost for non-profit purposes. This section describes our implementation that uses steady state genetic algorithm, the chromosome representation, objective function and the genetic operators employed.

Steady State Genetic Algorithm. The steady-state genetic algorithm given below uses overlapping populations. In each generation, the original population size is maintained by replacing a portion of the population by the newly generated individuals.

Chromosome Representation. This problem can be solved using either 1D or 2D representation of chromosome. GAlib supports both the representations. In this work we have used direct 2D representation where a row represents a personnel and a column represents one day of planning period. The cells contain the task. One of the problems associated with this representation is that it may generate infeasible solutions. We have tackled this by imposing penalty and knowledge augmented repair operator on such infeasible solutions.

Objective Function. The objective score of an individual is computed by assigning penalty costs (penalty points) for the violations of constraints. These penalty costs are specified according to the importance of the constraint. These values ensure that hard constraints are not violated in compensation for multiple soft constraints. The objective function for individual *t* is calculated as follows.

$$O(t)=\Sigma\ \alpha_i H_i\ + \Sigma\ \beta_j S_j\ . \tag{1}$$

where H_i and S_j represents number of hard and soft constraints violated by a solution for i^{th} hard constraint and j^{th} soft constraint respectively. α_i and β_j represents penalty costs for violation of hard constraint and soft constraints respectively. Objective values are negative numbers. Steady state Genetic algorithm does not work with negative objective values. Hence we have used sigma truncation scaling to convert them into positive fitness score.

Genetic Operators. Different selection, crossover and mutation applicable to given chromosome representation are tested. Roulette wheel selection is found to be effective than tournament selection. One point crossover with probability 0.8 gives better performance than two point crossover, multi-point crossover and uniform crossover. The swap mutation with probability 0.01 works better than flip mutation.

During the evolutionary process, some genetic operators may generate illegal individuals (that do not respect the problem constraints). To improve the quality of the solution and convergence speed it was necessary to introduce operator that uses problem specific knowledge. The idea is to apply repair operator which outperform the simple penalization of constraint violations. This operator is applied after standard genetic operators but before fitness calculation. This operator removes wrong task assignments in the solution. The operator first finds wrong task assignments. It then searches for a suitable task which is assigned to a part time employee on the same day and swaps the two assignments. It may be noted that, this operator does not necessarily generate a completely feasible solution. We have tested another version of this repair operator where the infeasibilities are completed removed. However, it did not give as good results as the partially feasibility maintaining repair operator.

5 Results and Discussion

We have solved the problem instance given by Philip and Post [8] with small variation. The laboratory has seven different tasks that should be assigned to personnel every day during the planning period. All the tasks are treated as week task that is same task must be allocated to personnel during the week. The task names are AF, CT, IN, KB, PB, PR and TF. Two persons are required to carry out tasks PB and TF; whereas, all other tasks require only one person. For remaining tasks single personnel is sufficient. The laboratory has 14 employees and works in two shifts, morning and afternoon. Two types of work regulations are under consideration, full time employees who work 40 hours per week and part time employees who work 16 or 24 hours in a week. We have assumed that each employee is available on all working days and that the part time employees will work only on limited number of days. Number of days for which the personnel can be assigned to work during the planning period is shown in Table 1. Thus, employees A, E, F, J, L and M will work on 3 days per week; whereas, employees C, D, I and N will work on 2 days per week. Employees B, G, H and K are fulltime employees.

Table 1 also shows history cost that is number of hours worked by employees on each task during last 15 weeks. History cost is treated as skill set for employees. The number of hours employee worked for any task represents expertise and willingness of employee for that job. Higher the history cost, more the expertise for that task.

Table 1. The History Cost (Skill Set) used by Philip and Post [8]

	$N^{\dagger}$	AF	CT	IN	KB	PB	PR	TF
A	3	64	88	-	64	104	64	96
B	5	-	160	-	80	136	72	124
C	2	64	-	-	32	48	36	68
D	2	-	-	-	-	80	48	104
E	3	-	-	300	-	-	-	-
F	3	48	32	-	32	44	24	44
G	5	-	-	-	92	152	104	176
H	5	88	128	-	80	128	48	104
I	2	56	-	-	32	56	32	64
J	2	-	-	300	-	-	-	-
K	5	88	120	-	40	88	40	120
L	3	56	56	-	32	64	32	64
M	3	72	-	-	32	80	32	64
N	2	56	-	-	40	84	48	68

† *n* represents number of days the employee can be assigned for work. n=5 represents a full-time employee whereas other values of n represent part-time employees. The values in this column are obtained from results given in [8].

Philip and Post [8] have solved this problem by using bipartite graph model followed by local search for further improvement in quality of solution. They have reported a good solution that has only two violations that is assignment of two or more tasks in a week to employees A and L.

Table 2. Task Schedule with Zero Violations

	MON	TUE	WED	THU	FRI
A	KB	KB	KB		
B	CT	CT	CT	CT	CT
C		TF		TF	
D			TF	TF	
E	IN		IN		IN
F	TF		TF		TF
G	PB	PB	PB	PB	PB
H	PB	PB	PB	PB	PB
I	AF			AF	
J		IN		IN	
K	PR	PR	PR	PR	PR
L	TF	TF			TF
M		AF	AF		AF
N				KB	KB

The proposed genetic algorithm has been applied to the problem instance and executed several times to test the effect of various parameters. The best task schedule obtained is shown in Table 2. One of the basic improvement obtained in this solution is that every employee has been assigned only a single task throughout the week.

To further compare the quality of our result with that obtained by Philip and Post, we have determined an expertise quality index calculated as shown below.

$$E = \sum_{e} \sum_{t} S_{et} \tag{2}$$

where, S_{et} is the skill score of an employee '*e*' for task '*t*'.

These values are simply obtained by assigning integer values (starting from 1) to task for which an employee has experience. For example, consider employee C who has experience on tasks KB, PR, PB, AT and TF (having 32, 36, 48, 64 and 68 hours history cost respectively). These tasks are assigned values as 1, 2, 3, 4 and 5 respectively. This index for our solution is 127 and 98 for solution given by Philip and Post. This clearly indicates the superiority of the genetic algorithmic approach.

The effect of repair operator is demonstrated in Fig. 1, where the best results in 10 runs (each started with random seed) are reported. In first 5000 generations genetic algorithm and hybrid genetic algorithm shows same convergence. Figure 2 shows objective values from generation number 5000 where we can see the fast convergence of hybrid genetic algorithm over standard genetic algorithm. The best initially generated random solution has objective value in the range of -5300 to -5900. The genetic algorithm without repair shows a good convergence speed in the initial part of the run that it moves from the region with objective value in between -5300 and 5900 to -200 to -500. However after 5000 generations, there is very little improvement which is a characteristic of genetic algorithm. Finally after 60000 generations we get a solution that still has -30 value (indicating soft constraint violations). On the other hand genetic algorithm with repair operator shows a considerable improvement in the solution. A solution having objective value -10 is obtained in less than 10000 generations only. However even after executing upto 60000 generations there was no further improvement.

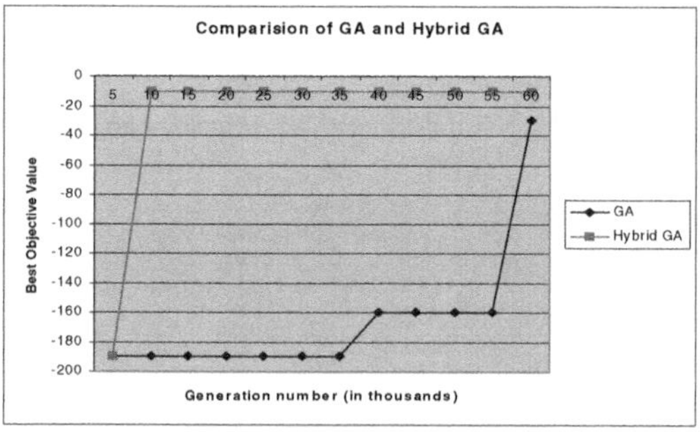

Fig. 1. Effect of repair operator in GA convergence

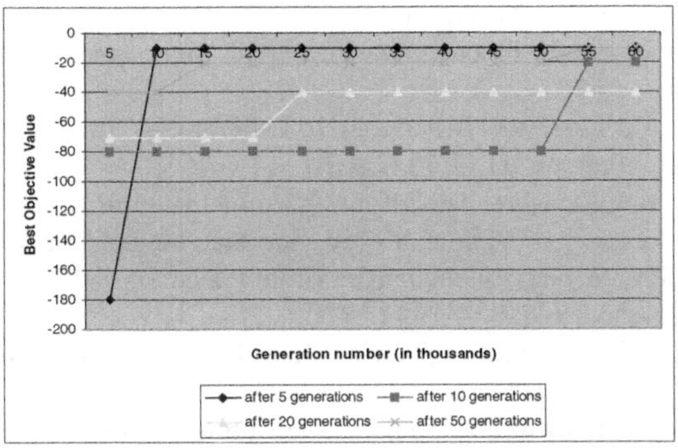

Fig. 2. Performance of hybrid genetic algorithm (GA + repair) with different repair activation frequencies

Proper balance between genetic algorithm and repair operator is necessary for better exploitation and exploration of the search space. Figure 2 shows performance of hybrid genetic algorithm (GA + repair) for different repair operator activation frequencies. Repair operator is applied to 10% of the population.

6 Conclusion

Results show that genetic algorithm is a powerful search technique to solve such multi-constrained scheduling problems. The genetic algorithm with standard selection, crossover and mutation performs well but is subject to premature convergence. To avoid these problems we have introduced a repair operator which helps genetic algorithm to speed up by making intelligent local moves. The proposed GA has given a significant improvement over the results presented in [8].

References

1. Causmaecker, P.D., Demeester, P., Berghe, G.V., Verbeke, B.: Analysis of Real-world Personnel Scheduling Problems. In: 5th Practice and Theory of Automated Timetabling (2004)
2. Burke, E.K., Cowling, P., Causmaecker, P.D., Berghe, G.V.: A Memetic Approach to the Nurse Rostering Problem. Applied Intelligence 15, 199–214 (2001)
3. Aickelin, U., White, P.: Building Better Nurse Scheduling Algorithms. J. Annals of Operations Research 128, 159–177 (2004)
4. Burke, E.K., Curtois, T., Post, G., Qu, R., Veltman, B.: A Hybrid Heuristic Ordering and Variable Neighbourhood Search for the Nurse Rostering Problem. Technical Report, Nottingham University (2005)
5. Özcan, E.: Memetic Algorithms for Nurse Rostering. In: Yolum, p., Güngör, T., Gürgen, F., Özturan, C. (eds.) ISCIS 2005. LNCS, vol. 3733, pp. 482–492. Springer, Heidelberg (2005)
6. Maenhout, B., Vanhoucke, M.: A Comparison and Hybridization of Crossover Operators for the Nurse Scheduling Problem. Working Papers of Faculty of Economics and Business Administration, Ghent University, Belgium (2006)

7. White, C.A., White, G.M.: Scheduling Doctors for Clinical Training Unit Rounds Using Tabu Optimization. In: 4th Practice and Theory of Automated Timetabling, pp. 120–128 (2002)
8. Franses, P., Post, G.: Personnel Scheduling in Laboratories. In: 4th Practice and Theory of Automated Timetabling, pp. 113–119 (2002)
9. Schmidt, M.: Solving Real-Life Time-Tabling Problems. In: Raś, Z.W., Skowron, A. (eds.) ISMIS 1999. LNCS, vol. 1609, pp. 648–656. Springer, Heidelberg (1999)
10. Li, J., Kwan, R.S.K.: A Fuzzy Genetic Algorithm for Driver Scheduling. European Journal of Operational Research 147, 334–344 (2003)
11. Li, J., Kwan, R.S.K.: A Self-Adjusting Algorithm for Driver Scheduling. Journal of Heuristics 11, 351–367 (2005)
12. Miyashita, T.: An Application of Immune Algorithms for Job-Shop Scheduling Problems. In: Proceedings of the 5th IEEE International Symposium on Assembly and Task Planning, France (2003)
13. Jensen, M.T.: Generating Robust and Flexible Job Shop Schedules Using Genetic Algorithms. IEEE Transactions on Evolutionary Computation 7, 275–288 (2003)
14. Ombuki, B.M., Ventresca, M.: Local Search Genetic Algorithms for the Job Shop Scheduling Problem. Applied Intelligence 21, 99–109 (2004)
15. Wilke, P., Grobner, M., Oster, N.: A Hybrid Genetic Algorithm for School Timetabling (2002)
16. Karova, M.: Solving Timetabling Problems Using Genetic Algorithms. In: 27th Int'l Spring Seminar on Electronic Technology. IEEE, Los Alamitos (2004)
17. Qu, R., Burke, E.K.: Hybrid Variable Neighborhood HyperHeuristics for Exam Timetabling Problems. In: 6th Metaheuristics International Conference (2005)
18. Goldberg, D.E.: Genetic Algorithms in Search, Optimization, and Machine Learning. Addison-Wesley, Reading (1989)
19. Grosan, C., Abraham, A.: Hybrid Evolutionary Algorithms. Methodologies, Architectures, and Reviews(2007)
20. Wren, A.: Scheduling, Timetabling and Rostering - A Special Relationship? In: 1st Practice and Theory of Automated Timetabling, pp. 46–75 (1996)
21. Wall, M.: GAlib: A C++ Library of Genetic Algorithm Components. Massachusetts Institute of Technology (1996)

Financial Market Prediction Using Feed Forward Neural Network

P.N. Kumar[1], G. Rahul Seshadri[2], A. Hariharan[3], V.P. Mohandas[4], and P. Balasubramanian[5]

[1,2,3] Dept. of CSE
pn_kumar@cb.amrita.edu, seshadri.rahul@gmail.com,
hariharan.anantharaman89@gmail.com
[4] Dept. of ECE
vp_mohandas@amrita.edu
[5] Amrita School of Business
Amrita Vishwa Vidyapeetham, Ettimadai, Coimbatore, Tamil Nadu, 641 105, India
bala@amrita.edu

Abstract. This paper outlines a methodology for aiding the decision making process for investment between two financial market assets (eg a risky asset versus a risk-free asset or between two risky assets itself), using neural network architecture. A Feed Forward Neural Network (FFNN) and a Radial Basis Function (RBF) Network has been evaluated. The model is employed for arriving at a decision as to where to invest in the next time step, given data from the current time step. The time step could be chosen on daily/weekly/monthly basis, based on the investment requirement. In this study, the FFNN has yielded good results over RBF. Consequently two such FFNN have been developed to enable us make a decision on investment in the next time step to decide between two risky assets. The prediction made by the two FFNN models has been validated from the actual market data.

Keywords: financial forecasting, risky assets, risk free assets, Feed Forward Neural Networks, Radial Basis Function Networks, minimum risk portfolio of two assets.

1 Introduction

Artificial Neural Networks (ANN) have earned themselves a unique position to model non-linear functions. In general, of all the AI techniques available, ANN deal best with uncertainty [1]. Like other forms of soft computing, ANN performs well in noisy data environments and has proved to exhibit a high tolerance to imprecision. These characteristics of ANN make them particularly suited to the arena of financial trading. The stock market represents a data source with an abundance of uncertainty and noise.

While ANN have been extensively studied [2,3] to perform a predictive analysis of the security prices, it may not be possible to model one general network that will fit every market and every security, hence models are built specific to markets and asset classes. Risky assets are those which do not have a guaranteed rate of return. An example of risky asset is stocks. Risk-free assets are those which give a return at a

K. Shah, V.R. Lakshmi Gorty, and A. Phirke (Eds.): ICTSM 2011, CCIS 145, pp. 77–84, 2011.

constant rate for eg, securities like Government bonds. Since there is always a particular amount of risk associated with a risky asset, an investor cannot be assured of making a profit. This creates a need for a choice between risky and risk-free assets in order to maximize the profits, while minimizing the risk at the same time. The historical market prices would provide us with a better idea about the market's behavior. This data is incorporated into a Feed-Forward Neural Network in order to predict the behavior of the market in the future. These have been modeled on the Bombay Stock Exchange (BSE) and the results are presented here.

2 Investment Decision: Risky Vs Risk Free Asset

2.1 Artificial Neural Network

An Artificial Neural Network (ANN) is a mathematical model or computational model that attempts to simulate the structure and/or functional aspects of biological neural networks. There are different types of Neural Networks. There are no standard rules available for determining the appropriate number of hidden layers and hidden neurons per layer. Smaller number of hidden nodes and hidden layers would render better generalization. A pyramid topology, which can be used to infer approximate numbers of hidden layers and hidden neurons, has been suggested by Shih [4]. Azoff [5] suggests that a network with one hidden layer and 2N + 1 hidden neurons is sufficient for N inputs, and states that the optimum number of hidden neurons and hidden layers is highly problem dependant. Gately [6] suggests setting the number of hidden nodes to be equal to the total of the number of inputs and outputs. Some researchers suggest training a great number of ANN with different configurations, and then select that configuration that performed best- Kim et al [7]. Finally, another reasonably popular method is used by some researchers such as Kim& Lee [8] and Versace et al[9], whereby genetic algorithms are used to select between possible networks given choices such as network type, architecture, activation functions, input selection and preprocessing. Another method Tan [1] , starts with a small number of hidden neurons and increase the number of hidden neurons gradually. A detailed comparative study can be seen at Vanstone [10]. For the purpose of this study, modeling has been attempted using a Feed-Forward Neural Network (FFNN) and a Radial Basis Function Network (RBF).

2.2 Modeling BSE Sensex

BSE Sensex, the most popular Indian stock index has been chosen for the study. The time step considered here is one month. BSE Sensex Index data pertaining to trading months starting from the year 2003 to 2008 is used for training the network. The duration is long enough and covers adequate market fluctuation. The one year monthly closing values of the Sensex for the year 2009 are used as the validation data set.

2.3 Model Architecture of FFNN-1

The functional form used is a FFNN (see Fig 1) with a single hidden unit with restricted inputs giving an output LeBaron[11,12,13]. The output is a simple function α (z_t, w_j). The equations given below define the network,

$$h_k = g_1(w_{0,k}\, z_{t,k} + w_{1,k}) \quad (1)$$

$$\alpha(z_t) = g_2\left(w_2 + \sum_{k=1}^{6} w_{3,k}\, h_k\right) \quad (2)$$

$$g_1(x) = \tanh(x) \quad (3)$$

$$g_2(x) = \tfrac{1}{2}(1+\tanh(x/2)) \quad (4)$$

where z_t is time t information and w_j are parameters. k takes values from 1 to 6 so that the weight array {w} consists of 19 parameters. The output from the intermediate neuron k is denoted h_k. The output from the network, α is a 0 or 1 which would suggest where to invest in the next time-step, a 0 indicating that risk-free asset would give higher returns for the particular time-step and a 1 indicating that investing in an Index Fund tracking the BSE would render higher returns.

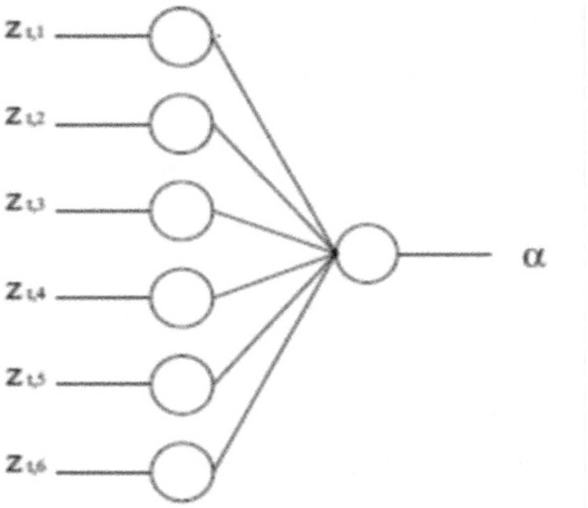

Fig. 1. FFNN-1: Decision between Risky Vs Risk Free Asset

2.4 Input Values to the FFNN-1: The Information Set

The information set consists of six items. These are designed to allow both passive and active, fundamental and technical trading strategies and combinations of these [11, 12]. Using just six items simplifies the decision making process by extracting potentially useful information from the large quantity of historic data The first three inputs are the returns on equity in the previous three time-steps, useful for technical trading. The fourth is a measure of how the current price differs from the rational-expectations price. The last two inputs measure the ratio between the current price and exponentially weighted moving averages of the price. Information set being:

$$z_{t,1} = r_t = \log((p_t + d_t)/p_{t-1}) \tag{5}$$

$$z_{t,2} = r_{t-1} \tag{6}$$

$$z_{t,3} = r_{t-2} \tag{7}$$

$$z_{t,4} = \log(r\, p_t / d_t) \tag{8}$$

$$z_{t,5} = \log(p_t / m_{1,t}) \tag{9}$$

$$z_{t,6} = \log(p_t / m_{2,t}) \tag{10}$$

Where p_t is the share price, d_t is the dividend paid, r is a constant and $m_{i,t}$ is the moving average given by

$$m_{i,t} = \rho_i\, m_{i,t-1} + (1 - \rho_i)\, p_t \tag{11}$$

with $\rho_1 = 0.8$ and $\rho_2 = 0.99$.

2.5 Results: Training and Testing FFNN-1

The MATLAB Neural Network Toolbox has been chosen for creating, training and testing the network. The FFNN-1 was trained with inputs from historical prices of BSE index, calculated taking monthly closing prices of BSE stock index from the year 2003 to 2008. The network was tested with data pertaining to the year 2009 and the results have been found to validate the market scenario. It has been found that the FFNN-1 with one hidden layer with six neurons has produced quite accurate results. The network prediction matched with the test data of 2009 market. The neural network thus establishes the functional dependency between the input parameters and the market behavior.

2.6 Radial Basis Function (RBF) Network

Radial Basis Function (RBF) network (Fig 2) is an artificial neural network that uses radial basis functions as activation functions. It is a linear combination of radial basis functions. They are used in function approximation, time series prediction, and control. RBF networks typically have three layers: an input layer, a hidden layer with a non-linear RBF activation function and a linear output layer. The output $\varphi : \mathbb{R}^n \rightarrow \mathbb{R}$, of the network is thus

$$\varphi(\mathbf{x}) = \sum_{i=1}^{N} a_i \rho(||\mathbf{x} - \mathbf{c}_i||) \tag{12}$$

where N is the number of neurons in the hidden layer, c_i is the center vector for neuron i, and a_i are the weights of the linear output neuron. In the basic form all inputs are connected to each hidden neuron. The norm is typically taken to be the Euclidean distance and the basis function is taken to be Gaussian.

$$\rho(\, ||\mathbf{x} - \mathbf{c}_i||\,) = \exp\left[-\beta\, ||\mathbf{x} - \mathbf{c}_i||^2\right] \tag{13}$$

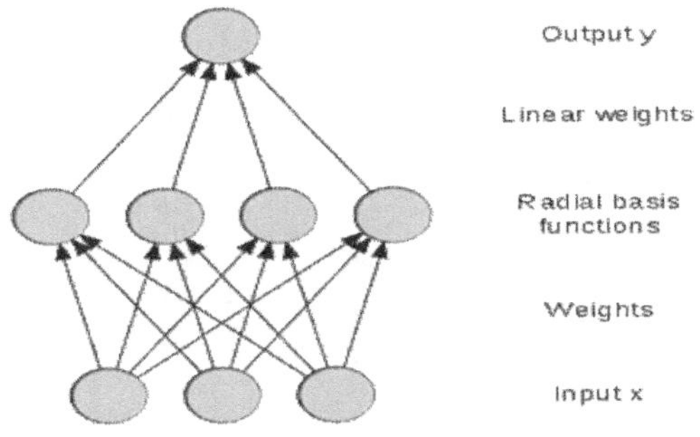

Fig. 2. Architecture of a radial basis function network

An input vector x is used as input to all radial basis functions, each with different parameters. The output of the network is a linear combination of the outputs from radial basis functions. The Gaussian basis functions are local in the sense that

$$\lim_{||x|| \to \infty} \rho(||\mathbf{x} - \mathbf{c}_i||) = 0 \tag{14}$$

i.e. changing parameters of one neuron has only a small effect for input values that are far away from the center of that neuron. RBF networks are universal approximators on a compact subset of $\mathbb{R}^n$. This means that a RBF network with enough hidden neurons can approximate any continuous function with arbitrary precision. The weights a_i, c_i and β are determined in a manner that optimizes the fit between φ and the data.

2.7 Results: Training and Testing RBF Network

The MATLAB Neural Network Toolbox has been chosen for creating, training and testing the RBF network. The network was trained with inputs from historical prices of BSE index, as was done above, taking monthly closing prices during the period 2003 to 2008. The network was tested with data pertaining to the year 2009. However, the RBF network did not yield results as was reflected in the market. This perhaps is attributable to the fact that the data presented to the network was close to each other, thereby resulting in improper clustering and inaccurate results.

3 Investment Decision: Between Two Risky Assets

3.1 The Concept

The above FFNN-1 model can be extended to enable comparison between two risky assets as well. The network architecture is modified accordingly, as given in Fig.3. This will give us the optimum proportion in which investment could be made between any two risky assets, and can be used as a guide for an informed investment decision between two stocks.

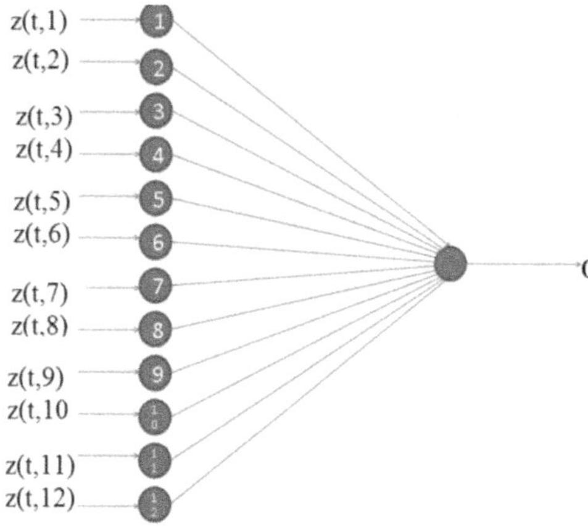

Fig. 3. FFNN-2: Decision between two risky assets

3.2 Model Architecture of FFNN-2

The network given in Fig.1 is modified to accommodate 12 inputs for this case (Fig.3). The same set of 6 inputs in respect of both the stocks is fed as input to the network as both stocks have risks associated with them. The functional form of the network is the same as the one discussed earlier. Historical prices of two stocks listed in the same index have been considered. The input sets are calculated for the same.

3.3 Training of FFNN-2

The targets for training the network was obtained using the Two-Asset case portfolio formula in (15)

$$\alpha = \frac{(\sigma_A)^2 - \rho_{AB}\ \sigma_A\ \sigma_B}{(\sigma_A)^2 + (\sigma_B)^2 - 2\ \rho_{AB}\ \sigma_A\ \sigma_B} \tag{15}$$

where α is the percentage invested in asset A, σ is the deviation and ρ is the correlation coefficient. Since this equation takes into account the deviation in prices of the stocks and the correlation between them, it gives the optimum proportion in which investment should be made between the two stocks, for minimal risk.

3.4 Results

Share prices of two listed BSE stocks viz TCS and Infosys were chosen to be compared. The input parameters and target values were calculated for prices from the period 2003-2008. The network was tested using data of the year 2009. The graph depicted in Fig.4 shows the percentage accuracy of the network's prediction with respect to the actual market value that emerges subsequently.

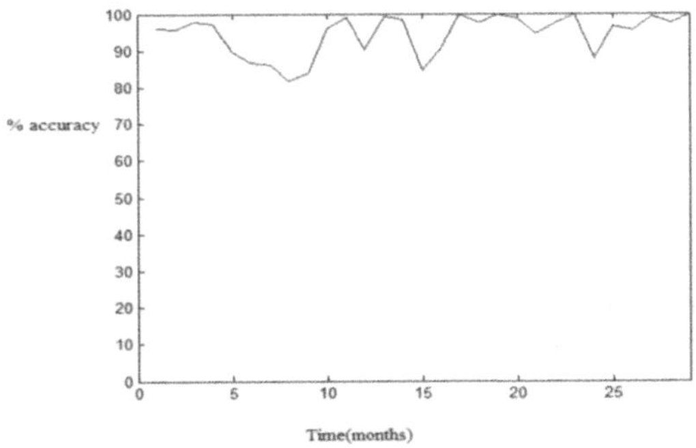

Fig. 4. Prediction: Selection between Two Risky Assets

4 Future Extension

The model presented above can be extended to choose between any two risky assets. This strategy can be effectively employed to compare between two mutual funds, two index funds or to arrive at an investment decision between two stock exchanges even. The limitation for this model is that it can be used for investment decision between only two assets. Further studies can explore the strategy for comparison of more than two risky assets. Whereas in FFNN-1 (in the case of Risky Vs Risk-free asset), the model only suggests a choice of where to invest, a possible extension might be to find out the proportion of wealth to be invested in Risky and Risk-free assets respectively.

5 Conclusion

Two neural network models (FFNN) have been designed, tested and validated from the BSE data to enable an investor make a decision at different time steps. FFNN with one hidden layer with six/twelve neurons has produced quite accurate results. However, the analysis carried out on a RBF network did not yield results reflected by the market. The model of FFNN-1 pertains to arriving at a decision between investment in a risky and a risk-free asset, whereas, FFNN-2 enables differentiation between two risky assets. Hence the models suggested by this paper can be used as a tool for an informed investment decision in the share markets. It is hoped that this can bring about a better investment strategy and help in achieving greater profits to investors.

References

1. Tan, C.N.W.: Artificial Neural Networks: Applications in Financial Distress Prediction and Foreign Exchange Trading. Wilberto Press, Gold Coast (2001)
2. White, H.: Economic Prediction using Neural Networks: The case of IBM Daily Stock Returns. In: Second Annual IEEE Conference on Neural Networks, pp. 451–458 (1988)
3. Oppenheimer, H.R., Schlarbaum, G.G.: Investing with Ben Graham: An Ex Ante Test of the Efficient Markets Hypothesis. Journal of Financial and Quantitative Analysis XVI(3), 341–360 (1981)

4. Shih, Y.: Neuralyst Users Guide: Cheshire Engineering Corporation (1994)
5. Azoff, M.E.: Neural Network Time Series Forecasting of Financial Markets. Wiley, Chichester (1994)
6. Gately, E.: Neural Networks for Financial Forecasting. Wiley, New York (1996)
7. Kim, J.-H., Park, S.-J., et al.: Stock Price Prediction using Back propagation Neural Network in KOSPI. In: International Conference on Artificial Intelligence IC-AI 2003, pp. 200–203 (2003)
8. Kim, K.J., Lee, W.B.: Stock market prediction using artificial neural networks with optimal feature transformation. Neural Computing and Applications 13(3), 255–260 (2004)
9. Versace, M., et al.: Predicting the exchange traded fund DIA with a combination of genetic algorithms and neural networks. Expert Systems with Applications 27(3), 417–425 (2005)
10. Bruce, V., Finnie, G.: An empirical methodology for developing stock market trading systems using artificial neural networks. Expert Systems with Applications: An International Journal 36(3) (2009)
11. LeBaron, B.: Empirical regularities from interacting long and short-memory investors in an agent based stock market. IEEE Transactions on Evolutionary Computation 5(5), 442–455 (2001)
12. LeBaron, B.: Evolution and time horizons in an agent based stock market. Macroeconomic Dynamics 5, 225–254 (2001)
13. LeBaron, B.: A builder's guide to agent based financial markets. Quantitative Finance 1(2), 254–261 (2001)
14. Markowitz, H.: Portfolio Selection. Journal of Finance, 77–91 (1952)

An Efficient Compression-Encryption Scheme for Batch-Image

Arup Kumar Pal, G. P. Biswas, and S. Mukhopadhyay

Department of Computer Science and Engineering
Indian School of Mines, Dhanbad-826004, India
arupkrpal@gmail.com, gp_biswas@yahoo.co.in,
msushanta2001@yahoo.co.in

Abstract. This paper presents an efficient scheme for image compression and encryption of a batch of images. Initially vector quantization is employed on a batch of images and it produces a common codebook and a batch of index-tables. Then all index-tables are fed to the encryption algorithm and the dimension of the codebook is further reduced by the principal component analysis. The proposed work gains high compression ratio as well as it supports speedy encryption-decryption process for a batch of images. The proposed scheme has been tested on a set of real images and simulation results show the improvement of compression ratio with acceptable image quality compare to other related scheme.

Keywords: Batch Image Encryption, Image Compression, Principal Component Analysis, Vector Quantization.

1 Introduction

With the rapid growth of usage of multimedia data like images, the demand for new approaches of their secured and speedy transmission over public channel is also rapidly increasing. Although the issues related to the multimedia data transmission and multimedia data security have been studied for a long time, but still there are several open issues like limited channel bandwidth, demand for faster and secure access of multimedia based applications. In general, before communication of confidential visual data, a suitable image compression [1] and image cryptosystem [2] are adopted for their efficient and secure transmission.

In literature [3-7], several integrated approaches of image compression and image encryption have been proposed. However all these schemes are designed for compression and encryption of a single image. Chen et al. [8] first pointed out that some application is used to send batch of images at one time. So they have devised a novel image compression-encryption scheme where at a time, a batch of images can be transmitted in secure and compressed form over public channel. In their proposed scheme initially they have trained a common VQ codebook for encoding all input images into corresponding index-tables. After VQ compression they have applied modified index-compression method for simultaneously compression and encryption of each index-table. In this paper we have also proposed a VQ [9-10] based integrated

K. Shah, V.R. Lakshmi Gorty, and A. Phirke (Eds.): ICTSM 2011, CCIS 145, pp. 85–90, 2011.

image compression and image encryption scheme for a batch of images. The proposed scheme can gain high compression and also ensure high security without using any conventional data encryption techniques.

The paper is organized as follows. After this introductory section, we present a brief overview of the VQ and PCA in section 2. The proposed algorithm is elaborated in section 3. Experimental results are given in section 4 to discuss the relative performance of the proposed scheme with other related schemes. In section 5 its security analysis is elaborated. Finally, conclusions are given in section 6.

2 Brief Concept of VQ and PCA

In this section we briefly review the working principal of VQ and PCA. Both VQ and PCA have lots of applications in image compression.

2.1 VQ for Image Compression

In VQ based image compression, initially image is decomposed into non-overlapping sub image blocks. Each sub block is then converted into one-dimension vector which is termed as training vector. From all these training vectors, a set of representative vectors are selected to represent the entire set of training vectors. The set of representative training vectors is called a codebook and each representative training vector is called codeword. Several algorithms have been proposed to construct a codebook. Among them the *LBG* algorithm [10] is one of the widely used codebook design methods. In *LBG* algorithm an initial codebook is chosen at random from the training vectors. Then the initial codebook is trained into an improved one by several iteration processes. After codebook design process, each codeword of the codebook is assigned a unique index value. In the next step i.e. encoding process, any arbitrary vector corresponding to a block from the image under consideration is replaced by the index of the most appropriate representative codeword. The matching is done based on the computation of minimum squared Euclidean distance between the input training vector and the codeword from the codebook. So after encoding process, an index-table is produced. The codebook and the index-table is nothing but the compressed form of the input image. In decoding process, the codebook which is available at the receiver end too, is employed to translate the index back to its corresponding codeword.

2.2 Principal Component Analysis

The principal component analysis (PCA) [11] is another popular and widely used technique for dimensionality reduction from a given data set. It aims at identifying the dependence structure within the data items. In PCA based data compression process, initially data is decomposed into non-overlapping blocks. After decomposition, each block is converted into a column vector. Then all vectors are concatenated to from a data matrix, X for performing PCA operation. Initially covariance matrix C of X is computed and then the eigenvalues (let there is total m number of eigenvalues) and respective eigenvectors are computed from $\boldsymbol{C}$ matrix. In PCA the compression is achieved when only first K number eigenvectors with largest eigenvalues are retained where $K<m$. The selected such eigenvectors are called feature matrix, V. Then the

coefficient matrix, Y is computed as follow, $Y = (V)^t X$. According to linear algebra, the original data matrix, X can be reconstruct as follow

$$X = (V)^{-1} Y = (V)^T Y \tag{1}$$

Since V is an orthogonal matrix and it can be found that $(V)^{-1} = (V)^t$.

3 The Proposed Technique

In the proposed scheme, the operations involved in the encoding and decoding process are described in the following subsections in details.

3.1 Encoding

In the scheme, first we have compressed all secret images using VQ then we perform the encryption operation only on index-tables. For codebook design, we have mainly followed the LBG algorithm. However the execution time of *LBG* algorithm is too high because it uses exhaustive search technique, which is computationally intensive. The searching technique is based on evaluating the minimum Euclidean distance between the input vector and the closest codeword. The computation of Euclidean distance can be avoided by using the *MCSS* function [12]. So after introducing *MCSS* function in the VQ process, the VQ process becomes faster compare to the conventional LBG algorithm. The trained codebook is used to encode all input images. In the next stage, the codebook is further compressed by PCA. Last step of the encoding process is to encrypt all index-tables. Initially, all index-tables are combined to form a single index-table. Then the confidentially of the index-table is done based on shuffling process using a secret pseudo random sequence (PNS). The shuffling process is used to replace an arrangement of an index-table accordingly to the sequence of the secret pseudo random number. The shuffling process for an index-table is shown as follow

$$T = \begin{pmatrix} p_1 & p_2 & \cdots & p_n \\ s_1 & s_2 & \cdots & s_n \end{pmatrix} \tag{2}$$

where p_i and s_i represent the original and random sequence position. The above shuffling process is faster and secure without using any conventional data encryption method. The schematic diagram of the encoding process for batch-image compression-encryption is shown in *Figure 1*.

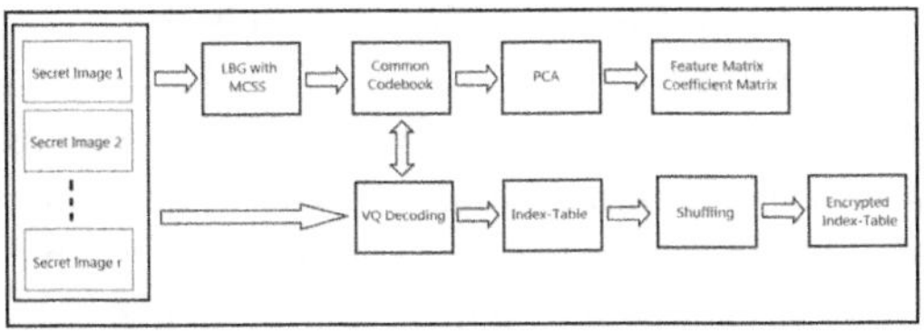

Fig. 1. The Schematic diagram of batch-image compression-encryption process

3.2 Decoding

In the decoding process, using Eq. 1, the receiver can obtain the codebook from the received feature matrix and coefficient matrix. Then the receiver will use same PNS which was used in encoding process. The inverse shuffling is employed for decryption the index-table and it is defined as follow

$$T^{-1} = \begin{pmatrix} s_1 & s_2 & \cdots & s_n \\ p_1 & p_2 & \cdots & p_n \end{pmatrix} \tag{3}$$

Since the shuffling and inverse shuffling process are based on same PNS. Instead of sending the PNS directly, both sender and receiver will generate same PNS using a PRNG and a seed. So in proposed scheme, seed will work as a cryptographic key. After decryption process, the secret images can be reconstructed by VQ decoding process. The schematic diagram of the decoding process for batch-image decompression-decryption is shown in *Figure 2*.

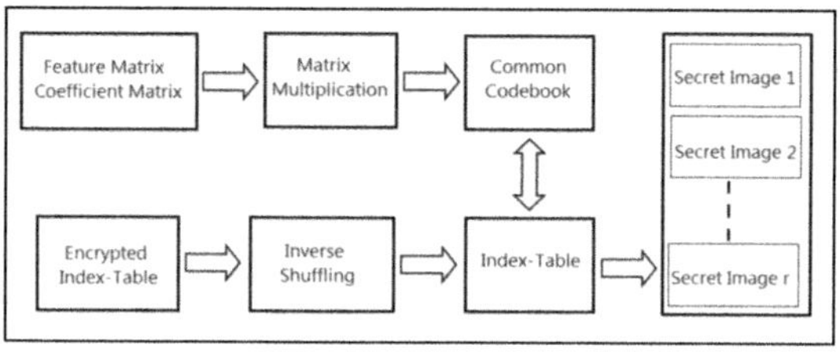

Fig. 2. The Schematic diagram of batch-image decompression-decryption process

4 Simulation Results

The proposed technique is applied on a set of standard images, but in this paper only the results for a batch of four images of size *512 × 512* pixels with 256 gray levels (shown in *Figure 3(a-d)*) are presented. In VQ process we have considered a common codebook of size *1024* with codeword of size *8×8* pixels. So after VQ process, it produces four index-tables of size *64 × 64* each. Since the physical size of the codebook is relatively huge so we compress the codebook further by applying PCA. In PCA, there is a tradeoff between the selection of the eigenvectors and the compression. In [13], authors have shown that the K $(4 \leq K \leq 8)$ number of eigenvectors is sufficient to represent the visual content. So in the simulation, we have restricted the selection of eigenvectors from 4 to 8. Table 1 presents the comparison with Chen et al. scheme [8] in terms of encrypted image size and the PSNR of the reconstructed images. From table 1, it is noted that the proposed scheme gives better PSNR value when $K=8$ but the size of the encrypted image is some extent high compare to Chen et al. scheme. But the scheme gives better compression than Chen et al.[8] scheme with acceptable image quality (as shown in Figure 3(e-h)) when $K=6$.

Table 1. Comparison between Chen et al. scheme [8] with the parameter TH=16 and the proposed scheme with K=4 to 8

Scheme	Original Image size	Encrypted image size	PSNR			
			Lena	Pepper	Airplane	Spaceman
Chen et al. scheme [8]	1024 KB	47.5KB	32.23	31.33	29.43	31.38
The Prop. when K=4	1024 KB	37KB	29.83	27.32	26.14	27.71
The Prop. when K=5	1024 KB	41.25KB	30.48	28.05	27.02	28.25
The Prop. when K=6	1024 KB	45.5KB	30.78	28.29	28.28	28.96
The Prop. when K=7	1024 KB	49.75KB	31.16	28.51	28.61	29.48
The Prop. when K=8	1024 KB	54KB	32.85	31.75	30.47	31.54

5 Security Analysis

The security of any encryption algorithm is measured by the size of the key space. When the size of the key space is large enough then the brute force attack becomes practically infeasible. In the proposed cryptosystem the secrecy depends on the selection of secret PNS. In our experiment we have taken a PNS of length *16384* (no of image × dimension of each index table). If attackers want to break the proposed cryptosystem then they have to know the PNS. The PNS guessing in our scheme is practically infeasible because there will be one optimal PNS out of total possible 16384! PNSs. Figures 3(i-l) show the decrypted images with wrong PNS. The images are completely non-traceable by human visual observation.

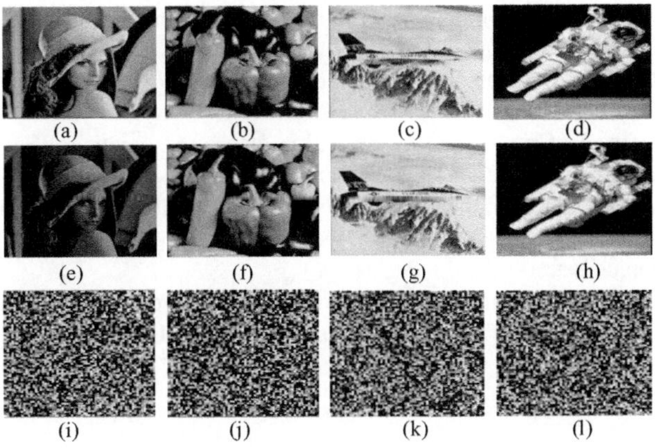

Fig. 3. (a-d) :The original test images of size 512×512 pixels; (e-f): The reconstructed images when K=6; (i-l)The decrypted images with wrong PNS

6 Conclusions

In this paper we have proposed a novel image compression-encryption scheme for a batch of images. A hybrid approach of VQ and PCA is presented for improving the compression ratio and the simple shuffling process is used to make the encryption process faster. In the scheme, we have achieved improved compression ratio with acceptable image quality. The scheme is also secured as evident from the security analysis. In conclusion, our proposed scheme is not only simple and effective in providing good quality compressed image, but also secure without using any convention data encryption algorithm.

References

1. Salomon, D.: Data Compression the Complete Reference. Springer international Edition (2005)
2. Lian, S.: Shiguo Lian: Multimedia content encryption techniques and applications. CRC Press, Boca Raton (2008)
3. Hwang, M.S., Chang, C.C., Chen, T.S.: A new encryption algorithm for image cryptosystems. The Journal of Systems and Software 58, 83–91 (2001)
4. Pal, A.K., Biswas, G.P., Mukhopadhyay, S.: Designing of High-Speed Image Cryptosystem Using VQ Generated Codebook and Index Table. In: 2010 International Conference on Recent Trends in Information, Telecommunication and Computing, pp. 39–43 (2010)
5. Chang, H.K., Liu, J.L.: A linear quadtree compression scheme for image encryption, Signal Processing. Image Communication, 279–290 (1997)
6. Chuanfeng, L., Qiangfu, Z.: Integration of data compression and cryptography: Another way to increase the information security. In: 21st International Conference on Advanced Information Networking and Applications Workshops, AINAW 2007 (2007)
7. Azman, H.K., Zurinahni, Z.: Enhance performance of secure image using wavelet compression. World Academy of Science, Engineering and Technology (2005)
8. Chen, T.H., Wu, C.S.: Compression-unimpaired batch-image encryption combining vector quantization and index compression. Information Sciences 180, 1690–1701 (2010)
9. Greso, A., Gray, R.M.: Vector Quantization and Signal Compression. Kluwer Academic Publishers, Boston (1991)
10. Linde, Y., Buzo, A., Gray, R.M.: An algorithm for vector quantizer design. IEEE Transactions on Communications 28, 84–95 (1980)
11. Jolliffe, I.T.: Principal Component Analysis, 2nd edn. Springer, Heidelberg (2002)
12. Cheng, S.M., Lo, K.T.: Fast clustering process for vector quantization codebook design. IEEE Electronics Letters 32, 311–312 (1996)
13. Kaarna, A., Zemcik, P., Kalviainen, P., Parkkinen, J.: Compression of multispectral remote sensing images using clustering and spectral reduction. IEEE Trans. GRS 38, 1073–1082 (2000)

Improving Performance of XML Web Services

Girish Tere[1] and Bharat Jadhav[2]

[1] Department of Computer Science, Shivaji University, Kolhapur – 416 004, India
girish.tere@gmail.com
[2] Department of Electronics and Computer Science, Y.C. Institute of Science, Satara – 415 001, India
btj21875@indiatimes.com

Abstract. XML is now worldwide standards for data definition. It is universal language for information exchange and has been used by many organizations for developing enterprise applications. With the widespread adoption of SOAP and Web services, XML-based processing, and parsing of XML documents in particular, is becoming a performance-critical aspect of business computing. There are connections between Formal languages, Automata theory and XML. Finite State Machines (FSM) provide a powerful way to describe dynamic behavior of systems and components. XML has many important features, including platform and language independence, flexibility, expressiveness, and extensibility. Thus, the combination of these characteristics with the interoperability trait of Web services is an attractive way to design distributed applications. To improve web service performance, we have parsed XML documents using Deterministic Finite Automata (DFA). DFA is constructed to efficiently parse XML documents containing SOAP messages by encoding the XML parser's states as a DFA. In this paper we discuss our simple example and performance results we obtained.

Keywords: Automata, DFA, SOAP, WSDL, XML documents.

1 Introduction

Use of Web services is mandatory in developing enterprise application. Producer of Web services need to publish Web service using WSDL and consumer of Web service need to search the needed Web service and use the Web service as per the contents of WSDL. WSDL documents are basically a contract for using Web service. WSDL plays an important role in the overall Web services architecture since it describes the complete contract for application. Web services communicate to each other as well as with client using WSDL contract defined in XML. This communication is done by exchanging SOAP messages. Many studies [6],[7],[8],[12],[15] have shown that the use of XML can lower performance. XML primarily uses UTF-8 as the representation format for data. Sending commonly used data structures via standard implementations of SOAP incurs severe performance overheads, making it difficult for applications to adopt Web services based grid middleware. Due to the widespread adoption of standards in Web services, it is critically important to investigate the impact on

K. Shah, V.R. Lakshmi Gorty, and A. Phirke (Eds.): ICTSM 2011, CCIS 145, pp. 91–98, 2011.

performance for the kinds of XML used in web applications. The flexibility and loose coupling of XML-based standards allows senders and receivers of XML data to independently deploy selected optimizations, according to the communication patterns and data structures in use [5],[13].

SOAP has plus points like transparency, expressiveness, platform and language independence, extensibility and robustness. SOAP is a popular choice as the common underlying protocol for interoperability between Web services. These features facilitate the use of SOAP in diverse applications with widely varying characteristics and requirements. Clients and Web service endpoints can also add optimizations in their implementations. The convergence of Web services standards has made SOAP important, requiring the evaluation of SOAP for data types and communication patterns used by grid applications. It is thus important to have a test framework to determine if a particular SOAP toolkit can meet the performance requirements of an application, or if some other communication protocol should be employed. Finite State Machines (FSM) provide a powerful way to describe dynamic behavior of systems and components. XML has many important features, including platform and language independence, flexibility, expressiveness, and extensibility [18],[19],[20]. Thus, the combination of these characteristics with the interoperability trait of Web services is an attractive way to design distributed applications. For processing of XML documents, we have used DFA based approach, as DFAs are executed faster on any computer [9].

2 Generating DFA Based Parser

Parsing is the process of reading a document and dissecting it into its elements and attributes, which can then be analyzed. In XML, parsing is done by an XML processor, the most fundamental building block of a Web application [18], [22]. This process is shown in Fig. 1. The XML processor parses and generates an XML document. The application uses an API to access objects that represent part of the XML document.

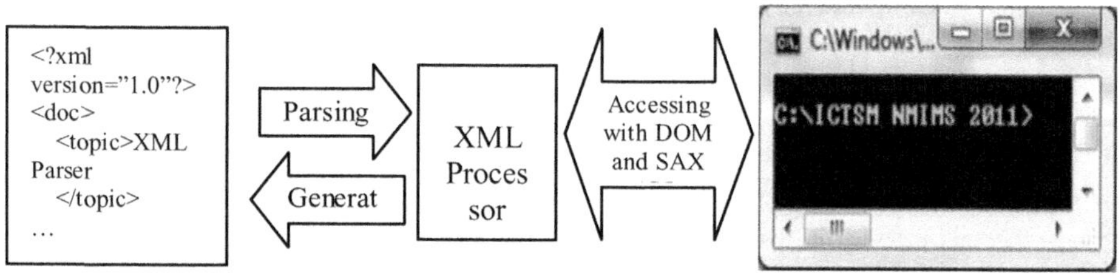

Fig. 1. XML Parsing Process

We developed WSDL Processor [2], which accepts WSDL or Schema and generates source codes for the implementation of a high-performance Web service as shown in Fig. 2. WSDL is an XML-based language for describing Web services and how to access them. A WSDL document is just a simple XML document. It contains set of definitions to describe a web service.

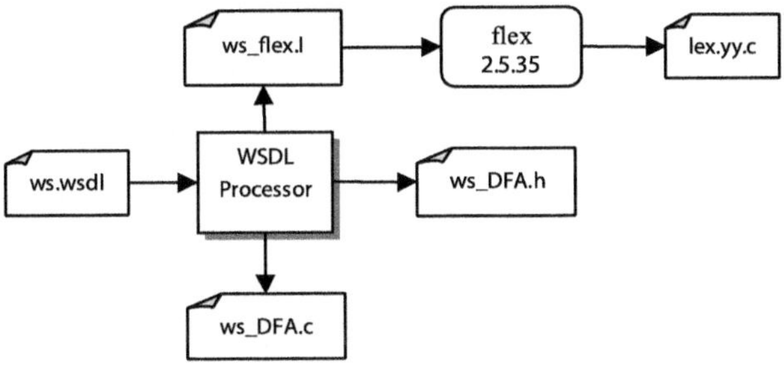

Fig. 2. Parser Genrator

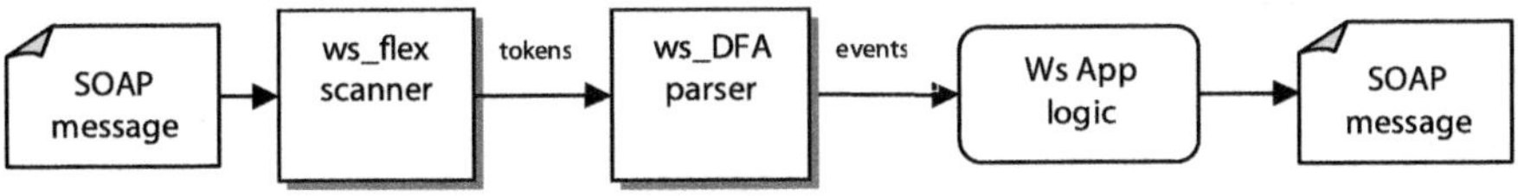

Fig. 3. Processing Web service

Many parsers are built using lexical analysis generators that tokenize input into logical chunks, called as tokens [1]. These tokens act as input for the parser. These generators generally take as input a collection of regular expressions matching each class of token and actions that are executed when those tokens are matched. This is explained in Fig. 3. Some of valid tokens are elements, attributes and data. flex is a commonly used scanner generator. flex transforms the collection of regular expressions into a Non-Deterministic Finite Automaton (NFA) which recognizes the union of the patterns[11]. This NFA is converted into a DFA and then reduced. The DFA is written out as a table of states of two dimensions, DFA state and character input. The input is scanned by reading each character in input order, looking up the current state and current input in the DFA table to find the next state. Any required action is executed when the state changes.

3 Application

Our example describes a parser for a `repeatMessage` service in detail. Fig. 4 shows the schema/DTD of the repeatMessage SOAP request message. The `repeatMessage` message element contains a child element `message` of type XSD string.

The wsdlProcessor tool generates the Flex description shown in Fig. 5. For this example, the PUSH and POP operations are simply keeping track of the node nesting level in the document. The level indicator is used for controlling the DFA transitions. The source code of the DFA generated by wsdlProcessor is shown in Fig. 6. The `yylex` function returns the next token from the Flex scanner shown in Fig. 5.

```
<schema targetNamespace="urn:repeatMessage"
xmlns:xsd="http://www.w3.org/2001/XMLSchema"
xmlns="http://www.w3.org/2001/XMLSchema">
<element name="repeatMessage">
<customType>
   <sequence>
<element name="message" type="xsd:string"/>
<any namespace="##any" lowerlimit="0"
upperlimit="100"/>
   </sequomence>
</customType>
</element>
</schema>
```

Fig. 4. The repeatMessage Message Schema

```
%{
#include "repeatMessageDFA.h"
#define PUSH level++;
#define POP level--;
%}
blank [ \t\v\n\f\r]*
name [^>/:= \t\v\n\f\r]+
qual {name}:|""
open <{qual}
close [^>]*>
data [^<]*
%%
{blank}                  // ignore white space
{open}"?"{close}         // ignore declaration
{open}"!"{close}         // ignore comment
{open}"/"{close}         POP
{open}"Header"{close}    PUSH return HEADER;
{open}"Body"{close}      PUSH return BODY;
{open}"repeatMessage"{close} PUSH return ELEMENT_repeatMessage;
{open}"message"{close}   PUSH return ELEMENT_message;
{open}{name}"/"{close}
{open}{name}{close}      PUSH
{data}                   return DATA;
<<EOF>>                  return EOF;
%%
```

Fig. 5. Flex Specification for repeatMessage

Fig. 6 shows the source code generated by WSDL Processor [2], [9], [14]. The `event("repeatMessage/message", yytext)` call returns an XPath expression and string data to the server application.

```
int repeatMessageDFA()
{
   int token, state = 0;
   while ((token = yylex()) != EOF)
   {
         switch (state) {
            case 0: if (token == BODY && level == 2)
                        state = 1;
                    break;
            case 1: if (token == ELEMENT_repeatMessage && level == 3)
                        state = 2;
                    break;
            case 2: if (token == ELEMENT_message && level == 4)
                        state = 3;
                    break;
            case 3: if (token != DATA || level != 4)
                        return error("Invalid input value");
                    event("repeatMessage/message", yytext);
                    state = 4;
                    break;
            case 4: if (token == EOF && level == 0)
                        return ACCEPT;
                    return error("Invalid message");
         }
   }
   return error("End of file");
}
```

Fig. 6. DFA for repeatMessage

4 Performance Analysis

Experiment were carried on Dell Inspiron Laptop with Intel Core 2 Duo, 4 GB RAM. We measured Web services performance on the same machine, i.e. client and server was installed on the same machine. The performance of the repeatMessage

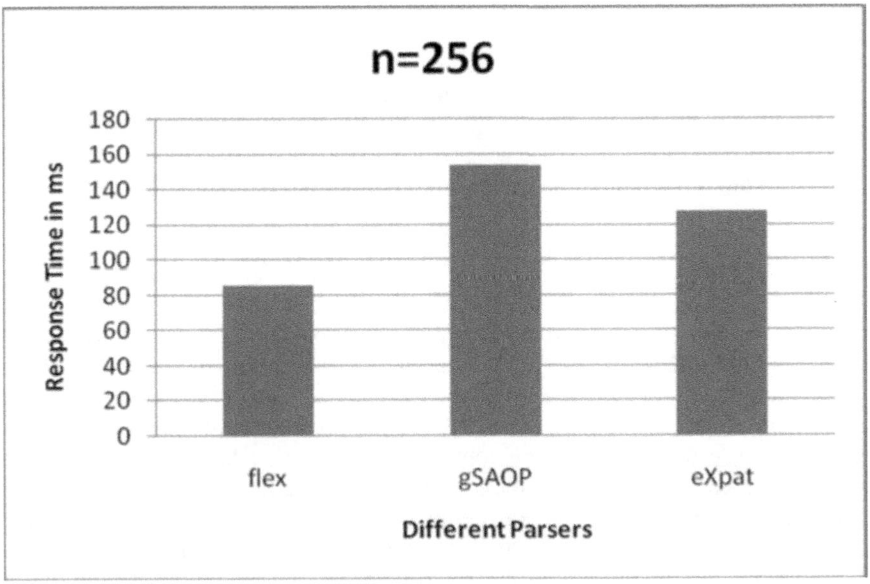

Fig. 7. Performance of parsing repeatMessage application with n=256

application is compared to the performance of parser built with gSOAP 2.5.1 and the performance of eXpat XML parser [10]. It is a stream-oriented parser written in C. The performance of the eXpat parser is considered to be good. The gSOAP toolkit is an efficient implementation of Web services standards in C and C++ [16], [17]. We changed the `message` string size n from 256 to 512 and repeated the experiment. Performance comparison of these parsers with varied n is shown in Fig. 7 and 8. Performance analysis of the result obtained show that the performance of DFA-based parser is better than other two parses considered, viz. gSOAP and eXpat.

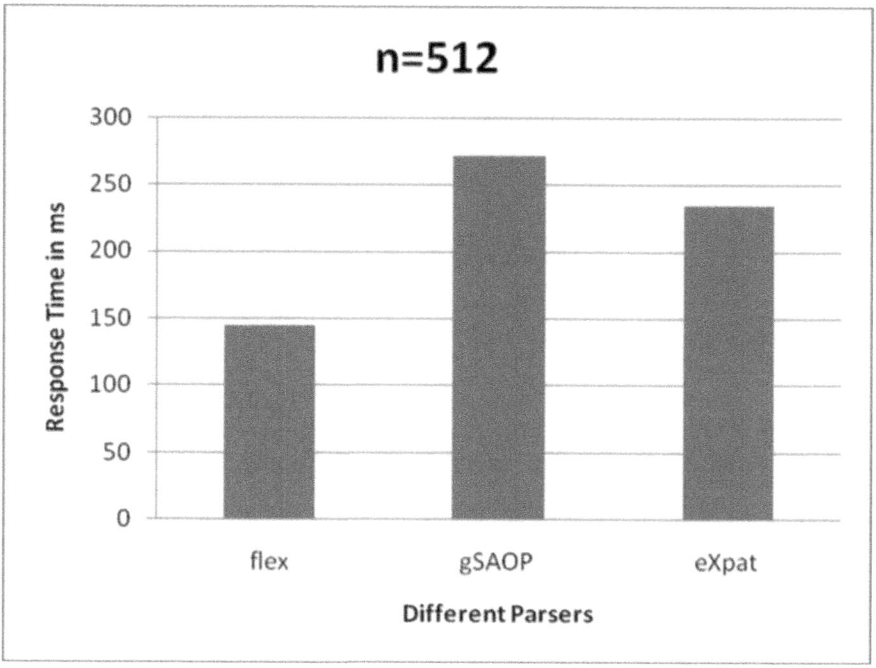

Fig. 8. Performance of parsing repeatMessage application with n=512

5 Conclusions

Many researchers tried to improve XML parsing. Web services interact with client by exchanging SOAP messages and can be used as per WSDL. Both SOAP and WSDL are basically XML document. Therefore to improve Web services performance, we tried to improve XML parsing work. XML parser has some finite states. Therefore, we used DFA to effectively reduce computational overheads for XML parsing of SOAP/XML messages. The DFA is generated by a code generator that takes a WSDL as input and generates codes for an optimized schema-specific SOAP/XML message parser. We observed that the performance of the DFA-based parser built with a scanner produced with Flex is better than the performance of parser built with gSOAP 2.5.1 and eXpat parser. However, we have not considered processing of XML namespaces. For developing this parser, we considered only ordered XML documents.

References

1. Aho, A., Sethi, R., Ullman, J.: Compilers: Principles, Techniques and Tools, 2nd edn. Addison-Wesley, Reading (2006)
2. Skonnard, A.: Understanding WSDL. Microsoft Corporation (2010), http://msdn.microsoft.com/enus/library/ms996486printe.aspx (accessed on August 30, 2010)
3. Apache Software Foundation, Xerces2 Java Parser, http://xml.apache.org/xerces2-j (accessed July 23, 2010)
4. Slominski, A.: XML Pull Parser version 2.1.8
5. Chan, C., Felber, P., Garofalakis, M., Rastogi, R.: Efficient Filtering of XML Documents with XPath Expressions. In: Proceedings of the International Conference on Data Engineering (2002)
6. Kohlhof, C., Steele, R.: Evaluating SOAP for high performance business applications: Real-time trading systems. In: Proceedings of the 2003 International WWW Conference, Budapest, Hungary (2003)
7. Davis, D., Parashar, M.: Latency performance of SOAP implementations. In: Proceedings of the 2nd IEEE International Symposium on Cluster Computing and the Grid (2002)
8. Chen, D., Wong, R.K.: Optimizing The Lazy DFA Approach for XML Stream Processing. In: The Fifteenth Australasian Database, Conference (ADC 2004), Dunedin, New Zealand, vol. 27 (2004)
9. Neven, F.: Automata theory for XML researchers. SIGMOD Record 31(3) (2002)
10. Clark, J.: eXpat XML parser, http://expat.sourceforge.net (accessed on September 16, 2010)
11. Chiu, K., Lu, W.: Compiler-based approach to schema-specific XML parsing. In: First International Workshop on High Performance XML Processing, May 17-22. ACM Press, New York (2004)
12. Chiu, K., Govindaraju, M., Bramley, R.: Investigating the limits of SOAP performance for scientific Computing. In: Proceedings of the 11th IEEE International Symposium on High-Performance Distributed Computing (2002)
13. Li, L., Niu, C., Chen, N., Wei, J.: High Performance Web Services Based on Service-Specific SOAP Processor. In: IEEE International Conference on Web Services (ICWS 2006), pp. 603–610 (2006)
14. Murata, M., Lee, D., Mani, M.: Taxonomy of XML schema languages using formal language theory. In: Extreme Markup Languages (2001)
15. van Engelen, R.: Pushing the SOAP envelope with Web services for scientific computing. In: Proceedings of the International Conference on Web Services (ICWS), Las Vegas, pp. 346–352 (2003)
16. van Engelen, R., Gallivan, K.: The gSOAP toolkit for web services and peer-to-peer computing networks. In: 2nd IEEE International Symposium on Cluster Computing and the Grid (2002)
17. van Engelen, R., Gupta, G., Pant, S.: Developing web services for C and C++. In: IEEE Internet Computing, pp. 53–61 (March 2003)
18. Green, T., Miklau, G., Onizuka, M., Suciu, D.: Processing XML Streams with Deterministic Automata. In: 9th International Conference on Database Theory, Siena, Italy, January 8-10 (2003)
19. Löwe, W.M., Noga, M.L., Gaul, T.S.: Foundations of Fast Communication via XML. Annals of Software Engineering 13(1-4), 357–359 (2002)
20. Zhang, W., van Engelen, R.A.: An Adaptive XML Parser for Developing High-Performance Web Services. In: Fourth IEEE International Conference on eScience, pp. 672–679 (2008)

21. Zhang, W., van Engelen, R.A.: High-Performance XML Parsing and Validation with Permutation Phrase Grammar Parsers. In: IEEE International Conference on Web Services, ICWS 2008, Beijing, pp. 286–294 (2008)
22. Martens, W., Niehren, J.: On the minimization of XML Schemas and tree automata for unranked trees. Journal of Computer and System Sciences 73(4) (June 2007)
23. XMLTK, The XML toolkit University of Washington (2002), `http://www.cs.washington.edu/homes/suciu/XMLTK/` (acessed on August 25, 2010)

Image Retrieval Using Texture Patterns Generated from Walsh-Hadamard Transform Matrix and Image Bitmaps

H.B. Kekre[1], Sudeep D. Thepade[2], and Varun K. Banura[3]

[1] Senior Professor
hbkekre@yahoo.com
[2] Ph.D. Research Scholar & Associate Professor
sudeepthepade@gmail.com
[3] B.Tech Student
Computer Engg. Dept., SVKM's NMIMS (Deemed-to-be University), Mumbai, India
varunkbanura@gmail.com

Abstract. The theme of the work presented here is texture pattern based image retrieval techniques using image bitmaps and Walsh-Hadamard transform. Different texture patterns namely '4-pattern', '16-pattern', '64-pattern' are generated using Walsh-Hadamard transform matrix and then compared with the bitmap of an image to generate the feature vector as the matching number of ones and minus ones per texture pattern. The proposed content based image retrieval (CBIR) techniques are tested on a generic image database having 1000 images spread across 11 categories. For each proposed CBIR technique 55 queries (randomly selected 5 per category) are fired on the image database. To compare the performance of image retrieval techniques average precision and recall of all the queries per image retrieval technique are computed. The results have shown improved performance (higher precision and recall values of crossover points) with the proposed methods compared to the colour averaging based image retrieval techniques. Further the performance of proposed image retrieval methods is enhanced using even image part. The proposed CBIR methods also show ameliorated performance with image bitmaps generated using tiling. In the discussed image retrieval methods, the combination of original and even image part for 16-pattern texture with image bitmaps generated using tiling gives the highest crossover point of precision and recall indicating better performance.

Keywords: CBIR, Walsh-Hadamard transform, Texture, Pattern, Bitmap.

1 Introduction

Today the information technology experts are facing technical challenges to store/transmit and index/manage image data effectively to make easy access to the image collections of tremendous size being generated due to large numbers of images generated from a variety of sources (digital camera, digital video, scanner, the internet etc.). The storage and transmission is taken care of by image compression [4,7,8]. The image indexing is studied in the perspective of image database [5,9,10,13,14] as one

K. Shah, V.R. Lakshmi Gorty, and A. Phirke (Eds.): ICTSM 2011, CCIS 145, pp. 99–106, 2011.

of the promising and important research area for researchers from disciplines like computer vision, image processing and database areas. The hunger of superior and quicker image retrieval techniques is increasing day by day. The significant applications for CBIR technology could be listed as art galleries [15,17], museums, archaeology [6], architecture design [11,16], geographic information systems [8], weather forecast [8,25], medical imaging [8,21], trademark databases [24,26], criminal investigations [27,28], image search on the Internet [12,22,23]. The paper attempts to provide better and faster image retrieval techniques.

1.1 Content Based Image Retrieval

For the first time Kato et.al. [7] described the experiments of automatic retrieval of images from a database by colour and shape feature using the terminology content based image retrieval (CBIR). The typical CBIR system performs two major tasks [19,20] as feature extraction (FE), where a set of features called feature vector is generated to accurately represent the content of each image in the database and similarity measurement (SM), where a distance between the query image and each image in the database using their feature vectors is used to retrieve the top "closest" images [19,20,29]. For feature extraction in CBIR there are mainly two approaches [8] feature extraction in spatial domain and feature extraction in transform domain. The feature extraction in spatial domain includes the CBIR techniques based on histograms [8], BTC [4,5,19], VQ [24,28,29]. The transform domain methods are widely used in image compression, as they give high energy compaction in transformed image [20,27]. So it is obvious to use images in transformed domain for feature extraction in CBIR [26]. But taking transform of image is time consuming. Reducing the size of feature vector using pure image pixel data in spatial domain and getting the improvement in performance of image retrieval is shown in [2] & [3]. But the problem of feature vector size still being dependent on image size persists in [2] & [3]. Here the query execution time is further reduced by decreasing the feature vector size further and making it independent of image size. Many current CBIR systems use the Euclidean distance [4-6,11-17] on the extracted feature set as a similarity measure. The Direct Euclidian Distance between image P and query image Q can be given as equation 1, where Vpi and Vqi are the feature vectors of image P and Query image Q respectively with size 'n'.

$$ED = \sqrt{\sum_{i=1}^{n} (Vpi - Vqi)^2} \tag{1}$$

2 Texture Patterns Using Walsh-Hadamard Transform Matrix

Using the Walsh-Hadamard transform assorted texture patterns namely 4-pattern, 16-pattern and 64-pattern are generated. To generate N^2 texture patterns, each column of the Walsh-Hadamard matrix [21,22,26] of size NxN is multiplied with every element of all possible columns of the same matrix (one column at a time to get one pattern). The texture patterns obtained are orthogonal in nature. Figure 1(a) shows a 2X2

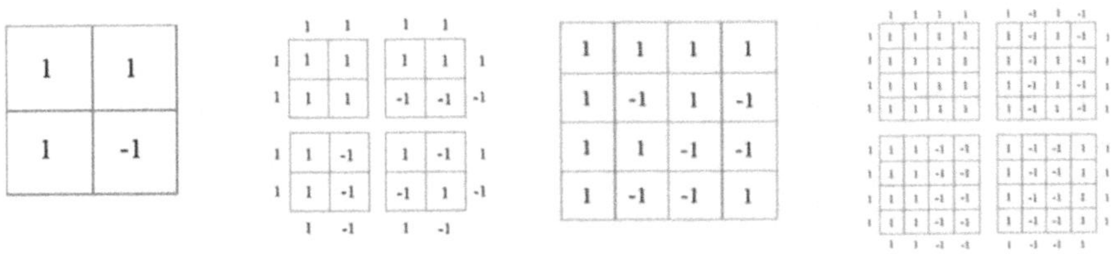

1(a) 1(b) 1(c) 1(d)

1(a). 2x2 Walsh-Hadamard transform matrix, **1(b).** '4-pattern' texture patterns, **1(c).** 4x4 Walsh-Hadamard transform matrix, **1(d).** First four '16-pattern' texture patterns

Fig. 1. Walsh-Hadamard Matrix Pattern Generation

Walsh-Hadamard matrix. The four texture patterns generated using this matrix are shown in figure 1(b). Similarly figure 1(d) shows first four texture patterns (out of total 16) generated using 4X4 Walsh-Hadamard matrix shown in figure 1(c).

3 Generation of Image Bitmaps

Image bitmaps of colour image are generated using three independent red (R), green (G) and blue (B) components of image to calculate three different thresholds. Let X={R(i,j),G(i,j),B(i,j)} where i=1,2,....m and j=1,2,....,n; be an m×n color image in RGB space. Let the thresholds be TR, TG and TB, which could be computed as per the equation given below as 2.

$$TR = \frac{1}{m*n}\sum_{i=1}^{m}\sum_{j=1}^{n} R(i,j), \quad TG = \frac{1}{m*n}\sum_{i=1}^{m}\sum_{j=1}^{n} G(i,j), \quad TB = \frac{1}{m*n}\sum_{i=1}^{m}\sum_{j=1}^{n} B(i,j) \tag{2}$$

Here three binary bitmaps will be computed as BMr, BMg and BMb. If a pixel in each component (R, G, and B) is greater than or equal to the respective threshold, the corresponding pixel position of the bitmap will have a value of 1 otherwise it will have a value of -1.

$$BMr(i,j) = \begin{cases} 1, & if.R(i,j) >= TR \\ -1, & if.R(i,j) < TR \end{cases}, \quad BMg(i,j) = \begin{cases} 1, & if.G(i,j) >= TG \\ -1, & if.G(i,j) < TG \end{cases}, \quad BMb(i,j) = \begin{cases} 1, & if.B(i,j) >= TB \\ -1, & if.B(i,j) < TB \end{cases} \tag{3}$$

To generate tiled bitmaps, the image is divided into four non-overlapping equal quadrants and the average of each quadrant is considered to generate the respective tile of the image bitmap.

4 Proposed CBIR Methods

After generating bitmap of the image, to generate feature vectors the bitmap of each image is compared with the generated texture patterns to find matching number of ones and minus ones. The size of the feature vector of image is given by equation 4.

$$\text{Feature vector size}=2*3*(\text{no. of considered texture-pattern}) \tag{4}$$

Using three assorted texture pattern set along with original and original-even image, total six novel feature vector generation methods can be used resulting into six new image retrieval techniques. Colour averaging based CBIR methods [1,2,3] are considered to compare the performance of proposed CBIR techniques. In the proposed CBIR techniques the combination of original and even part of images give better results than original image alone [1,2]. The result of the original-even feature vector is further ameliorated using image bitmaps generated by tiling. The main advantage of proposed CBIR methods is reduced time complexity for query execution due to reduced size of feature vector resulting into faster image retrieval with better performance. Also the feature vector size is independent of image size in proposed CBIR methods.

Table 1. Feature vector size of discussed image retrieval techniques

CBIR Technique	**RCM**	**FDM**	**RCFDM**	**4-Pattern**	**16-Pattern**	**64-Pattern**
Feature Vector Size for NxN image	2N	2N-1	4N-1	8	32	128

5 Implementation

The implementation of the discussed CBIR techniques is done in MATLAB 7.0 using a computer with Intel Core 2 Duo Processor T8100 (2.1GHz) and 2 GB RAM. The CBIR techniques are tested on the augmented Wang image database [18] of 1000 variable size images spread across 11 categories of human being, animals, natural scenery and manmade things, etc. To assess the retrieval effectiveness, we have used the precision and recall as statistical comparison parameters [4,5] for the proposed CBIR techniques.

6 Results and Discussion

For testing the performance of each proposed CBIR method, 55 queries (randomly selected 5 from each category) are fired on the image database. The feature vector of query image and database image are matched using the Euclidian distance. The average precision and recall values are found for all the proposed CBIR methods. The intersection of precision and recall values gives the crossover point. The crossover point of precision and recall is computed for all the proposed CBIR methods. The one with higher value of crossover point indicates better performance.

Figure 2 shows the performance comparison of proposed CBIR methods with the colour averaging based CBIR methods [1,2,3]. It is observed that the '4-pattern' texture based image retrieval gives the worst performance. The '16-pattern' texture based image retrieval has the highest crossover point thus indicating better performance. Figure 3 shows performance comparison of the proposed CBIR methods with the combination of original and even image part. It is observed that the combination of original and even image part gives better performance than the original alone. Here the '16-pattern' texture based image retrieval gives the highest crossover point with the combination of original and even image. Figure 4 shows

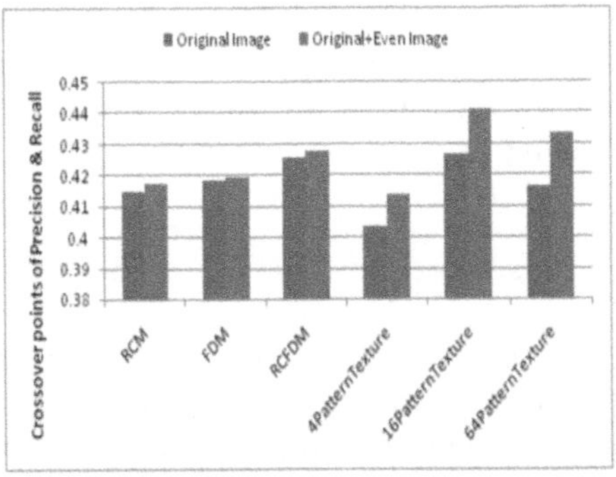

Fig. 2. Performance comparison of proposed CBIR methods with the colour averaging based CBIR methods

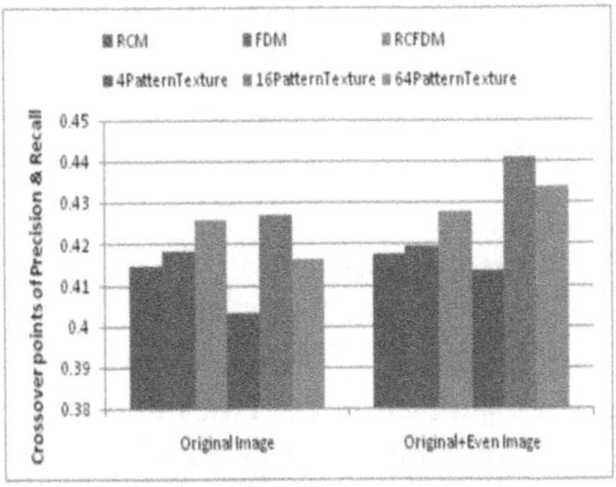

Fig. 3. Performance comparison of the proposed CBIR methods with the combination of original and even image part

amelioration of the '16-pattern' texture based image retrieval using tiled bitmaps. The '16-pattern' texture gives the highest crossover point of precision and recall for the 4-tile bitmaps. On further increasing the number of tiles in the bitmap, the results start deteriorating. Figure 5 shows the performance comparison of the '16-pattern' texture based image retrieval using tiled bitmaps with the combination of original and even image part. It is observed that the combination of original and even part image gives better performance than original alone. The '16-pattern' texture based image retrieval using 4-tile bitmaps with the combination of original and even part image gives the highest crossover point among the proposed CBIR methods thus indicating better performance. From comparison of colour averaging based image retrieval techniques [1,2,3] with the proposed CBIR methods it is observed that the 4-pattern texture gives the worst result. However increasing the number of texture patterns helps in improving the performance of proposed image retrieval method. Feature extraction using the combination of original image with even part of image further improves the performance of the proposed texture pattern based CBIR. Increasing the number of texture patterns helps in performance amelioration up to certain extent only

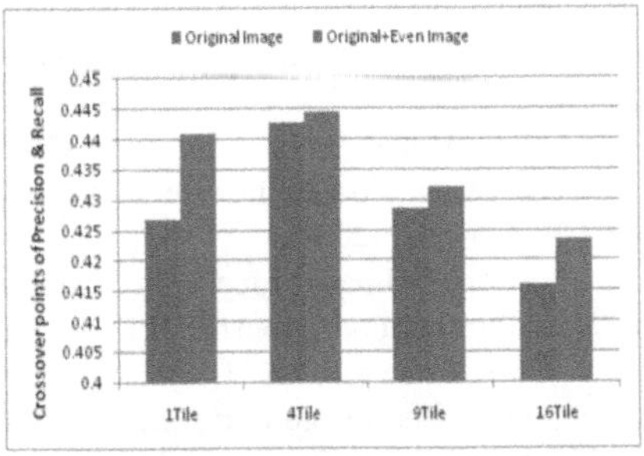

Fig. 4. Amelioration of the '16-pattern' texture based image retrieval using tiled bitmaps

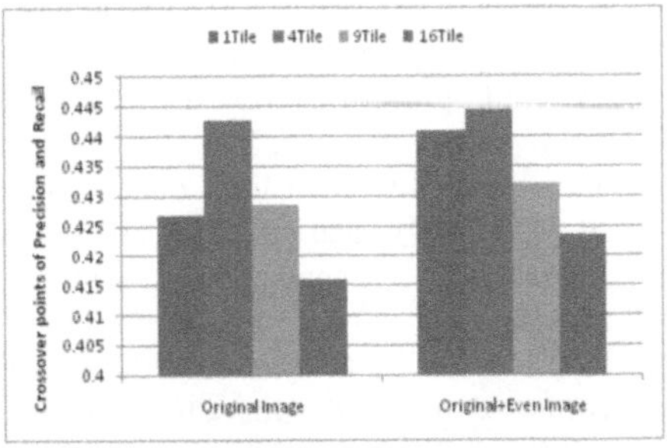

Fig. 5. Performance comparison of the '16-pattern' texture based image retrieval using tiled bitmaps with the combination of original and even image part

(16-pattern texture) beyond which the performance is degraded. The result of the 16-pattern texture is further enhanced by using image bitmaps generated using tiling. However tiled bitmaps can improve the performance up to a certain level (4 tiles) beyond which the performance starts deteriorating.

7 Conclusion

As compared to conventional feature extraction like colour averaging, the performance of image retrieval can be improved using texture patterns generated from Walsh-Hadamard transform and image bitmaps. Among the various texture patterns used for content based image retrieval, "16-pattern" texture patterns give the best result with the combination of original image and even part of image using "4-Tile" image bitmaps. The precision-recall crossover point of this 16-pattern texture is higher than that of RCM, RCFDM [1,2] colour averaging based image retrieval techniques . Moreover, it is observed that the performance of proposed CBIR method improves with increasing number of texture patterns up to a certain level. The image bitmaps generated using tiling also help in ameliorating the performance up to a certain level.

References

1. Kekre, H.B., Thepade, S.D., Banura, V.K.: Augmentation of Colour Averaging Based Image Retrieval Techniques using Even part of Images and Amalgamation of feature vectors. Int. Journal of Engg. Science and Tech (IJEST) 2(10) (2010), http://www.ijest.info
2. Kekre, H.B., Thepade, S.D., Banura, V.K.: Amelioration of Colour Averaging Based Image Retrieval Techniques using Even and Odd parts of Images. Int/ Journal of Engineering Science and Technology (IJEST) 2(9) (2010)
3. Kekre, H.B., Thepade, S.D., Maloo, A.: Query by Image Content Using Colour Averaging Techniques. Int. Journal of Engineering Science and Technology (IJEST) 2(6), 1612–1622 (2010), http://www.ijest.info
4. Kekre, H.B., Thepade, S.D.: Boosting Block Truncation Coding using Kekre's LUV Color Space for Image Retrieval. WASET International Journal of Electrical, Computer and System Engineering (IJECSE) 2(3), 172–180 (Summer 2008)
5. Kekre, H.B., Thepade, S.D.: Image Retrieval using Augmented Block Truncation Coding Techniques. In: ACM Int. Conf. on Advances in Computing, Communication and Control (ICAC3 2009), January 23-24, pp. 384–390. FCRCE, Mumbai (2009)
6. Kekre, H.B., Thepade, S.D.: Scaling Invariant Fusion of Image Pieces in Panorama Making and Novel Image Blending Technique. International Journal on Imaging (IJI) 1(A08), 31–46 (Autumn 2008), http://www.ceser.res.in/iji.html
7. Hirata, K., Kato, T.: Query by visual example – content-based image retrieval. In: Pirotte, A., Delobel, C., Gottlob, G. (eds.) EDBT 1992. LNCS, vol. 580, pp. 56–71. Springer, Heidelberg (1992)
8. Kekre, H.B., Thepade, S.D.: Rendering Futuristic Image Retrieval System. In: National Conference on Enhancements in Computer, Communication and Information Technology, EC2IT 2009, March 20-21, K.J.S. COE Mumbai 77 (2009)

9. Do, M.N., Vetterli, M.: Wavelet-Based Texture Retrieval Using Generalized Gaussian Density and Kullback-Leibler Distance. IEEE Transactions on Image Processing 11(2), 146–158 (2002)
10. Prasad, B.G., Biswas, K.K., Gupta, S.K.: Region –based image retrieval using integrated color, shape, and location index. Int. Journal on Computer Vision and Image Understanding Special Issue: Colour for Image Indexing and Retrieval 94(1-3), 193–233 (2004)
11. Kekre, H.B., Thepade, S.D.: Creating the Color Panoramic View using Medley of Grayscale and Color Partial Images. WASET Int. Journal of Electrical, Computer and System Engg (IJECSE) 2(3) (2008), http://www.waset.org/ijecse/v2/v2-3-26.pdf
12. Edvardsen, S.: Classification of Images using color, CBIR Distance Measures and Genetic Programming. Ph.D. Thesis, Master of science in Informatics, Norwegian university of science and Tech., Dept. of computer and Info. science (June 2006)
13. Kekre, H.B., Sarode, T., Thepade, S.D.: DCT Applied to Row Mean and Column Vectors in Fingerprint Identification. In: International Conference on Computer Networks and Security (ICCNS), September 27-28. VIT, Pune (2008)
14. Pan, Z., Kotani, K., Ohmi, T.: Enhanced fast encoding method for vector quantization by finding an optimally-ordered Walsh transform kernel. In: ICIP 2005, IEEE International Conference, September 2005, vol. 1, pp. I -573-6 (2005)
15. Kekre, H.B., Thepade, S.D.: Improving 'Color to Gray and Back' using Kekre's LUV Color Space. In: IEEE Int. Advanced Computing Conference 2009 (IACC 2009), Thapar University, Patiala, INDIA, March 6-7 (2009), Is uploaded at online at IEEE Xplore
16. Kekre, H.B., Thepade, S.D.: Image Blending in Vista Creation using Kekre's LUV Color Space. In: SPIT-IEEE Colloquium & Int. Conf., SPIT, Mumbai (February 2008)
17. Kekre, H.B., Thepade, S.D.: Color Traits Transfer to Grayscale Images. In: IEEE First International Conference on Emerging Trends in Engg. & Technology (ICETET 2008), G.H.Raisoni COE, Nagpur, INDIA (2008), Uploaded on online IEEE Xplore
18. http://wang.ist.psu.edu/docs/related/Image.orig (Last referred on September 23, 2008)
19. Kekre, H.B., Thepade, S.D.: Using YUV Color Space to Hoist the Performance of Block Truncation Coding for Image Retrieval. In: IEEE Int. Advanced Computing Conference 2009 (IACC 2009), Thapar University, Patiala, INDIA (March 2009)
20. Kekre, H.B., Thepade, S.D., Athawale, A., Shah, A., Verlekar, P., Shirke, S.: Energy Compaction and Image Splitting for Image Retrieval using Kekre Transform over Row and Column Feature Vectors. Int. Journal of Computer Science and Network Security (IJCSNS) 10(1) (January 2010), http://www.IJCSNS.org
21. Kekre, H.B., Thepade, S.D., Athawale, A., Shah, A., Verlekar, P., Shirke, S.: Walsh Transform over Row Mean and Column Mean using Image Fragmentation and Energy Compaction for Image Retrieval. Int. Journal on Computer Science and Engineering (IJCSE) 2S(1) (January 2010), http://www.enggjournals.com/ijcse
22. Kekre, H.B., Thepade, S.D.: Image Retrieval using Color-Texture Features Extracted from Walshlet Pyramid. ICGST Int. Journal on Graphics, Vision and Image Processing (GVIP) 10(1), 9–18 (2010)
23. Kekre, H.B., Thepade, S.D.: Color Based Image Retrieval using Amendment Block Truncation Coding with YCbCr Color Space. Int. Journal on Imaging (IJI) 2(A09), 2–14 (Autumn 2009), http://www.ceser.res.in/iji.html

24. Kekre, H.B., Sarode, T., Thepade, S.D.: Color-Texture Feature based Image Retrieval using DCT applied on Kekre's Median Codebook. Int. Journal on Imaging (IJI) 2(A09), 55–65 (Autumn 2009), http://www.ceser.res.in/iji.html
25. Kekre, H.B., Thepade, S.D., Maloo, A.: Performance Comparison for Face Recognition using PCA, DCT & Walsh Transform of Row Mean and Column Mean. ICGST Int. Journal on Graphics, Vision and Image Processing (GVIP) 10(II), 9–18 (2010), http://209.61.248.177/gvip/Volume10/Issue2
26. Kekre, H.B., Thepade, S.D.: Improving the Performance of Image Retrieval using Partial Coefficients of Transformed Image. International Journal of Information Retrieval, Serials Publications 2(1), 72–79 (2009)
27. Kekre, H.B., Thepade, S.D., Athawale, A., Shah, A., Verlekar, P., Shirke, S.: Performance Evaluation of Image Retrieval using Energy Compaction and Image Tiling over DCT Row Mean and DCT Column Mean. In: Springer-Int. Conf. on Contours of Computing Technology (Thinkquest-2010), BGIT, Mumbai (March 2010)
28. Kekre, H.B., Sarode, T.K., Thepade, S.D., Suryavanshi, V.: Improved Texture Feature Based Image Retrieval using Kekre's Fast Codebook Generation Algorithm. In: Springer-Int. Conf. on Contours of Computing Technology (Thinkquest-2010), BGIT, Mumbai (March 2010)
29. Kekre, H.B., Sarode, T.K., Thepade, S.D.: Image Retrieval by Kekre's Transform Applied on Each Row of Walsh Transformed VQ Codebook. In: Invited at ACM-Int. Conf. and Workshop on Emerging Trends in Tech (ICWET 2010),TCET, Mumbai (February 2010); The paper is uploaded on online ACM Portal

Cache Enabled MVC for Latency Reduction for Data Display on Mobile

Sylvan Lobo, Kushal Gore, Prashant Gotarne, C. R. Karthik, Pankaj Doke, and Sanjay Kimbahune

Tata Consultancy Services Limited, Innovations Lab, Mumbai, India
{sylvan.lobo,kushal.gore,prashant.gotarne,cr.karthik, pankaj.doke,sanjay.kimbahune}@tcs.com

Abstract. Data display on a mobile device is not the same as on a desktop system. There are constraints of memory and screen size, which make it difficult to display large a amount of data. We have proposed a simple but novel caching and data replenishment mechanism through which a large amount of data can be displayed efficiently on a mobile device without slowing down the mobile.

Keywords: Caching, Mobile, Model-View-Controller, MVC.

1 Introduction

With mobile devices getting more powerful day by day, the complexity of applications developed is also increasing. Instead of simple stand-alone applications or games, we observe client-server applications that allow users to generate and browse through a lot of data - e.g. micro-blogging and social networking applications. Such systems have immense data owing to the huge user space generating data daily. All this data resides on remote server systems. The challenge here is to download relevant data to the mobile device and present it to the user seamlessly without slowing down the device to display and manage this immense data.

Backend services have a large quantum of memory and processing power, which mobile devices do not. The ratio of the capability of a backend system to that of a mobile device being large, there needs to be some transformation or mapping from the large scale system to a miniature system. We need to create an illusion of a large scale system on the mobile device. The challenge is how to create this illusion given the constraints of a mobile device. Unlike personal computer systems on a wired network, mobile devices have network bandwidth and connectivity issues. Memory and storage capacity is also limited. Neither do they enjoy the luxury of the large screen real estate and enormous amounts of memory and persistent storage that simplest desktop systems have.

In this paper, we describe how we can create such an illusion of a large-scale system on a mobile, by using the concept of caching. We discuss a 2-level caching mechanism fused with the Model-View-Controller (MVC) [1] design pattern, to display a seemingly infinite list of data on a resource constrained mobile without slowing

K. Shah, V.R. Lakshmi Gorty, and A. Phirke (Eds.): ICTSM 2011, CCIS 145, pp. 107–114, 2011.

it down. We have used this approach in the development of GappaGoshti™, a mobile-based social networking platform [2][3].

The rest of the paper is organized as follows: In section 2 we review the existing approaches for displaying a large amount of content on mobile devices. In section 3 we introduce our approach, which is based on MVC with an integrated caching mechanism. Section 4 and 5 state our observations and conclusions.

2 Existing Approaches

Browsing voluminous content on a mobile device can be an unpleasant experience owing to the limited processing power, memory, form factor, uncertain quality of wireless service and frequent interruptions such as telephone calls. [4] There are a few noted approaches to deal with displaying data on constrained devices.

Thierry Violleau and Ray Ortigas [4] discuss a solution where the client-side data model replicates the server-side data model. It uses MVC, Facade and Proxy patterns [5] to manage offline and online mode of the application. The mobile application session is prone to interruption. Therefore, the data changes done by the user or data received from a previous web call is stored in the client side model. When the network is available, the client is synchronized with the server to bring the data up to date. This solution facilitates an improved response time and reduced bandwidth usage.

Another approach present is a client side database to maintain model data. In this approach, the database along with the database engine resides on the mobile phone itself. Whenever the application needs data, instead of sending a web request to the server, it sends a database query to the database residing on the handset. This reduces number of web calls to the server and helps to improve the performance by decreasing the response time. A disadvantage of this approach is that we may have stale data on the handset. However, this is acceptable when data changes less frequently. The database on the mobile can be synchronized with the database on the server at regular time intervals to address this disadvantage [6].

Another well known related concept is the Data Access Object (DAO) pattern [7]. Microsoft discusses a mechanism to cache multiple records in a Recordset. A Recordset object represents the records in a base table or the records that result from running a query. If operations like searching, scrolling are used more often, you can use DAO to cache records. When a request for a record is made, the database engine downloads the record from the server only when the record is not present in local cache [8].

3 Our Approach

There is considerable latency when data is accessed from secondary or remote storage as compared to primary storage. This is where a cache comes handy. Caching is a mechanism that stores a partial copy of data so that future requests for data can be served instantly, rather than suffering latency to fetch data from the original source. The cache is comparably faster. The cache is more useful if it smartly stores relevant data that is most probable to be requested. This prevents cache misses that force you to interact with the original data source, leading to latency. A good cache leads to a faster system performance.

In our architecture, a large amount of list based data resides on a remote server. A mobile client system needs to scroll through this seemingly endless amount of content on the mobile device without any performance issues.

We have proposed an adaptation of MVC pattern where we combine MVC with the multi level caching architecture typically seen on microprocessors. That is to say, we have two levels of caching where, a smaller cache, which is the L1 cache, is maintained at the View part of the MVC. Another considerably bigger cache, which is the L2 cache, is maintained at the Model part of MVC.

The idea behind two level caching is to minimize latency and improve the hit rate. In case of microprocessors, the L1 cache is smaller and closer to the processor than the L2 cache. This helps in faster access and hence low latency. The L2 cache, on the other hand, is larger and resides further from the processor. Its role is to improve the hit rate. When there is a cache miss in the L1 cache, data is expected to be found in the L2 cache.

In our case, the L1 and L2 cache reside in the same memory. The cost involved and associated impacts are hence in terms of object creation and destruction. Another artifact of the nature of mobile application in our case is the frequency of usage of objects. Objects in L1 are rich and are used at the most once or twice. Hence we need to be judicious in their creation else it might lead to memory fragmentation due to frequent releases. In the worst case, closure of application can happen due to highly fragmented memory. Hence the L1 cache in the view is small and consequently contains low number of high level objects.

The L2 cache in the model is larger and contains raw objects. There is computation time involved in transforming raw model data to high level objects required by the view. This computational latency is what is saved by the existence of L1 cache. A small amount of data from the L2 cache is transformed to high level objects and loaded into the L1 cache much before it is accessed. (This mechanism of pre-fetching data is done with the help of a windowing concept which is discussed later in section 3.2). Hence, when the user interacts with the L1 cache, there is no latency of object creation involved. Another benefit of the L1 cache is that being small it can help in reducing memory fragmentation. The content in the L1 cache is prone to have a short life. The objects would be released frequently compared to the models L2 cache data. Frequent release of memory can lead to memory fragmentation.

Subsequent sub-sections of this section will delve into the details of this approach of using MVC with a caching mechanism to reduce latency in displaying content on mobile device.

3.1 Architecture

We have conceptualized the architecture of the application to be constituent of certain components that go hand in hand with the pertinent parts of the MVC architecture. First, we have the L2 cache combined with the Model. It consists of a list (*ModelList*) of *Nodes*. This is basic content fetched from the remote server. Then, we have the L1 cache merged with the View component. This is made up of a *ViewList*, which is

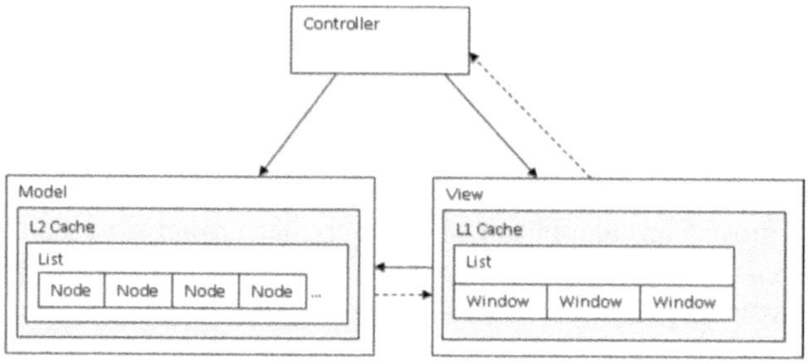

Fig. 1. The caching components merged into the caching architecture. The Model contains a list of Nodes that contain subset of raw data fetched from the remote server. The View contains a subset of the Model's data in a list with three Windows to scroll over the list.

managed using three *Windows* i.e. blocks. The list in the View also transforms the Model content to richer user viewable objects.

The Model represents the entities and their relationships in the system. We structurally store the model data in a list, in a raw format. The object representation of this data could be in a much richer format in the View. This structural representation of the data in the list is a snap shot of the massive data in the remote storage. The list is a doubly linked list of ***Nodes*** that contain data and references to the next and previous *Nodes*. The list so obtained is termed ***ModelList*** and makes up the L2 cache. The idea behind the L2 cache is to improve the cache hit rate of the system. As the model contains raw data, it can hold a considerably large amount of data. Thus there is a low chance of cache-miss. In case the requested data is not present, then fresh data needs to be fetched from the remote server.

The View presents the Model's data in a way that the user can interact with it. It maintains a subset of the Model's list – ***ViewList***. The list acts as the L1 cache, which is smaller than the Model's cache and some transformations are applied to create richer objects than the Model's raw data. The existence of the L1 cache improves the latency because the list is small, and its content is pre-fetched and converted to high level objects much before use, thus improving the performance of scrolling.

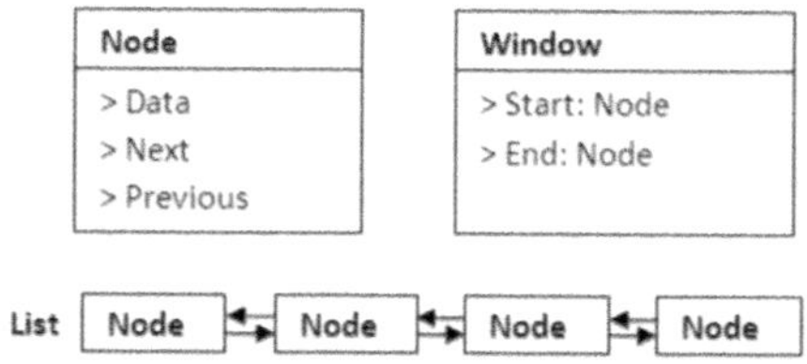

Fig. 2. Basic components of the caching mechanism: Nodes, Lists and Window

In order to manage the prefetching of objects in the *ViewList*, we have conceived a windowing concept. The list is managed as three ***Windows***: *PreviousWindow*, *CurrentWindow* and *NextWindow*. A *Window* is essentially a viewport into the data. It is demarcated by *start* and *end* references. The *start* of the *PreviousWindow* references the first *Node* in the list. The *end* of the *NextWindow* references the last *Node* in the list. The *CurrentWindow* has the data which is displayed to the user. As the user scrolls, and the *Window* changes from current to next. At this point, the *Windows* are updated and new data from the Model is populated into the *ViewList*. The windows, hence, let you seamlessly scroll through the list, and manage the cache content. The windows in effect enable the L1 cache to be populated with content before a cache-miss. The detailed working is explained in the next section.

3.2 Detailed Approach

First, data is downloaded from the remote server into the L2 cache, i.e. the Model. Then a portion of this raw data is transformed into richer objects for the View, and a new list is created for the View, which is the L1 cache. This list is accessed via the *Windows*. While scrolling before running out of L1 data, more data is fetched from the L2 cache. If there is no further data in the L2 cache, fresh data is downloaded from the remote server. This system is highlighted in Figure 3.

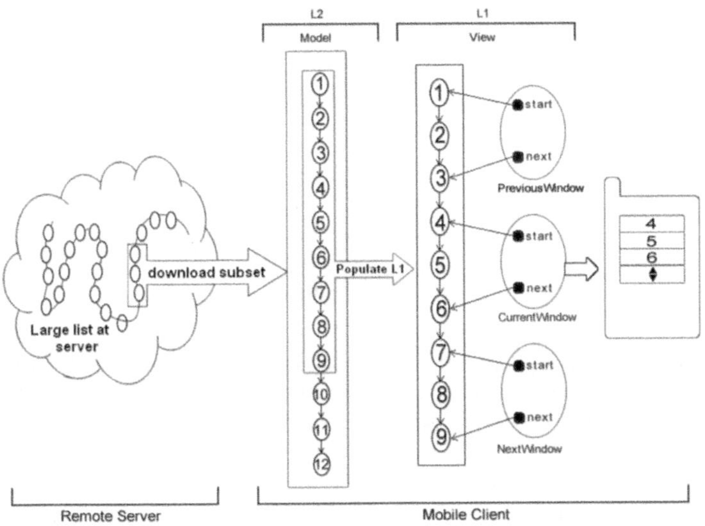

Fig. 3. This figure highlights that a subset of the remote server's data is fetched by the Model, into a list. A subset of the Model's list is present in the View's list. This list is scrolled with the help of windows.

When the user scrolls from the *CurrentWindow* to the next, a series of events get activated. The Nodes referenced by *PreviousWindow* are deleted. New data is fetched from the model and appended to the ViewList. The Windows are shifted downwards.

To explain further, upon scrolling from the *CurrentWindow* to the *NextWindow*, the *PreviousWindow* is deleted. This means that the *Nodes* managed by the *PreviousWindow* in the *ViewList* are released. The windows are then shifted downward. The original *CurrentWindow* now becomes the new *PreviousWindow*. The original

NextWindow now becomes the current. New content is required to be appended to the list in order to be referenced by the new *NextWindow*. A notification is sent to the View to populate a new *Window* with data from the L2. This new *Window* becomes the new *NextWindow*. Hence, while the user is working in the *CurrentWindow*, the *NextWindow* is already populated, thus allowing hassle free scrolling to the user. This mechanism is explained in Figure 4.

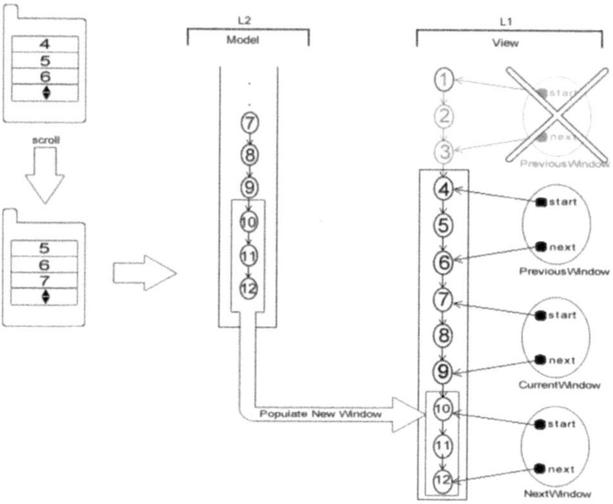

Fig. 4. The user scrolls from the 6th item to the 7th. This causes a shift in windows, from the current to the next. This leads to a series of activities where the previous window is released, the windows are shifted down and new data is populated from the model to the view's list.

Similar activities take place when the user scrolls up. The *NextWindow* is deleted and the *CurrentWindow* becomes the next and the *PreviousWindow* becomes the current. In addition, a notification is sent to the View to populate a new window with data from the L2. This new window becomes the new *PreviousWindow*.

There are special cases for the beginning and end of the Model list. Initially, the *PreviousWindow* is null and at the end, the *NextWindow* is null. When the Model list is exhausted, it is replaced by fresh data from the remote server, i.e. the next set of *Nodes*. This is the only point where the user will need to wait for data.

4 Observations

We have tested this approach by applying the mechanism in a social networking mobile application that we have developed. The application is used to create and browse through a large database consisting of posts. We have found the mechanism of scrolling very useful, where we could scroll through seemingly endless lists without any noticeable delays.

Consider T_{System} = f(T, R) where T_{System} is the System Latency, T is time taken for computation and network data fetch and R is the amount of memory available with the handset. We state that the latency of the system will decrease if we could increase R or reduce T. However, since T is a parameter over which we may not have significant control (telecom network), we can only tune the value of R. Hence,

Min(T_{System}) can be achieved by ensuring Max(R). However, while doing so, we have to ensure that the numbers of objects created are minimal since object creation and destruction on a mobile handset is a costly exercise resulting in memory fragmentation and quick battery discharge.

Based on our experiments conducted on Motorazr handset (15MB) RAM, if we observed that with 30% of the RAM, we could fetch 2000 objects. At 10%, we could fetch around 600.

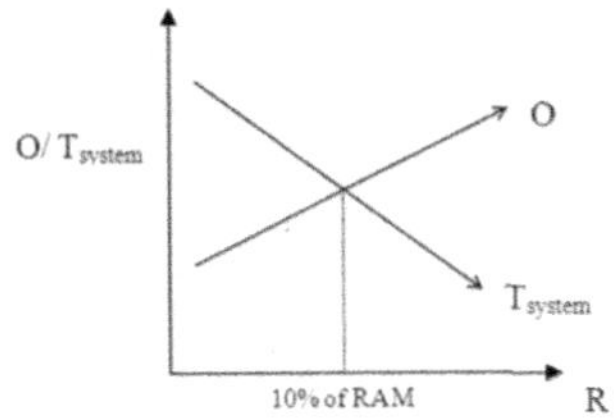

Fig. 5. Graph of the number of objects fetched (O), Latency of the system (Tsystem) vs RAM(R)

From the graph we observe that the optimal amount of RAM to be 10%. This is the minimal amount of useful memory for object storage, thereafter the benefits accrued have costly tradeoffs. Also, consider:

W_s = Size of window, i.e. the number of objects that can be displayed on one screen. We compute this value based on the handset display form factor, metrics of the fonts used as $W_s = HeightOfScreen / (HeightOfFont + Padding)$. This value happens to be 3 in our case.

N_w = the number of windows
N_{l1} = the number of items in L1 cache = N_w x W_s

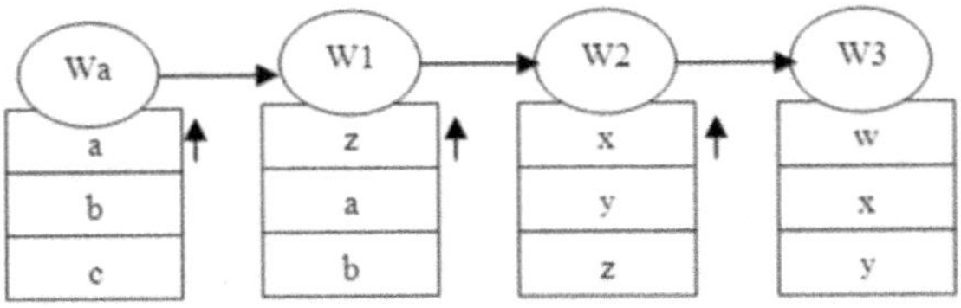

Fig. 6. Illustration to show that the probability of scrolling from W_a to windows W_1, $W_2$and W_3 keeps decreasing

$P(W_aW_1)$ – Probability of going from W_a to W_1
$P(W_aW_2)$ – Probability of going from W_a to W_2
$P(W_aW_3)$ – Probability of going from W_a to W_3
Event W_aW_2 – Scroll from W_a to W_2
Event W_aW_1 – Scroll from W_a to W_1

$P(W_aW_2) < P(W_aW_1)$ that is, the probability of going to W_2 from W_a is less than the probability of going to w1 from W_a.

$K = P(W_2 \mid W_aW_1) < p(W_aW_1)$

We have empirically determined that when the number of windows (N_w) exceeds 3, the value of K decreases significantly.

5 Conclusion and Future Work

A two level caching mechanism fused with the MVC architecture provides a seamless hassle free experience browsing through immense lists on a mobile device, creating an illusion of a large scale system on a miniature system.

As future work, we would be looking into a predictive caching mechanism instead of the current mechanism of updating the cache upon exhaustion. The server could also send push notifications over User Datagram Protocol (UDP) about fresh data, which the client could then fetch, as required.

Acknowledgments. We thank Arun Pande, Ananth Krishnan, Hiten Panchal, Devanuj and the Development team at TCS Innovations Lab, for their invaluable support, feedback and efforts.

References

1. Krasner, G.E., Pope, S.T.: A Description of the Model-View-Controller User Interface Paradigm in the Smalltalk-80 System. ParcPlace Systems (1988)
2. Pande, A., Kimbahune, S., Doke, P., Chunduru, D.: GappaGoshti: Multimedia based Mobile phone Solution for Social Networking for Rural masses. In: Third International Conference of Pattern Recognition and Machine Intelligence (PReMI 2009) (2009)
3. Lobo, S., Doke, P., Kimbahune, S.: GappaGoshti™ – A Social Networking Platform for Information Dissemination in the Rural World. In: Proceedings of the 6th Nordic Conference on Human-Computer Interaction, pp. 727–730. ACM Press, New York (2010)
4. Supporting Disconnected Operation in Wireless Enterprise Applications, `http://java.sun.com/blueprints/earlyaccess/wireless/disconnected/disconnected.pdf`
5. Gamma, E., Helm, R., Johnson, R., Vlissides, J.: Design Patterns: Elements of Reusable Object-Oriented Software (1994)
6. Gore, K., Kabra, P., Kimbahune, S., Doke, P.: Databases on Mobile - Porting SQLite to BREW. In: National/Asia Pacific Regional Conference in ICTM (Innovation in ICT an Technology Management) (2010)
7. Core J2EE Patterns - Data Access Objects. Sun Microsystems, `http://java.sun.com/blueprints/corej2eepatterns/Patterns/DataAccessObject.html`
8. DAO Recordset: Caching Multiple Records for Performance, `http://msdn.microsoft.com/en-us/library/aa984793VS.71.aspx`

An Approach to Optimize Fuzzy Time-Interval Sequential Patterns Using Multi-objective Genetic Algorithm

Sunita Mahajan[1] and Alpa Reshamwala[2]

[1] Principal,
Institute of Computer Science, M.E.T, Bandra, Mumbai
[2] Assistant Professor, Research Scholar
Computer Department, MPSTME, SVKM's NMIMS University, Mumbai

Abstract. Sequential pattern mining, which discovers frequent subsequences as patterns in a sequence database, is an important data-mining problem with broad applications. From these discovered sequential patterns, we can discover the order of the patterns; however, they cannot tell us the time intervals between successive patterns. Accordingly, Chen *et al.* have proposed a fuzzy time-interval (FTI) sequential pattern mining algorithms, which reveals the time intervals between successive patterns [12][13]. In this paper, we contributed to the ongoing research on FTI sequential pattern mining by proposing a multi objective Genetic Algorithm (GA) based method. Fuzzy solves the sharp boundary problem and the refinement ability of GA helps to find the global optimum FTI sequential patterns. Our approach uses two measures as objectives, namely: Confidence and Coverage to prune the traditional Apriori algorithm. The main objective is to achieve maximum confidence and maximum coverage in the FTI sequential patterns. The paper defines the confidence of the FTI sequences, which is not yet defined in the previous researches. The main advantage of the proposed algorithm is the use of fuzzy genetic approach to discover optimized sequences in the network traffic data to classify and detect intrusion.

Keywords: Data mining, fuzzy sets, sequence data, time interval, genetic algorithm, intrusion detection system.

1 Introduction

Data mining extracts implicit, previously unknown and potentially useful information from databases. Many approaches have been proposed to extract information, and mining sequential patterns is one of the most important ones [1][2][3].

The security of computer network plays a strategic role in modern computer systems with the widespread use of network. Intrusion Detection Systems (IDS) are effective security tools, placing inside a protected network and looking for known or potential threats in network traffic and/or audit data recorded by hosts. The problem of intrusion detection has been studied extensively in computer security [4][5][6], and has received a lot of attention in machine learning and data mining [7].Intrusion detection techniques can be categorized into misuse detection, for example, IDIOT [8] and STAT [9], and anomaly detection, for example, IDES [10]. Lee and Stolfo [11]

K. Shah, V.R. Lakshmi Gorty, and A. Phirke (Eds.): ICTSM 2011, CCIS 145, pp. 115–120, 2011.

discuss data mining approaches for intrusion detection. With the improvement of intrusion detection means, sometime it is difficult to judge whether an isolated sequence of audit event belongs to intrusion or not, so the sequence pattern algorithms of data mining techniques are applied in intrusion detection and intrusion pattern rules are found by learning the frequent episodes. A frequent episode is a set of events that occur frequently within a time window.

However, the discovered sequential patterns, cannot tell us the time gaps between successive patterns. Accordingly, Chen *et al.* have proposed a generalization of sequential patterns, called time-interval sequential patterns, which reveals not only the order of patterns, but also the time intervals between successive patterns [12]. Two efficient algorithms, FTI-Apriori algorithm and the FTI-PrefixSpan algorithm, are developed by Chen *et al* for mining FTI sequential patterns [13].

2 Related Work

The problem of mining sequential patterns was first introduced by Agarwal and Srikant [1] which discovers patterns that occur frequently in a sequence database. After mid 1990's, following Agrawal and Srikant [1], many scholar provided more efficient algorithms [15][16][17][18].

Existing approaches to find appropriate sequential patterns in time related data are mainly classified into two approaches. In the first approach developed by Agarwal and Srikant [14], the algorithm extends the well-known Apriori algorithm. The later, uses a pattern growth approach [15], employs the same idea used by the Prefix-Span algorithm.

In the algorithm [17], Shrikant and Agrawal, specified the maximum interval (max-interval), the minimum interval (min-interval) and the sliding time window size (windowsize). Moreover, they cannot find a pattern whose interval between any two sequences is not in the range of the window-size.

To address the intervals between successive patterns in sequence database, Chen *et al.* have proposed a generalization of sequential patterns, called time-interval sequential patterns, which reveals not only the order of patterns, but also the time intervals between successive patterns [12]. An extension of the algorithm developed by Chen *et al* [12], to solve the problem of sharp boundaries to provide a smooth transition between members and non-members of a set, is addressed in Chen *et al* [13]. The sharp boundary problems can be solved by the concept of fuzzy sets.

Anrong *et al* [20], addresses application of sequential pattern in intrusion detection by refining the pattern rules and reducing redundant rules. Zhou *et al*, Prasad *et al* and Yunwu in [21][22][23], incorporates fuzzy logic and GA for Intrusion Detection System. Hence, Saggar *et al* [19], has addressed optimization of the Apriori algorithm by using GA.

This paper focuses on pruning the traditional Apriori algorithm for pattern mining, to mine frequent sequential pattern using FTI sequential pattern in network audit data to classify and detect intrusion. A pruning algorithm is proposed with the use of GA's global search to prune the FTI sequential pattern mining, Apriori based algorithm to detect high accuracy for intrusions and optimize the network intrusion classifier.

3 Theory

In all the definitions for **n** patterns in **S** sequences with **sid** as the sequence-id in a network traffic pattern **T** is represented as **<sid, S>**.

3.1 Sequence Pattern

A pattern-set is a non-empty set of patterns. A sequence is an ordered list of pattern-set. Without loss of generality, we assume that the set of patterns is mapped to a set of contiguous integers. We denote a pattern-set **a** as $(a_1a_2...a_n)$, where a_j is a pattern. We denote a sequence **S** by $< s_1s_2...s_n>$, where s_j is a pattern-set.

A sequence $< a_1a_2...a_n >$ is contained in another sequence $< b_1b_2...b_m >$ if there exist integers $i_1 < i_2 < ... < i_n$ such that $a_1 \subseteq bi_1$, $a_2 \subseteq bi_2$, ..., $a_n \subseteq bi_n$.

3.2 Time Interval Sequence Pattern

A sequence **ST** is represented as $((a_1,t_1),(a_2,t_2), (a_3,t_3), ..., (a_n,t_n))$, where a_j is an pattern and t_j stands for the time at which a_j occurs, $1 \leq j \leq n$, and $t_{j-1} \leq t_j$, for $2 \leq j \leq n$. In the sequence, if patterns occur at the same time, they are ordered alphabetically. The time interval values can be calculated as $ti_j = | t_{j+1} - t_j|$, where $j = 1,2,...,n-1$.

3.3 Fuzzy Time Interval Sequence Pattern

A Fuzzy set can be defined as - If *U* is a collection of objects denoted generically by *x*, then a *fuzzy set A* in *U* is defined as a set of ordered pairs:

$$A= \{(x, \mu_A(x))| x \in U\},$$

where, $\mu_A(x)$ is the membership function and *U* is the universe of discourse.

Suppose we want to represent a time interval by using three linguistic terms: *Short* (*S*), *Middle* (*M*), and *Long* (*L*) within a month. Their membership functions can be represented as in [13].

Let $A=\{a_1a_2...a_n\}$, be the set of patterns and $LT = \{ lt_j / j= 1,2,...,l\}$ be a set of all linguistic terms. A sequence $\alpha = (b_1, lg_1, b_2, lg_2,..., b_{r-1}, lg_{r-1}, b_r)$ is a FTI sequence if $b_i \in A$ and $lg_i \in LT$ for $1 \leq i \leq r-1$ and $b_r \in A$.

If a sequence α is contained in with degree γ, then we call α a FTI subsequence of **S** with degree γ. The total number of patterns in a FTI sequence α is referred to as the *length* of the sequence. A FTI sequence whose length is *k* is referred to as a fuzzy *k*-time-interval sequence.

$$\text{Support}_S(\alpha) = \sum_{(sid,s)\ in\ S} \frac{\gamma(\alpha,s)}{|S|} \quad (1)$$

Support of a FTI sequence α is given by equation (1). Given a sequence database and *min_sup*, the goal of FTI sequential pattern mining is to determine in the sequence database all the FTI subsequences whose supports are more than or equal to *min_sup*.

$$\text{support\_count}_S(\alpha) = \{(sid,s) \mid (sid,s) \in S \quad \alpha \text{ is contained in } s\} \quad (2)$$

$$\text{Confidence}_S(\alpha_1 \Rightarrow \alpha_2) = \frac{\text{support\_count}(\alpha1 \cup \alpha2)}{\text{support\_count}(\alpha1)} \quad (3)$$

Confidence of FTI sequence α_1 on α_2 is given by equation (3), which uses the *support_count* defined in equation (2).

3.4 Genetic Algorithms

Genetic algorithm is a domain-independent method that genetically breeds a population of computer programs to solve a problem. Specifically, GA iteratively transforms a population of computer programs into a new generation of programs by applying analogs of naturally occurring genetic operations. The genetic operations include crossover (sexual recombination), mutation, reproduction, gene duplication, and gene deletion.

4 Proposed Algorithm

A pruning algorithm is proposed, to prune the FTI sequential pattern mining. The proposed algorithm uses two measures as objectives, namely: confidence and coverage to prune the traditional Apriori algorithm. In the algorithm defined,
L_k - the set of all frequent k-FTI sequences,
C_k – the set of candidate k-FTI sequences.

Input: Sequence Database S, Minimum Support *min_sup*, and Linguistic Terms LT
Output: The complete set of FTI sequential patterns.
C_1= find all patterns in S. // *all distinct patterns in database S*
$L_1 = \{c \in C_1 | \left(\frac{c.count}{|S|}\right) \geq min\_sup\}$ // *candidate generated with support ≥ min_sup*
for (k=2; L_k-1≠ Ø ; k++) {
C_k= new candidates generated from L_{k-1} // *all distinct patterns from previous pass*
    for each $p_1 \in L_{k-1}$ { // *first pattern of the FTI sequence s*
        for each $p_2 \in L_{k-1}$ { // *second pattern of the FTI sequence s*
            If (k=2) {
                for each *ltd* ∈ LT // *ltd is Fuzzy Linguistic variables*
                c= p_1* ltd* p_2; // *all possible combination of patterns*
                add c to C_k;
            }
        }
    }
    if (k>2)
    Build the fuzzy candidate tree from C_k;
    for each sequence s ∈ S
        Traverse the fuzzy candidate tree and accumulate the supports;
$L_k = \{c \in C_k | \left(\frac{c.count}{|S|}\right) \geq min\_sup\}$
}
P = randomly generate population from UL_k; // *all candidate FTI sequences s ∈ S, with support ≥ min_sup*
While (not termination condition)
    Apply GA operators on individuals of P. // *candidate FTI sequences form the individual*
    Evaluate fitness criteria on individuals of P.
return L'_k; // *L'_k is candidate FTI sequence which satisfies the fitness criteria*

The algorithm proceeds in phases, in the first phase, for k=1, L_1 is found from C_1: Clearly, C_1 can be generated by listing all distinct patterns in databases. To determine the supports of all patterns in C_k, a tree structure, called fuzzy candidate tree, is used as a basis which forms the individuals *I* in the second phase. Basically, the candidate tree is similar to the prefix tree adopted in previous research [14]. The major difference lies in that the traditional approach connects each tree branch with an item name, whereas in the new approach two components are attached— a pattern name and a linguistic term. In the second phase, having obtained L_k, population is generated randomly of the frequent FTI sequential patterns. On the individuals, GA operators are applied, for example, crossover, mutation, reproduction, gene duplication, and gene deletion. A multi objective fitness function is defined with the parameters to maximum confidence and maximum coverage of the individual.

Maximum coverage as given in equation (9), is considered to get the largest frequent FTI sequence and to achieve accuracy maximum confidence is taken as a parameter as given by equation (8).
The fitness function is defined as in equation (7).

$$\text{Fitness (I)} = F_{\text{confidence}}(I) * F_{\text{coverage}}(I) \tag{7}$$

where,

$$F_{\text{confidence}}(I) = \text{Confidence}_S(\alpha_1 => \alpha_2) \tag{8}$$

$$F_{\text{coverage}}(I) = \frac{\textit{number of patterns in the individual}}{\textit{Total number of patterns in the Database}} \tag{9}$$

Thus the above algorithm prunes the traditional Apriori algorithm for pattern mining, to mine frequent sequential pattern using FTI sequential pattern in network audit data to classify and detect intrusion.

5 Conclusion

In this paper, we contributed to the ongoing research on FTI sequential pattern mining by proposing a multi objective GA based method for mining FTI sequential patterns. Our approach uses two measures as objectives, namely: confidence and coverage to prune the traditional Apriori algorithm. The main objective is to achieve maximum confidence and maximum coverage in the FTI sequential patterns. The paper defines the confidence of the FTI sequences, which is not yet defined in the previous researches. The main advantage of the proposed algorithm is the use of fuzzy genetic approach to discover optimized sequences in the network traffic data to classify and detect intrusion. Fuzzy solves the sharp boundary problem and the refinement ability of GA helps to find the global optimum FTI sequential patterns and copes better with attribute interaction than the greedy rule induction algorithms

References

1. Agrawal, R., Srikant, R.: Mining sequential patterns. In: Proc. Int. Conf. Data Engineering, pp. 3–14 (1995)
2. Chen, Y.L., Chen, S.S., Hsu, P.Y.: Mining hybrid sequential patterns and sequential rules. Inf. Syst. 27(5), 345–362 (2002)
3. Han, J., Kamber, M.: Data Mining: Concepts and Techniques. Academic, New York (2001)
4. Murali, A., Rao, M.: A survey on intrusion detection approaches. In: The First International Conference on Information and Communication Technologies, pp. 23–240 (2005)

5. Nong, Y., Qiang, C., Borror, C.M.: EWMA forecast of normal system activity for computer intrusion Detection. IEEE Trans., Reliab. 53(4), 557–566 (2004)
6. Axelsson, S.: Intrusion detection systems: a survey and taxonomy. Technical report no. 99–15, Department of Computer Engineering. Chalmers University of Technology, Sewden (2000)
7. Tian, J.F., Fu, Y., Wang, J.-L.: Intrusion detection combining multiple decision trees by fuzzy logic. In: Sixth International Conference on Parallel and Distributed Computing. Application an Technologies, December 5-8, pp. 256–258 (2005)
8. Kumar, S., Spafford, E.H.: A software architecture to support misuse intrusion detection. In: Proceedings of the 18th National Information Security Conference, pp. 194–204 (1995)
9. Ilgun, K., Kemmerer, R.A., Porras, P.A.: State transition analysis: A rule-based intrusion detection approach. IEEE Transactions on Software Engineering 21, 181–199 (1995)
10. Lunt, T., Tamaru, A., Gilham, F., Jagannathan, R., Neumann, P., Javitz, H., Valdes, A., Garver, T.: A real-time intrusion detection expert system (IDES)-final technical report, Technical report, Computer Science Laboratory, SRI International, Melo Park, California (February 1992)
11. Lee, W., Stolfo, S.J.: Data mining approaches for intrusion detection. In: Proceedings of the 7th USENIX Security Symposium, pp. 26–29 (1998)
12. Chen, Y.L., Chiang, M.C., Ko, M.T.: Discovering time-interval sequential patterns in sequence databases. Expert Syst. Applicat. 25(3), 343–354 (2003)
13. Tony, Y.-L., Huang, C.-K.: Discovering Fuzzy Time-Interval Sequential Patterns in Sequence Databases. IEEE Transactions on Systems, Man, and Cybernetics-Part B: Cybernetics 35, 959–972 (2005)
14. Agrawal, R., Srikant, R.: Fast algorithms for mining association rules. In: Proc. Int. Conf. Very Large Data Bases, pp. 487–499 (1994)
15. Pei, J., Han, J., Pinto, H., Chen, Q., Dayal, U., Hsu, M.-C.: PrefixSpan: Mining sequential patterns efficiently by prefix-projected pattern growth. In: Proceedings of 2001 International Conference on Data Engineering, pp. 215–224 (2001)
16. Han, J., Pei, J., Mortazavi-Asl, B., Chen, Q., Dayal, U., Hsu, M.-C.: FreeSpan: Frequent pattern-projected sequential pattern mining. In: Proceedings of 2000 International Conference on Knowledge Discovery and Data Mining, pp. 355–359 (2000)
17. Srikant, R., Agrawal, R.: Mining sequential patterns: Generalizations and performance improvements. In: Proceedings of the 5th International Conference on Extending Database Technology, pp. 3–17 (1996)
18. Zaki, M.J.: SPADE: An efficient algorithm for mining frequent sequences. Machine Learning Journal 42(1/2), 31–60 (2001)
19. Saggar, M., Agrawal, A.K., Lad, A.: Optimization of Association Rule Mining using Improved Genetic Algorithms. In: IEEE International Conference on Systems, Man and Cybernetics, pp. 3725–3729 (2004)
20. Anrong, X., Shijie, H., Shiguang, J., Weihe, C.: Application of Sequential Patterns Based on User's Interest in Intrusion Detection. In: Proceedings of 2008 IEEE International Symposium on IT in Medicine and Education, pp. 1089–1093 (2008)
21. Zhou, Y., Fang, J., Yu, D.: Research on Fuzzy Genetics-Based Rule Classifier in Intrusion Detection System. In: International Conference on Intelligent Computation Technology and Automation, pp. 914–919 (2008)
22. Prasad, G.V.S.N.R.V., Dhanalakshmi, Y., Vijaya Kumar, V., Ramesh Babu, I.: Modeling An Intrusion Detection System Using Data Mining And Genetic Algorithms Based On Fuzzy Logic. IJCSNS International Journal of Computer Science and Network Security 8(7), 319–325 (2008)
23. Yunwu, W.: Using Fuzzy Expert System Based on Genetic Algorithms for Intrusion Detection System. In: International Forum on Information Technology and Applications, pp. 221–224 (2009)

Signaling Architectures for Efficient Resource Utilization in NGN End-to-End QoS

Seema A. Ladhe[1] and Satish R. Devane[2]

[1] Asst.Professor, Dept of Information Technology,
[2] Professor, Dept of Computer Engineering,
Ramrao Adik Institute of Technology, Nerul, Navi Mumbai, India
{seema,satish}@rait.ac.in

Abstract. With increase in popularity of internet on multimedia applications, it is a key challenge in front of Service Providers to provide required Quality of Service (QoS) to their customers. Customer strongly demands for guaranteed services to their service providers. The existing network need separate protocol and methods for services VoIP, multimedia and data, as increase in the demand for multimedia services on existing network raises the issues of QoS. Next Generation Network (NGN) is a packet based network in which service related functions are independent from underlying transport-related technologies. End-to-End QoS is one of the additional functionality provided in NGN. Issues related to the resource utilization in existing network are not completely solved and same are extended to the NGN. This paper presents signaling architectures for achieving End-to-End QoS using single-phase and two-phase scheme. After comparison of these two, paper concludes that the resources can be utilized efficiently using two-phase scheme.

Keywords: Next Generation Network (NGN), Quality of Service (QoS), RACF.

1 Introduction

An intense competition is expected in future in the area of network as the explosion of the internet and popularity of internet multimedia services. Today, customer's expectations are to have access to their preferred facilities and services irrespective of type of network and their geographical location. Their primary expectations are good coverage of internet, mobility, portability, simplicity and value for money. Today's network scenario, where voice, data, mobile and internet uses separate network eg. PSTN, MPLS IP VPN, Internet etc. for each network service. With the huge increase in real time and high priority traffic on the internet, network operator demand for a service independent network architecture, which can facilitate to integrate new services easily, where to have converged multi-service platform to deliver all services. Quality of Service issues to satisfy the consumer requirements is not yet solved completely in present IP network and hence End-to-End QoS with best utilization of resources are the key problems in researchers view. It is the key challenge in front of

K. Shah, V.R. Lakshmi Gorty, and A. Phirke (Eds.): ICTSM 2011, CCIS 145, pp. 121–127, 2011.

service providers to provide best services to their customers with desired Quality of Service (QoS).

Next Generation Network (NGN) also refers as packet based network [1], [3] is a powerful emerging platform to provide variety of services like Voice, Video and Data. NGN is designed to provide a single multiservice network which encompasses all the elements of existing telecommunication networks and hence increases the service delivery capabilities of networks. The attractive feature of NGN is QoS enabled transport technologies in which service-related function is independent of the underlying transport-related technologies, support for a wide range of data and multimedia services, unfettered access by users to different service providers etc [1], [3]. Various organizations like International Telecommunication Union Telecommunication Standardization Sector (ITU-T), Telecoms & Internet Converged Services & Protocols for Advanced Networks (TISPAN) are working on Next Generation Network. One of the tasks of the ITU-T organization is to develop a global architecture supporting end-to-end services to be used on future specifications. The key points covered by study groups are NGN requirements, the general reference model, functional requirements and architecture of the NGN, evolution to NGN, services requirements and capabilities, functional architecture and mobility, quality of service (QoS) control aspects , security capability, migration of current networks into NGN (e.g. PSTN, ISDN etc.), future packet based network requirements etc.

The key feature of NGN is to provide End-to-End QoS [2]. The issues in present network for improving QoS with best utilization of resources will also exist in the NGN. Resource utilization is possible using single-phase scheme, two-phase scheme and three phase scheme. This paper presents RACF signaling architectures for resource utilization in End-to-End QoS using single-phase scheme and two-phase scheme. This paper also shows that efficient resource utilization is possible using two phase scheme.

2 Ends-to-End Quality of Service in NGN

ITU-T study group 12 is working on NGN QoS and it's focus is on performance, quality of service and quality of experience [2]. Quality of Service is collective effect of service performance which determines the degree of satisfaction of a user of the service [Recommendation E.800] [2], [8]. Network Performance (NP) is measured in terms of parameters which are meaningful to the network provider and are used for the purpose of system design, configuration, operation and maintenance [Recommendation I.350] [2]. Quality of Experience (QoE) [7] is the overall acceptability of an application or service, as perceived subjectively by the end-user. QoE in NGN includes the complete end-to-end system effects. For network provider, QoS provides a valuable framework but does not specify performance requirements for particular network technologies [2]. NP determines the user observed QoS, but it does not necessarily describe that quality in a way that is meaningful to users [2]. QoE is depending upon user actions and subjective opinions [7].

End to End QoS gives performance guarantees on the resources that end user use. ITU-T introduces RACF architecture in NGN for supporting End-to-End QoS. RACF hides the details of transport network to the service layer to support the separation of service control from transport function [9]. It ensures that there are sufficient resources available to guarantee the desired level of QoS. The functional architecture of the NGN contains Access Network, Network Attachment Control Functions (NACF), Service Stratum, RACF and Transport Functions, Customer Premises Equipment (CPE). Functional entities in NGN are divided in two parts, Service Stratum and Transport Stratum which is functionally connected through RACF [6],[9].

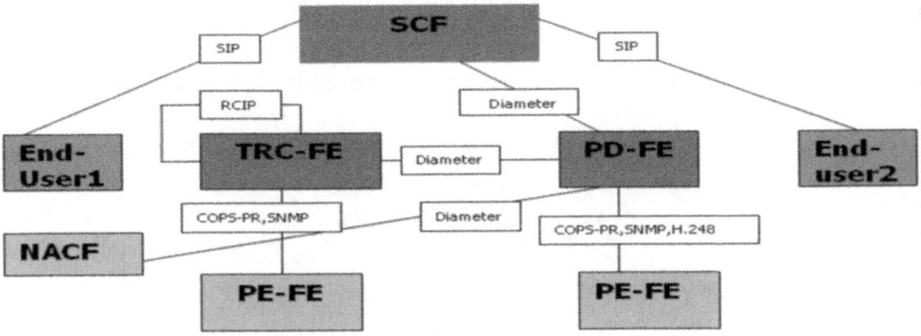

Fig. 1. RACF Protocol Architecture

Service stratum consists of Service Control Functions (SCF) and RACF consist of Transport Resource Control Functional Entity (TRC-FE), Policy Decision Functional Entity (PD-FE) and Transport Functions consist of Policy Enforcement Functional Entity (PE-FE). Figure 1 represents the RACF protocol architecture with the protocols used for the interfaces between end points of RACF architecture. End user1 sends service request to SCF. The SCF sends resource availability check message to PD-FE. The main functionality of PD-FE is to make policy decisions and TRC-FE is to determine availability of network resources. The PE-FE, transport layer functional entity is the gateway at the boundary of different packet networks, e.g., edge routers and/or between the CPE and access networks. Dynamic QoS is enforced in PE-FE. PE-FE guarantees dynamic QoS. Table 1 represents the interface between end points and protocols recommended for each interface in RACF [10].

Table 1. Interface and Protocol recommendations in RACF

End-Points	Interface	Protocol
SCF, PD-FE	Rs	Diameter
PD-FE,TRC-FE	Rt	Diameter
Between TRC-FE	Rp	RCIP
TRC-FE, PE-FE	Rc	COPS-PR, SNMP
PD-FE , PE-FE	Rw	COPS-PR, SNMP, H.248
PD-FE, NACF	Ru	Diameter
PD-FE–PD-FE (intra domain)	Rd	Not yet selected
PD-FE–PD-FE (inter domain)	Ri	Not yet selected

3 Signaling Architectures for End-to-End Quality of Service

RACF supports three different schemes of resource allocation and utilization: single-phase, two-phase and three-phase. Single-phase scheme does authorization, reservation and commitment in a single step while two-phase scheme do the authorization and reservation in first phase and followed by commitment in another phase. In three-phase scheme authorization, reservation and commitment are performed in three steps sequentially [4], [5].

3.1 Resource Utilization in End-to-End QoS Using Single-Phase Scheme

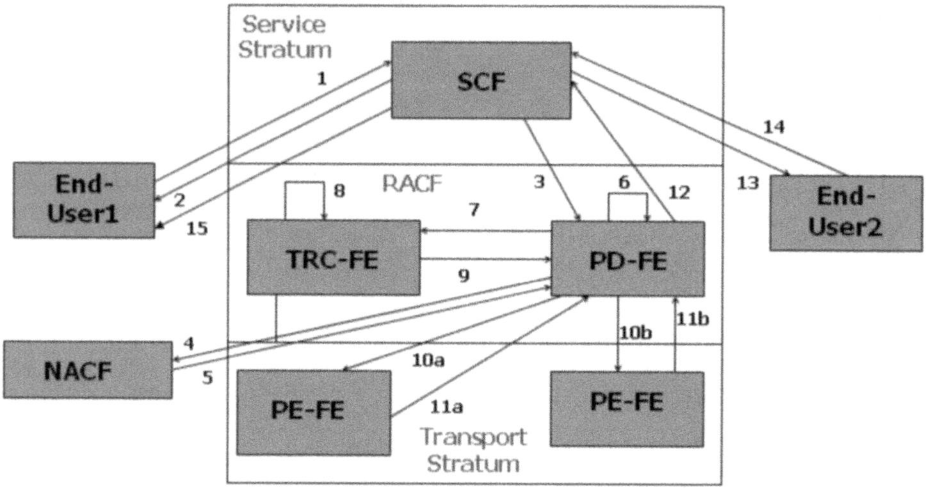

Fig. 2. RACF Signaling Architecture for Single-Phase Scheme

Figure 2 describes End-to-End QoS signaling call flow for single-phase scheme between End-User1/CPE1 and End-User2/CPE2. The sequence of events for single-phase scheme is listed below.

Event 1: End-User1 sends service request by issuing INVITE command to Service Control Functions (SCF).
Event 2: After receiving service request message, Service Control Function authenticates the service and sends an acknowledgement to EndUser1. It has Service Description Parameters (SDP) describing the details about the multimedia sessions (codec, etc) that the EndUser1 wishes to establish.
Event 3: For the service requested by End-User1, Service Control Function sends request for the resource reservation and commitment to PD-FE via Rs interface for which DIAMETER (Q.3301.1) protocol is recommended. The SCF sends AAR command with option Resource-Reservation-Mode=2 (Q.3301.1) indicating that authorization, reservation and commitment steps should be performed in a single step.

Event 4: PD-FE sends request for network-level authentication, transport subscription to NACF via Ru interface.
Event 5: NACF provides transport subscription to PD-FE via Ru interface. NACF uses DIAMETER protocol to provide network-level authentication, manage the IP address space of the access network, and authenticate access sessions.
Event 6: Taking into account the request for resource reservation from SCF and transport subscription information from NACF, PD-FE decides policy.
Event 7: PD-FE sends resource availability check message to TRC-FE via Rt interface. Diameter Protocol (Q.3305.1) is recommended for Rt interface.
Event 8: TRC-FE is responsible for collection and maintenance of the transport network topology and resource status information. Based on topology, connectivity, availability of network and node resources, transport subscription information in access networks, and other network information, the TRC-FE authorize resource admission control of the transport network.
Event 9: TRC-FE informs resource availability status to PD-FE via Rt interface.
Event 10a and 10b: PD-FE confirms resource reservation and commitment to PE-FE in transport layer by issuing a command which is based on COPS-PR protocol. The command is issued through DEC message and it should have the Decision Flags: Command-Code=Install option indicating that configuration should be installed.
Event 11a and 11b: PE-FE informs PD-FE about successful resource reservation and commitment.
Event 12: PD-FE informs SCF that resource reservation and commitment have been performed successfully
Event 13: Service request is forwarded to End-User 2/CPE2.
Event 14: Acknowledgement for the service request is send by End-User2 to SCF.
Event 15: SCF forwards acknowledgement received from End-User2/CPE2 to End-User1/CPE1. Session establishment is done between End-User1 and End-User2.

Single-phase scheme is mainly used for allocation and utilization of resources in single step. Reservation and commitment steps are initiated immediately after receiving service request from End-User1. SCF send request for resource authorization, reservation and commitment in single step to RACF. SCF does not forward the service request of End-User1 to End-User2 until the requested resources are committed. Once requested resources are committed, these resources can't be used for further signaling or services. Reservation and commitment is done simultaneously in single-phase scheme, which results in poor utilization of resources.

3.2 Resource Utilization in End-to-End QoS Using Two-Phase Scheme

Figure 3 describes End-to-End QoS signaling call flow for two-phase scheme between End-User1/CPE1 and End-User2/CPE2. Event 1 to event 9 is same as single-phase scheme as described in section 3.1 except even 3. The events for two-phase scheme are listed below.

Event 3: For the service requested by End-User1, SCF sends request only for the resource reservation to PD-FE via Rs interface for which DIAMETER (Q.3301.1) protocol is recommended. The SCF sends AAR command with option

Resource-Reservation-Mode=1 indicating that only authorization and reservation should be performed in a single step.

Event 10a and 10b: PD-FE confirms resource reservation to PE-FE in transport layer by issuing a command which is based on COPS-PR protocol. The command is issued through DEC message.

Event 11a and 11b: PE-FE informs PD-FE about successful resource reservation.

Event 12: PD-FE informs SCF about successful resource reservation.

Event 13: Service request is forwarded to End-User2.

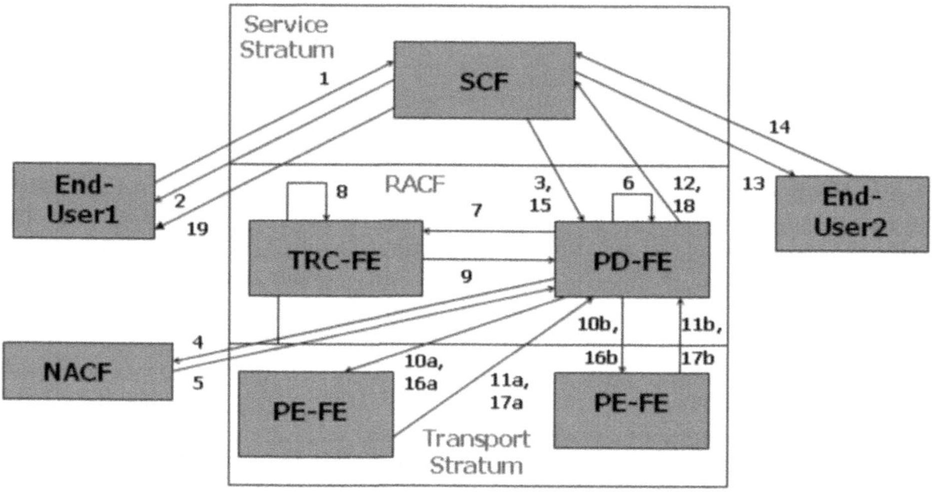

Fig. 3. RACF Signaling Architecture for Two-Phase Scheme

Event 14: Acknowledgement of the service request is send by End-User2 to SCF.

Event 15: SCF sends request for the resource commitment to PD-FE via Rs interface. The request is based on the DIAMETER protocol. The SCF sends AAR command with option Resource-Reservation-Mode=3 indicating that only commitment should be performed.

Event 16a and 16b: PD-FE confirms resource commitment to PE-FE in transport layer by issuing a command which is based on COPS-PR protocol. The command is issued through DEC message and it should have the Decision Flags: Command-Code=Install option indicating that configuration should be installed.

Event 17a and 17b: PE-FE informs PD-FE about successful resource commitment.

Event 18: PD-FE informs SCF about successful resource commitment through AAA command.

Event 19: After confirming successful resource commitment, SCF forwards acknowledgement of End-User2 to End-User1/CPE1. Session establishment is done between End-User1 and End-User2.

Referring above scenario of two-phase scheme, authorization and reservation is done in first phase (event 1 to 14) followed by commitment in second phase (event 15 to 19). SCF sends request only for resource authorization and reservation to RACF after receiving service request from End-User1. After confirmation of resource reservation, request is forwarded to End-User2. Resource commitment is done after receiving acknowledgement of service request from End-User2.

Efficient resource utilization can be achieve if the commitment is deferred as much as possible since reserved but uncommitted resources can be used for signaling as well as for services. Two-phase scheme initiates commitment after acknowledgement is received from End-User2 and since commitment is deferred, this results in efficient resource utilization.

4 Conclusion

NGN will provide most successful environment for not only internet users but to all type of communication network with End-to-End QoS between end applications or between end users. The key challenges in NGN are resource management and End-to-End QoS. This paper presented the RACF signaling architectures which represent the resource utilization in End-to-End QoS using single-phase scheme and two-phase scheme. From signaling architectures of single-phase scheme and two-phase scheme, resources can be utilized in efficient way in two-phase scheme as compared to single-phase scheme since resources are committed after receiving acknowledgement from End-User2. Even most efficient resource utilization can be possible using three-phase scheme without affecting the QoS.

References

1. International Telecommunication Union ITU-T, NGN FG Proceedings Part I, http://www.itu.int/en/ITU-T/gsi/ngn/Pages/default.aspx
2. International Telecommunication Union ITU-T, NGN FG Proceedings Part II, http://www.itu.int/en/ITU-T/gsi/ngn/Pages/default.aspx
3. ITU-T Rec. Y., General overview of NGN (2004), http://www.itu.int/en/ITU-T/gsi/ngn/Pages/default.aspx
4. ITU-T., Resource and admission control functions in Next Generation Networks (2006), http://www.itu.int/en/ITU-T/gsi/ngn/Pages/default.aspx
5. Pirhadi, M., Seyed, M., Safavi, H.: Analysis of Call Set-up Delay for Different Resource Control Schemes in Next Generation Networks. World Applied Sciences Journal 7 (Special Issue of Computer & IT), 129–137 (2009) ISSN
6. Draft Recommendation Q.3300: Architectural framework for the Q.33xx series of Recommendations, http://www.itu.int/en/ITU-T/gsi/ngn/Pages/default.aspx
7. Kamaljit, I., Lakhtaria: Enhancing QOS And QOE in IMS Enabled Next Generation Networks. International Journal on Applications of Graph Theory in Wireless Ad Hoc Networks and Sensor Networks (GRAPH-HOC) 2(2) (2010)
8. Jongtae, S.: Overview of ITU-T NGN QoS Control. IEEE Communications Magazine (2007)
9. ETSI ES: Resource and Admission Control Sub-system (RACS), Functional Architecture. 282 003 v1.1.1 (2006)
10. Reference Specification for Next Generation Networks (NGN) Framework. Issue 1 (2007), http://www.ida.gov.sg/home/index.aspx

Application of Software in Mathematical Bioscience for Modelling and Simulation of the Behaviour of Multiple Interactive Microbial Populations

B. Sivaprakash[1], T. Karunanithi[2], and S. Jayalakshmi[3]

[1] Assistant Professor, Department of Chemical Engineering, Annamalai University, Annamalainagar – 608002
shiv_bhaskar@yahoo.com
[2] Principal, Srinivasa Engineering College, Bommidi
[3] Associate Professor, CAS in Marine Biology, Annamalai University

Abstract. Engineering problems are tedious and time consuming to solve manually due to higher complexity. Though the biological systems are very complex they are susceptible to engineering analysis. This forms the basis for bioprocess modeling optimization and simulation. In the microbial world mixed culture operations play a vital role. In such case interaction among them decides the output of the system and five patterns of interactions (namely neutralism, **amensalism**, competition, commensalism, mutualism, predation and parasitism) are observed so far. In the present work an innovative and unified approach is developed to characterize these patterns of interactions among microorganisms for two species interaction. The models were derived from experimental data in batch mode using CFTOOL kit and the differential equations were solved using ODE SOLVER in MATLAB 7.1. The simulation for continuous operations was carried out using C++ software and the interpretations are obtained using surface plots drawn using MINITAB.

Keywords: Interaction, Chemostat, **Dilution** rate, cftool, ode solver, **surface** plot, MATLAB.

1 Introduction

Mathematical Bioscience is the application of mathematical modelling and mathematical techniques to get insight into the problems of biosciences. Biochemical reactions involve microorganisms either as single species or multi species [1]. In multi species system interactions between microbial populations can be recognized as negative and positive [3,5,6,7] and the patterns are neutralism, amensalism, competition, commensalism, mutualism, predation and parasitism. The behaviour and performance of the mixed species system depends on the type of interaction among them [9]. The present investigation aims to develop a new methodology to describe the mixed microbial population kinetics incorporating these interaction phenomena in terms of mathematical equations and to solve them using computer. This is done by obtaining the growth equations for interacting microorganisms in batch scale and extending

K. Shah, V.R. Lakshmi Gorty, and A. Phirke (Eds.): ICTSM 2011, CCIS 145, pp. 128–137, 2011.

them to continuous process using chemostat model. The objective is to determine the stable operating conditions of a chemostat without washout of any of the species.

2 Mathematical Models

Mathematical model in bioprocesses is defined as an equation that relates the influence of various parameters to cell growth (x) and consumption of the nutrients (s) by the microorganisms. The analysis of multiple interactive microbial populations involves mathematical modelling to quantify the several interactive effects in describing the growth rates of the interacting species in mixed culture. In the present study the logistic model is used for cell growth and the interaction effects are given as specific functions depending on the interaction mechanism. The growth kinetics of microorganisms in the logarithmic phase which attains stationary phase is well explained by the classical logistic model [4].

$$\frac{dx_i}{dt} = k_i x_i (1 - \beta_i x_i) \tag{1}$$

where x_i is the cell concentration and k_i and β_i are logistic constants. Since our study is limited to two species interaction, the general equations for describing the growth rates in associated form for the two species (x_1 and x_2) and the balances for substrates (s_1 and s_2) in batch case are given by equations (2) and (3).

$$\frac{dx_1}{dt} = a_{11} f_{11}(x_1, s_1) + a_{12} f_{12}(x_1, x_2) \tag{2}$$

$$\frac{dx_2}{dt} = a_{22} f_{22}(x_2, s_2) + a_{21} f_{21}(x_1, x_2) \tag{3}$$

where f_{11} and f_{22} represent the pure culture growth patterns of species 1 and 2 respectively and f_{12} and f_{21} give the interactive effect of species 2 on 1 and 1 on 2 respectively. All possible combinations of interactions namely neutralism, amensalism, competition, commensalism, mutualism, predation and parasitism defined in this manner are shown in Table 1[4].

Table 1. Classification of pairwise interactions based on the sign of the entries a_{12} and a_{21}

		Effect of species 2 on species 1 (sign of a_{12})		
		-	0	+
Effect of species 1 on species 2 (sign of a_{21})	-	-- competition	-0 amensalism	-+ predation
	0	0-amensalism	00 neutralism	0+commensalism
	+	+-predation	+0commensalism	++mutualism

In an amensal relationship, the growth of one species is inhibited by the presence of another [5]. Competition represents a negative relationship between two populations in which both populations are adversely affected with respect to their survival and growth [3, 6]. In a commensal relationship, one population benefits while the other remains unaffected [8]. In mutualistic interaction, growth of both species is enhanced by the presence of each other [2]. In predation one species forms the food for the other [7].

2.1 Chemostat

A chemostat is a well-stirred vessel of volume fed with the medium that contains nutrients required by the microorganisms. An important feature of chemostat cultivation is the dilution rate (D), defined as the volume (F) of nutrient medium supplied per unit time divided by the volume (V) of the culture. During chemostat cultivation, equilibrium is established (steady state) at which the growth rate of the cells equals the dilution rate. The growth rate of the microorganisms is purely based on the dilution rate. The concentration of the cells (biomass) in the chemostat is dependent on the concentration of the growth-limiting nutrient in the medium feed.

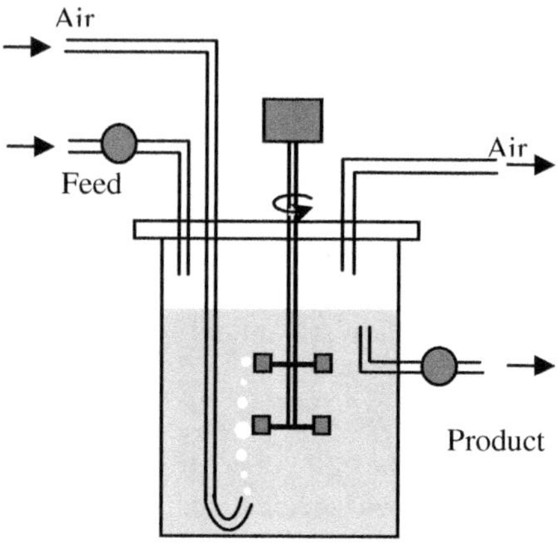

Fig. 1. Schematic representation of chemostat

3 Modeling and Discussions

3.1 Amensalism

System: *Pseudomonas aeruginosa (1) and Micrococcus luteus (2)*

The experimental studies on the pure and mixed culture operations of the bacterial system *Pseudomonas aeruginosa* and *Micrococcus luteus* proved the existence of amensal interaction against luteus species. The experiments were carried out with various initial substrate concentrations S_0 (0, 100, 200, 300 and 400 mg/ml) to study its influence on the pure and mixed culture growth. Suitable models were developed to predict the pure and associated growth rates based on equations (2) and (3) and extended to continuous culture (4) and (5). For pure culture $a_{21} = 0$.

$$\frac{dx_1}{dt} = -Dx_1 + k_1x_1(1-\beta_1x_1) \tag{4}$$

$$\frac{dx_2}{dt} = -Dx_2 + k_2x_2(1-\beta_2x_2) - a_{21}x_1x_2 \tag{5}$$

The β values change with respect to the dosage level of the glucose showing decreasing trend with increasing glucose concentrations for both the species and is expressed by the cubic polynomial and the equations for β_1 and β_2 in terms of s_0

$$\beta_1 = 0.03167s_0^3 - 0.008786s_0^2 - 0.003652s_0 + 0.003602 \tag{6}$$

$$\beta_2 = 0.2367s_0^3 - 0.09886s_0^2 + 0.0006762s_0 + 0.005337 \tag{7}$$

Table 2. Logistic constants for *Pseudomonas aeruginosa* and *Micrococcus luteus*

Initial glucose concentration (mg/ml)	k_1 (h^{-1})	β_1 (millions cfu/ml)$^{-1}$	R^2	k_2 (h^{-1})	β_2 (millions cfu/ml)$^{-1}$	R^2
0	0.2264	3.59×10^{-3}	0.9967	0.1751	5.34×10^{-3}	0.9794
100	0.2338	3.23×10^{-3}	0.9935	0.2098	4.64×10^{-3}	0.9871
200	0.2443	2.70×10^{-3}	0.9852	0.2104	3.43×10^{-3}	0.9747
300	0.2734	2.62×10^{-3}	0.9782	0.2413	3.02×10^{-3}	0.9637
400	0.2305	2.75×10^{-3}	0.9891	0.1960	4.94×10^{-3}	0.9902

The interaction parameter too varies with respect to the initial substrate concentration as a cubic polynomial.

$$a_{21} = -0.009367 s_0^3 + 0.003922 s_0^2 + 0.00032 s_0 + 0.000297 \tag{8}$$

The system of equations (4) to (8) are solved using ODE solver in MATLAB 7.1. It was observed that the model was able to fit the experimental data with greater accuracy with R^2 > (Table 2).The comparison between experimental and predicted values of the growth of the two species from the above equations is presented in figures 2 and 3 respectively.

$$x_{1s} = \left(1 - \frac{D}{k_1}\right)\frac{1}{\beta_1} \tag{9}$$

$$x_{2s} = \left(1 - \frac{(D + a_{21}x_{1s})}{k_2}\right)\frac{1}{\beta_2} \tag{10}$$

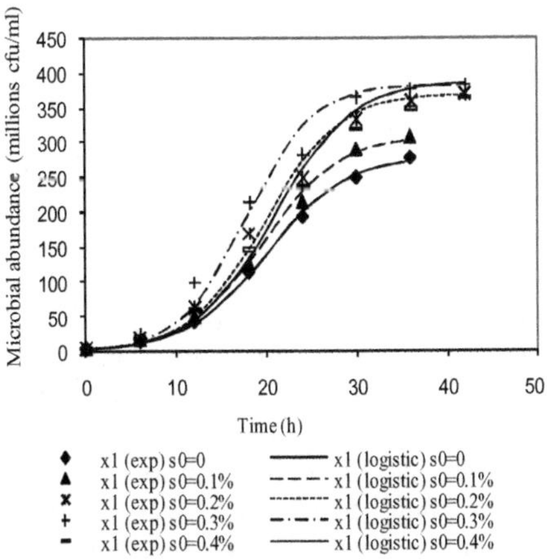

Fig. 2. Growth curve of *Pseudomonas auereginosa* in mixed culture with varying substrate concentrations

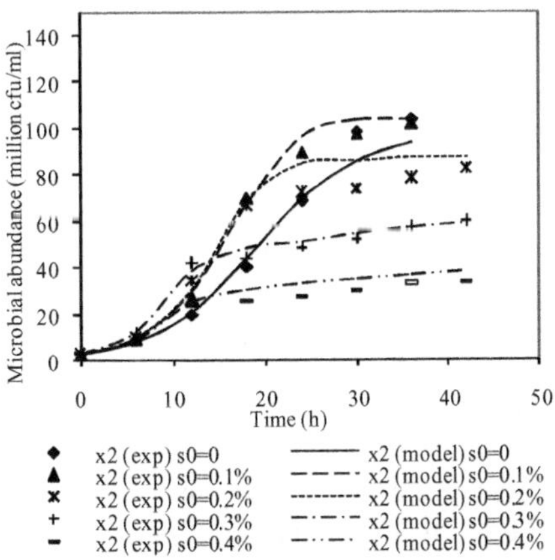

Fig. 3. Growth curve of *Micrococcus Luteus* in mixed culture with varying substrate concentrations

Table 3. Range of the operating conditions for coexistence of both the species in a chemostat

Initial glucose concentration (mg/ml)	Maximum limit for D (h^{-1})	Range of the steady state concentrations			
		x_{1s} (million cfu/ml)		x_{2s} (million cfu/ml)	
		Minimum	Maximum	Minimum	Maximum
0	0.146	98.92	278.55	0.357	100.6351
100	0.185	64.62	309.59	0.005	93.49
200	0.156	133.87	370.37	0.106	83.28
300	0.156	164.51	381.68	0.043	59.82
400	0.113	183.37	363.63	0.145	34.55

The chemostat model equations are used to simulate the range of dilution rates to be maintained for coexistence of both the species for the various initial concentrations and the respective steady state microbial concentrations that would be obtained (Table 3). The simulation was carried out in C++ language.

3.2 Competition

System: *Escherichia coli* (1) and *Staphylococcus aureus* (2)

The competitive interaction of the systems *E.coli* and *S. aureus* 255 and *E.coli* and *S. aureus* 261 based on the literature data [9] for three different temperatures viz. 15, 30 and 44 °C were modelled and given by equations (11) and (12). Under pure culture the second terms in each of these equations become zero.

$$\frac{dx_1}{dt} = -Dx_1 + k_1x_1(1-\beta_1x_1) - a_{12}x_2^2 \tag{11}$$

$$\frac{dx_2}{dt} = -Dx_2 + k_2x_2(1-\beta_2x_2) - a_{21}x_1^2x_2 \tag{12}$$

The values of the logistic constants for the pure culture growth are evaluated using cftool option in MATLAB 7.1 and are presented in Table 4 for *E .coli, S. aureus* (255) and *S. aureus* (261 The error percentages in all the cases are less than 7%. It can be inferred from the values of the logistic constants (Table 4) that, though β values did not change considerably, 'k' values show increasing trend with increasing temperature. Though the stationary phase numbers of the species did not change with temperature, the rate at which it is attained increases with the temperature. Similar observations are found with the *S. aureus* 255 and 261 in their pure culture growths. In mixed culture case the function f_{12} is second order whereas f_{21} is of third order. It was inferred that the competitive effect is independent of temperature for the coliform, whereas for *S aureus* the suppression is more at low and high temperatures (15 and 44 °C) than at 30 °C. The interaction parameters are found to vary with respect to temperature and are presented in Table 5. The comparison between experimental and predicted growth rates for the two sets at various temperature conditions specified are represented in figures 4 to 9. The simulation study on chemostat model showed that continuous culture of these two systems is not possible.

Table 4. Logistic constants and R^2 for *E .coli, S. aureus* (255) and *S. aureus* (261)

Species	Temperature (°C)	k (h^{-1})	β ($numbers^{-1}$)	R^2
	15	0.108	2.93×10^{-9}	0.9635
E .coli	30	0.598	3.49×10^{-9}	0.9935
	44	1.252	3.04×10^{-9}	0.9889
	15	0.0194	3.58×10^{-9}	0.9838
S. aureus (255)	30	0.1173	3.78×10^{-9}	0.9949
	44	0.4275	3.24×10^{-9}	0.9906
	15	0.0313	3.15×10^{-9}	0.9913
S. aureus (261)	30	0.2083	3.62×10^{-9}	0.9897
	44	1.002	3.45×10^{-9}	0.9536

Table 5. Interaction parameters for different temperatures

Temperature (°C)	*E.coli - S. aureus* 255		*E.coli - S. aureus* 261	
	a_{12} (1/number hour)	a_{21} ($1/number^2$ hour)	a_{12} (1/number hour)	a_{21} ($1/number^2$ hour)
15	9.13×10^{-4}	1.82×10^{-3}	6.49×10^{-4}	3.92×10^{-3}
30	8.76×10^{-4}	8.69×10^{-4}	6.12×10^{-4}	3.11×10^{-3}
44	7.64×10^{-4}	2.30×10^{-3}	5.72×10^{-4}	4.49×10^{-3}

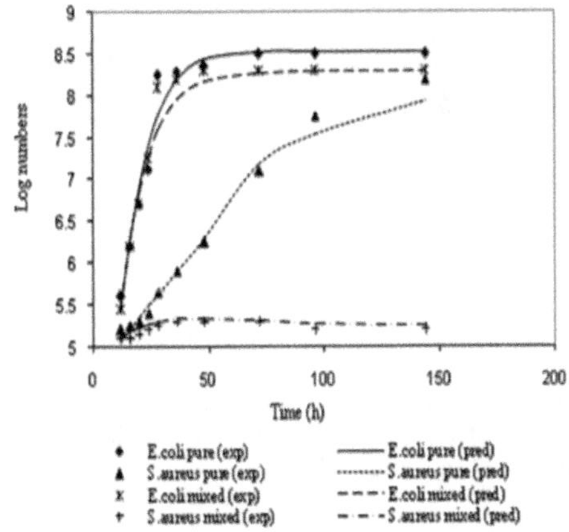

Fig. 4. Experimental and predicted values of the growth of *E.coli* and *S. aureus* 255 in pure and mixed culture at 15 °C

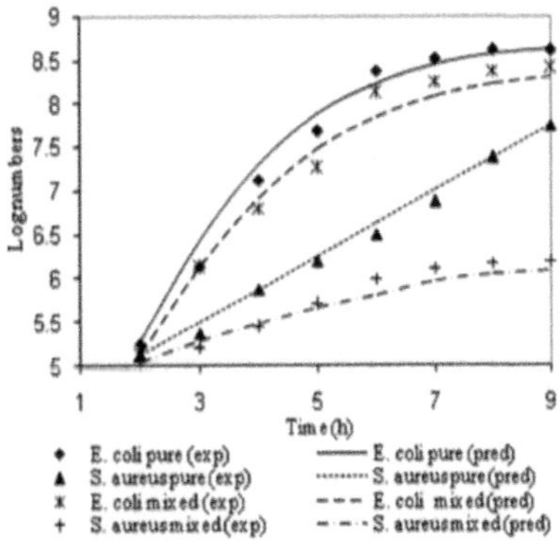

Fig. 5. Experimental and predicted values of the growth of *E.coli* and *S. aureus* 255 in pure and mixed culture at 30 °C

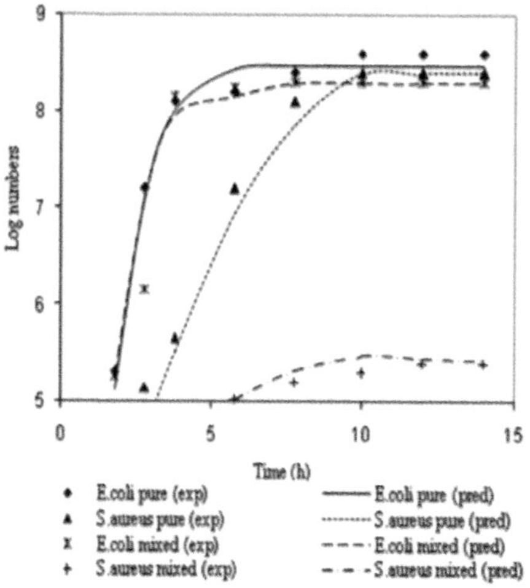

Fig. 6. Experimental and predicted values of the growth of *E.coli* and *S. aureus* 255 in pure and mixed culture at 44 °C

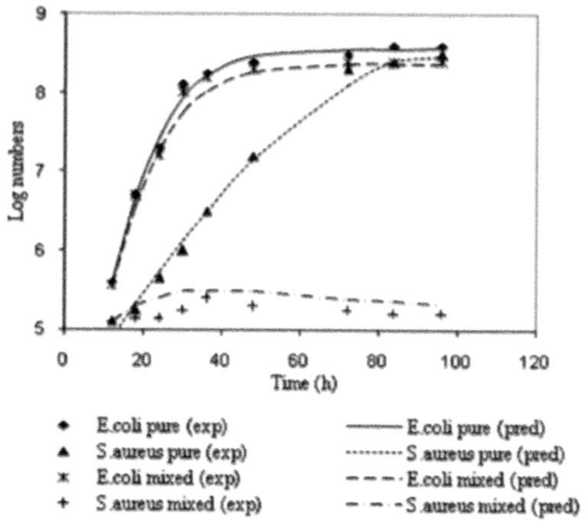

Fig. 7. Experimental and predicted values of the growth of *E.coli* and *S. aureus* 261 in pure and mixed culture at 15 °C

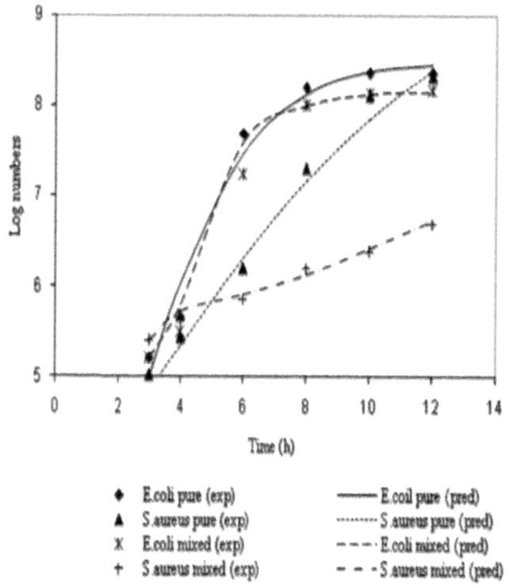

Fig. 8. Experimental and predicted values of the growth of *E.coli* and *S. aureus* 261 in pure and mixed culture at 30 °C

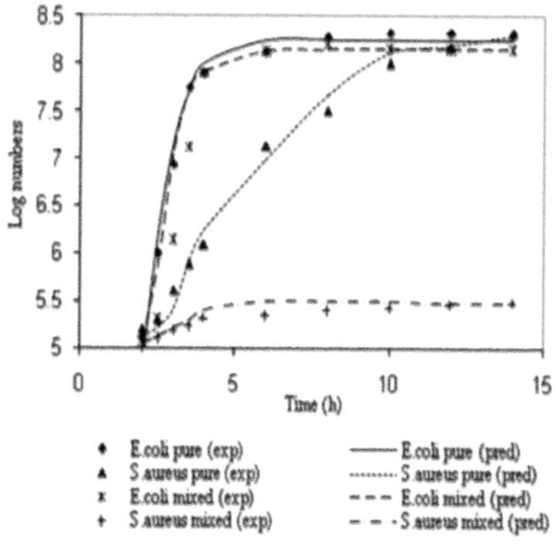

Fig. 9. Experimental and predicted values of the growth of *E.coli* and *S. aureus* 261 in pure and mixed culture at 44 °C

3.3 Commensalism

System: *Streptococcus thermophilus* (1) and *Lactobacillus bulgaricus* (2)

Modelling studies of the commensal pattern of interaction for the system *Streptococcus thermophilus* and *Lactobacillus bulgaricus* in batch case (D = 0) using the literature data [8] revealed that the interaction function for the species 1 by 2 is of first order.

$$\frac{dx_1}{dt} = -Dx_1 + k_1x_1(1-\beta_1x_1) + a_{12}x_2 \tag{13}$$

$$\frac{dx_2}{dt} = -Dx_2 + k_2x_2(1-\beta_2x_2) \tag{14}$$

These equations were able to fit the experimental growth with reasonable accuracy. The logistic constants along with the regression coefficients are given in Table 6. The interaction parameter a_{12} is calculated as 10.281 (1/h).

Table 6. Values of the logistic constants

Species	k (h^{-1})	β (ml/numbers)	R^2
Streptococcus thermophilus	4.224	1.00 x 10^{-8}	0.9892
Lactobacillus bulgaricus	0.52	1.41 x 10^{-9}	0.9885

The simulation studies for continuous culture was carried out by solving equations (13) and (14) for steady state and given as equations (15) and (16).

$$x_{1s} = \frac{-(D-k_1)+\sqrt{(D-k_1)^2+4k_1\beta_1(a_{12}x_{2s})}}{2k_1\beta_1} \tag{15}$$

$$x_{2s} = \left(1-\frac{D}{k_2}\right)\frac{1}{\beta_2} \tag{16}$$

These were solved using C++ software and the simulation result is given as surface plot (Fig.10) using MINITAB software package. The results indicate that the dilution

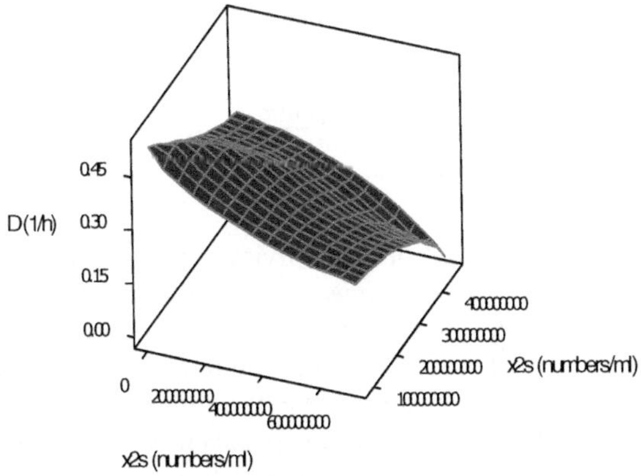

Fig. 10. Surface Plot of Dilution rate vs steady state concentrations of *S. thermophilus* and *L. bulgaricus* in mixed culture

rate in the chemostat should not exceed 0.52 h^{-1} for coexistence of both the species. Also it was inferred that when the dilution rate is maintained below 0.26 h^{-1} the yield of bacillus species is predominant and in the range 0.26 > D > 0.52 the coccal species' yield is predominant. Thus when the species 1 or the product from it is desired, the chemostat should be operated in the dilution rate below 0.26 h^{-1} and if the second species is desired, a dilution rate greater than 0.26 h^{-1} (but lesser than 0.52 h^{-1}) should be maintained.

3.4 Mutualism

System: *Geotrichum candidum* (1) and *Penicillium camembertii* (2)

The mutualistic interaction between the two fungal populations *Geotrichum candidum* and *Penicillium camembertii* was modelled using the experimental data available from literature [2].

$$\frac{dx_1}{dt} = -D(x_1) + k_1 x_1 (1 - \beta_1 x_1) + a_{12} \frac{x_2}{x_1 + x_2} \tag{17}$$

$$\frac{dx_2}{dt} = -D(x_2) + k_2 x_2 (1 - \beta_2 x_2) + a_{21} \frac{x_1}{x_1 + x_2} \tag{18}$$

The interaction parameters a_{12} and a_{21} become zero for pure culture and D is zero for batch case in the above equations. The logistic constants and the regression coefficients are given in Table 7. The mutualistic effect is more pronounced for the second species than the first. This is evident from the values of the interaction constants (a_{12} = 400.34 cfu/ml h and a_{21} = 4.5 x 10^4 cfu/ml h).

Table 7. Parameters of the Logistic model

Species	Logistic constants		R^2
	k (1/h)	β (ml/cfu)	
Geotrichum candidum	0.0572	5.01 x 10^{-8}	0.9932
Penicillium camembertii	0.0295	7.94 x 10^{-8}	0.9983

The simulation studies are carried out by solving equations (13) and (14) for steady state conditions and the result are given as surface plot (Fig.11) obtained using MINITAB. It is inferred that that the former species shows relatively higher decrement with the increasing dilution rate when compared to the latter. The decrement trend is found to be sharp for the candidum species for the entire range of the dilution rates obtained from simulation whereas for the penicillium species it is sharp till 0.03 h^{-1} and beyond this it attains a minimum (non zero) value. It is observed from the simulation data that coexistence of both the species can be attained only when the dilution rate of the chemostat is maintained lesser than 0.0576 h^{-1} above which *G. candidum* becomes extinct.

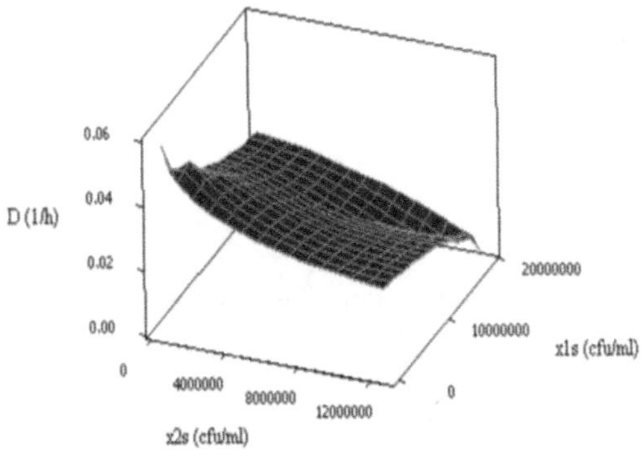

Fig. 11. Surface Plot of Dilution rate vs steady state concentrations of *G. candidum* and *P. camembertii* in mixed culture

References

1. Atlas, R.M., Bartha, R.: Microbial Ecology. Pearson Education, New Delhi (2005)
2. Aziza, M., Couriol, C., Amrane, A., Boutrou, R.: Evidences for synergistic effects of Geotrichum candidum on Penicillium camembertii growing on cheese juice. Enzyme and Microbial Technology 37, 218–224 (2005)
3. Baltzis, B.C., Fredrickson, A.G.: Competition of two suspension-feeding protozoan populations for a growing bacterial population. Microbial Ecology (1984)
4. Bailey, E., James, D.F.: Ollis: Biochemical Engineering Fundamentals. McGraw-Hill, New Delhi (1986)
5. Chapuis, C., Flandrois, J.P.: Mathematical model of the interactions between Micrococcus spp. and Pseudomonas aeruginosa on agar surface. Journal of Applied Bacteriology 77, 727–732
6. Fredrickson, A.G., Stephanopoulos, G.: Microbial competition. Science 213, 972–979 (1981)
7. Gause, G.F.: Experimental demonstration of Volterra's periodic oscillation in the numbers of animals. Journal of Experimental Biology 12, 44–48 (1935)
8. Moon, J., Nancy, R.G.W.: Commensalism and Competition in mixed cultures of Lactobacillus bulgaricus and Streptococcus thermophilus. Journal of Milk Food Technology 39, 337–341 (1975)
9. Oberhofer, T.H., Frazier, W.C.: Competition of Staphylococcus aureus with other organisms. Journal of Milk Food Technology 24, 172–175 (1961)

Effect of Different Target Decomposition Techniques on Classification Accuracy for Polarimetric SAR Data

Varsha Turkar and Y.S. Rao

Centre of Studies in Resources Engineering, IIT, Bombay, Powai, Mumbai-400 076, India
Mobile: 9920038965
varshaturkar@iitb.ac.in, ysrao@csre.iitb.ac.in

Abstract. Effect of different decomposition techniques on the classification accuracy for polarimetric SAR data is studied. We applied different target decomposition techniques (both coherent and incoherent) on ALOS-PALSAR data over Sunderban, West Bengal district of India and later **classified** the data using various classification techniques. The same training areas are used for classification of decomposed images. A comparative study has indicated that Van Zyl decomposition gives better classification accuracy than other decomposition techniques. It is observed that among the different classifiers applied, Maximum Likelihood Classifier gives highest accuracy.

Keywords: Radar polarimetry, synthetic aperture radar, speckle, target decomposition, terrain classification.

1 Introduction

Classification of polarimetric SAR images has become a very important topic after the availability of Polarimetric SAR images through ENVISAT ASAR, ALOS PALSAR, SIR-C and Radarsat-2. Classification is the task of assigning a set of given data elements to a given set of labels or classes such that the cost of assigning the data element to a class is minimum. Radar polarimetry is a well established technique for classification of land use features. It has also become an important and cost-effective tool for wetland investigation and researches. Wetlands are under pressure due to high demand for land development for housing and agriculture. Most of mangrove forests were cleared for settlements, agriculture and fire wood. It is important to manage the wetlands and conserve them for the benefit of the society, flora and fauna. Remote sensing is very useful tool to map the wetlands and classify them. Synthetic aperture radar technology is an advantage over optical remote sensing due to microwave penetration through vegetation and interaction with water under vegetation. The major steps of image classification may include determination of a suitable classification system, selection of training samples, image preprocessing and feature extraction, and selection of suitable classification approaches, post-classification processing and accuracy assessment.

In this paper different target decomposition techniques are used before applying classification techniques. Same training sites are used for all the decompositions to do the comparative study of the results. The objective of Target decomposition (TD)

K. Shah, V.R. Lakshmi Gorty, and A. Phirke (Eds.): ICTSM 2011, CCIS 145, pp. 138–145, 2011.

theory is to express the average scattering mechanism as the sum of independent elements to associate a physical mechanism with each component. There are two types of TD. One is Coherent (CTD) and other is Incoherent (ICTD). CTD was developed to characterize completely polarized scattered waves for which fully polarimetric information is contained in the scattering matrix. The CTD can be used only to study coherent targets also known as point or pure targets. Man made objects are the example of pure targets. Pauli, Krogager, Cameron are the Coherent type of decomposition. In this paper we have used Krogager decomposition. Krogager is also known as SDH decomposition because the scattering matrix can be represented as the combination of the response of sphere, diplane and helix. Helix scattering is a general scattering mechanism which appears in an urban area whereas disappears for almost all natural distributed scattering. It can distinguish man-made target from natural targets well but can not divide one type of man-made target from another kind. The scattering matrix is only able to characterize coherent or pure scatterers. However the matrix can not characterize distributed scatterers (natural targets). ICTD was developed to characterize distributed scatterers. In this paper we have used Freeman, Van Zyl and Yamaguchi which are the types of ICTD. Freeman and Van Zyl has three types of scattering mechanisms namely volume, double bounce and surface or single bounce. Yamaguchi 4- component has one additional scattering mechanism that is helix. Helix scattering often appears in complex urban areas where as disappears in almost all natural distributed scenarios.

2 Polarimetric Decomposition

The target decomposition was first introduced by Chandrasekhar (1960) [1] and later applied to polarized microwave by Huynen (1970) [2]. Coherent decomposition theorems use [S] matrices. It considers a matrix [S] as a linear combination of several other scatterers. A method for coherent target decomposition was presented by Krogager, 1988 [3]. His approach was based on the observation that any complex, symmetric scattering matrix can be decomposed into three components, as if the scattering were due to a sphere, a diplane and a right or left rotating helix. A three-component scattering model for polarimetric SAR data is proposed by Freeman and Durden (1998) [4]. Cloude et.al. (1995, 1996) [5] have suggested the H-A-alpha target decomposition theorem. Van Zyl (1989) [6] describes the use of an imaging radar polarimeteric data for unsupervised classification of scattering behavior by comparing the polarization properties of each pixel in an image to that of simple classes of scattering such as even number of reflections, odd number of reflections, and diffuse scattering. Coherent target decomposition methods can only be applied to coherent scattering. Generally, the scattered wave is partially polarized and the user might be interested in the extraction of geophysical parameters from an area that exhibits significant natural variability in the scattering properties (Van Zyl, 1992) [7]. For different target decomposition methods Alberga et al. (2004) [8] has applied Minimum Distance, Maximum Likelihood and Parallelepiped classifier. A four-component scattering model is proposed by Yamaguichi et al., 2005 [9] to decompose polarimetric synthetic aperture radar images. Circular polarization power is added as the fourth component to the three component scattering model which describes surface, double bounce, and volume scattering. This circular polarization term is added to take into account of the co-pol and the cross-pol correlations which generally appear in complex urban area scatttering and disappear for natural distributed scatterer. A comparison of polarimetric target decomposition methods is proposed by Zhang et.al. (2008) [10]. Results show that among many target decomposition algorithms, the coherent and incoherent formulations are quite comparable in distinguishing natural targets and man made buildings. Pauli decomposition, Cameron decomposition and Freeman decomposition are suitable for the detection of natural targets. On the other hand, SDH decomposition, OEC decomposition, and Four-component model, in particular, are very useful for man-made target extraction. The Touzi decomposition is investigated for wetland characterization [11]. A target scattering decomposition was investigated by Touzi et. al. (2009) [12], for wetland classification. The Touzi decomposition, which permits a roll-invariant target scattering decomposition, leads to the characterization of wetland classes in terms of unique target parameters. Ballester-Berman et. al. (2010) [13] proposed a procedure for exporting the Freeman–Durden PolSAR TD concept to PolInSAR data. The formulation of the Freeman–Durden decomposition has been adapted to PolInSAR in order to jointly retrieve not only the magnitude but also the interferometric phases (related to the vertical locations) of the direct (odd-bounce), double-bounce, and volume scattering mechanisms.

3 Test Sites and Data Sources

ALOS PALSAR data in fully polarimetric mode was acquired over Sunderban, West Bengal district of India on May 21, 2007. Close to the area covering Gangetic plains with agriculture is also acquired by ALOS PALSAR on May 21, 2008.

Sunderban is a part of the world's largest mangrove forest. It is located in southern West Bengal. Lothian Island Wildlife Sanctuary lies south of Sunderbans in South 24 Parganas District, West Bengal. This 2,585-sq km park is the world's largest mangrove forest. Named after the Sundari trees, once found in large quantities here, the park features an extremely diverse array of vegetation and plant life, as well as houses an astounding variety of wildlife. It is the home of the endangered Royal Bengal tiger. In addition to the tigers, the sanctuary is also home to other wild and marine life, including wild boars, macaques, jungle cats, chitals, monkeys, Olive Ridley turtles, dolphins, sea snakes, king cobras and estuarine crocodiles to name a few.

4 Data Processing

The target decomposition techniques are applied with the help of PolSARPro software on ALOS-PALSAR Sunderban area. The software creates files one file for each scattering mechanism for example for Freeman decomposition it creates three files namely Freeman_dbl.bin, Freeman_vol.bin, Freeman_odd.bin for double bounce, volume scattering and odd bounce scattering respectively. These three files are separately processed using ENVI software. Further the three files are combined to form one file: Freeman_dbl_vol_odd. This file is further processed by different classifiers namely Minimum Distance Classifier (MDC), Maximum Likelihood Classifier (MLC), Parallelepiped and Mahalanobis. The chosen algorithms implement quite general image classification methods and are not specifically intended for SAR data; hence they are not the optimal tools for analyzing them. Since these classifiers are the simpler approaches and are available to wide array of potential users, it would be the method of choice. The same procedure is done for all the decompositions.

5 Results

After applying the decompositions, different classification techniques are applied using ENVI. Table 1 show the classification accuracy of various features for three separate components of Freeman and Van Zyl decomposition namely Double, Volume and Odd or Single bounce. From figures 1, 2 and the table no 1 it is clearly seen that Van Zyl decomposition gives better accuracy than Freeman decomposition. For double bounce scattering Van Zyl gives 20% more accuracy than that of Freeman. From the results of both the decompositions, it is clearly seen that wetland gives odd bounce scattering and mangrove gives volume scattering. Figure 1 and 2 shows the classified images using Freeman and Van Zyl decomposition respectively. Figure 1 (a) and figure 2 (a) are the decomposed images before classification. In this red color shows double bounce, green shows volume scattering and blue shows odd bounce scattering. Figure 1 and 2 (b)-(e) shows the classified images after applying different classifiers namely Minimum Distance (MDC), Mahalanobis, Parallelepiped and Maximum Likelihood (MLC) respectively. Table 2 (a)-(e) shows confusion matrices for ALOS/PALSAR Sunderban area computed after applying Minimum distance classifier for different decompositions. Table 3 gives comparison of classification accuracies for different decompositions for different classifiers.

After comparing the results with other decomposition techniques like Freeman, Krogager, Yamaguchi-3 and -4 components it is found that Van Zyl decomposition gives the best results than all these decomposition techniques. The classification accuracy for mangrove is highest (93.55%) in case of Van Zyl which is much more than all other decomposition. The results are verified by using different classifiers like Mahalanobis, Parallelepiped, MDC and MLC. The result for ML classifier is the best among all the classifiers.

Table 1. Classification accuracy for double, volume and odd bounce scattering. (a) Freeman (b) Van Zyl.

	Double	Volume	Odd
Accuracy	52.80%	61.94%	41.72%
Mangrove	12.83	81.07	26.62
Water	86.79	34.59	28.21
Agri-bare	21.2	98.08	83.17
Veg+village	57.3	83.02	25.48
Wetland	1.42	55.38	92.02

(a)

	Double	Volume	Odd
Accuracy	71.51%	63.23%	46.14%
Mangrove	74	90.05	70.94
Water	79.83	37.39	34.7
Agri-bare	71.38	85.28	47.26
Veg+village	63.46	78.97	26.02
Wetland	13.78	79.59	99.94

(b)

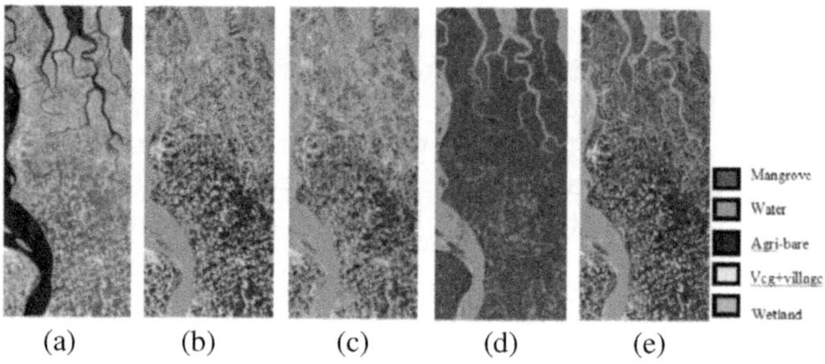

Fig. 1. Classified images for Freeman decomposition (a) Freeman_dbl_vol_odd_db (before classification) Red = double bounce, Green = volume scattering, Blue = odd or single bounce (b) Minimum distance classified image (c) Mahalanobis classified image (d) Parallelepiped classified image (e) Maximum likelihood classified image

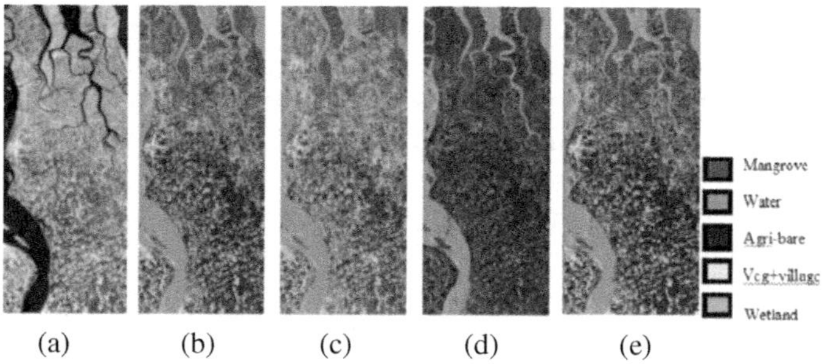

Fig. 2. Classified images for Van Zyl decomposition (a) Van Zyl3_dbl_vol_odd_db (before classification) Red = double bounce, Green = volume scattering, Blue = odd or single bounce (b) Minimum distance classified image (c) Mahalanobis classified image (d) Parallelepiped classified image (e) Maximum likelihood classified image

Table 2. Confusion matrices for ALOS-PALSAR Sunderban area

Class	Mangrove	Water	Agri-bare	Veg+Village	Wetland	Total
Mangrove	64.98	0	0	11.39	0.26	12.88
Water	0	96.17	1.13	0	0.64	44.07
Agri-bare	0.59	3.81	98.85	0.24	0.07	20.82
Veg+Village	34.43	0	0.02	88.37	0	16.66
Wetland	0	0.02	0	0	99.03	5.56

(a) Freeman

Class	Mangrove	Water	Agri-bare	Veg+Village	Wetland	Total
Mangrove	93.55	0	0.15	8.37	0.06	17.6
Water	0	99.9	2.49	0	0.58	46.03
Agri-bare	0.06	0.06	97.36	0.06	0.26	18.72
Veg+Village	6.38	0	0	91.57	0	12.07
Wetland	0	0.04	0	0	99.1	5.58

(b) Van Zyl

Table 2. (*continued*)

Class	Mangrove	Water	Agri-bare	Veg+Village	Wetland	Total
Mangrove	40	9.16	4.65	4.32	1.74	12.76
Water	13.16	54.45	28.19	9.4	1.55	33.75
Agri-bare	10.48	33.28	35.03	13.75	0.32	25.39
Veg+Village	36.35	1.21	32.01	72.5	0	21.79
Wetland	0.02	1.89	0.11	0.03	96.39	6.3

(c) Krogager

Class	Mangrove	Water	Agri-bare	Veg+Village	Wetland	Total
Mangrove	56.52	0	1.17	19.16	0.06	12.53
Water	0	91.15	11.76	0.03	3.03	43.96
Agri-bare	7.55	7.86	86.07	5.44	0.13	22.08
Veg+Village	33.69	0	0.94	74.86	0	15.09
Wetland	2.24	0.98	0.06	0.51	96.78	6.35

(d) Yamaguchi_3

Class	Mangrove	Water	Agri-bare	Veg+Village	Wetland	Total
Mangrove	9.81	0	0	8.16	3.86	2.93
Water	5.57	86.85	23.89	2.48	49.26	48.2
Agri-bare	66.76	9.8	69.96	71.41	0.26	38.25
Veg+Village	9.04	3.35	6.14	13.12	3.54	6.07
Wetland	8.83	0	0	4.84	43.08	4.56

(e)Yamaguchi_4

Table 3. Classification accuracies for different decompositions for different classifier

Classifier	Freeman	Van Zyl3	Krogager	Yamaguchi3	Yamaguchi4
Minimum Distance	90.39%	97.25%	52.68%	82.41%	48.51%
Mahalanobis	84.53%	85.64%	72.34%	77.14%	60.87%
Parallelepiped	74.50%	89.27%	34.16%	69.22%	46.91%
Maximum Likelihood	96.83%	97.84%	80.98%	85.44%	71.52%

As discussed before ICTD is giving better accuracy for this type of data set since it has more natural or distributed scatterers than pure scatterers. Yamaguchi 4 gives poor classification accuracy. The 4th helix scattering component *Pc* in Yamaguchi 4 decomposition becomes minor contribution for the natural distributed target area [14].

6 Conclusion

Different target decomposition techniques have been applied on ALOS-PALSAR Sunderban data set for finding classification accuracy. It is observed that among

Freeman, Van Zyl, Krogager, Yamaguchi 3-components and 4- components decomposition Van Zyl gives the best results. The Van Zyl decomposition gives better classification accuracy for mangroves (93.55%) while Freeman decomposition gives almost 30% less than Van Zyl. The overall classification accuracy obtained by Van Zyl is 97.8%.

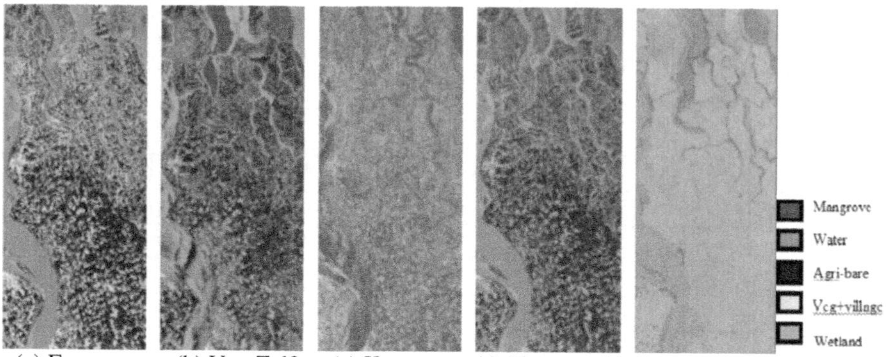

(a) Freeman (b) Van Zyl3 (c) Krogager (d) Yamaguchi_3 (e) Yamaguchi_4

Fig. 3. Minimum distance classified images

Acknowledgements

ALOS-PALSAR data were obtained through announcement opportunity of JAXA, Japan under project no. 381.

References

1. Chandrasekhar, S.: Radiative transfer. Dover, New York (1960)
2. Huynen, J.R.: Phenomenological theory of radar targets, PhD thesis, Technical University of Delft, the Netherlands (1970)
3. Krogager, E.: Decomposition of the Sinclair matrix into fundamental components with applications to high resolution radar target imaging. In: Proceedings of the NATO-ARW, Bad Windsheim, West Germany, pp. 18–24 (1988)
4. Freeman, A., Durden, S.L.: A Three-Component Scattering Model for Polarimetric SAR Data. IEEE TGRS 36(3) (1998)
5. Cloude, S.R., Pottier, E.: A review of target decomposition theorems in radar polarimetry. IEEE Transactions on Geoscience and Remote Sensing 34(2), 498–518 (1996)
6. Van Zyl, J.J.: Unsupervised Classification of Scattering Behavior Using Radar Polarimetry Data. IEEE Transactions on Geoscience and Remote Sensing 27(1), 37–45 (1989)
7. Van Zyl, J.J.: Application of Cloude's target decomposition theorem to polarimetric imaging radar data. In: Mott, H., Boerner, W.M. (ed.) Radar Polarimetry (1992)
8. Alberga, V., Krogager, E., Chandra, M., Wanielik, G.: Potential of coherent decompositions in SAR polarimetry and interferometry. In: International Geoscience and Remote Sensing Symposium (2004)

9. Yamaguchi, Y., Moriyama, T., Ishido, M., Yamada, H.: Four-component scattering model for polariemtric SAR image decomposition. IEEE TGRS 43(8) (2005)
10. Zhang, L., Zhang, J., Zou, B., Zhang, Y.: Comparison of Methods for Target Detection and Applications Using Polarimetric SAR Image. Piers online 4(1) (2008)
11. Touzi, R., Deschamps, A., Rother, G.: Scattering type phase for wetland classification using C-band polarimetric SAR. In: IGARSS (2008)
12. Touzi, R., Deschamps, A., Rother, G.: Phase of Target Scattering for Wetland Characterization Using Polarimetric C-Band SAR. IEEE Transactions on Geoscience and Remote Sensing 47(9), 3241–3261 (2009)
13. Berman, D.B., Lopez-Sanchez, J., Juan, M.: Applying the Freeman–Durden Decomposition Concept to Polarimetric SAR Interferometry. IEEE Transactions on Geoscience and Remote Sensing 48(1), 466–479 (2010)
14. Sato, R., Yamaguchi, Y., Yamada, H.: Polarimetric scattering feature estimation for accurate vegetation area classification. In: Proc. IGARSS, vol. 2, pp. 888–891 (2009)

Audio Steganography Using Differential Phase Encoding

Nikhil Parab, Mark Nathan, and K.T. Talele

Sardar Patel Institute of Technology, Munshi Nagar, Andheri (West), Mumbai-400058, India
{parabnik17,marknathan.147}@gmail.com, kttalele@spit.ac.in

Abstract. This paper proposes a novel technique of camouflaging a block of binary data in an audio file for secured covert communication. The phase of the cover signal is used as a medium for steganography. Cover signal may be a song or a speech signal. Instead of modifying the absolute phase values of selected component frequencies, the difference between the phase values of the selected component frequencies and their adjacent frequencies of the cover signal are modified by a small amount to hide data bits .The change is so minute that it hardly alters the phase spectrum and is imperceptible to human auditory system(HAS). This technique provides robustness, noise insensitivity, high data capacity and almost accurate data recovery from the cover signal.

Keywords: Steganography, component frequencies, cover signal, phase.

1 Introduction

Steganography is the art and science of writing hidden messages in such a way that no one, apart from the sender and intended recipient, suspects the existence of the message; a form of security through obscurity. Steganography means "covered writing" in Greek. A steganography system, in general, is expected to meet three key requirements, namely, imperceptibility of embedding, correct recovery of embedded information, and large payload. Audio steganography, or information hiding in audio signals, is gaining widespread importance for secure communication of information such as covert battlefield data and banking transactions via open audio channels. Audio Steganography consists of a cover audio signal and a data to be camouflaged. Cover signal is the one in which secret data to be transmitted is blended in such a manner that it is imperceptible to Human Auditory System (HAS).

Communication between two parties over long distances has always been subject to interception. This led to the development of cryptography schemes. Cryptography is a process where the sensitive data is encrypted and sent over the channel to the desired receiver. It achieves security mainly through a process of making the message unintelligible so that those who do not possess necessary keys cannot decrypt the message. Though cryptography can hide the content of the message, the existence of a cryptographic communication in progress cannot be hidden from a third party. If the third party discovers the cryptographic communication, they might be able to decipher the message. Also, it creates suspicion about the presence of valuable data. Steganography has many advantages over cryptography. In steganography, data is camouflaged in such a way that it is very difficult to detect the presence of any

K. Shah, V.R. Lakshmi Gorty, and A. Phirke (Eds.): ICTSM 2011, CCIS 145, pp. 146–151, 2011.

sensitive hidden data. Thus, the messages do not attract attention to themselves. The stego signal has an outward appearance almost identical to that of the original signal.Therefore, whereas cryptography protects the contents of a message, steganography can be said to protect both messages and communicating parties. Several steganography methods by indirectly exploiting various limitations of HAS have been proposed with varying degrees of success[1-5]. These methods alter the audio signal by a small amount so that the change is imperceptible to HAS thereby, avoiding suspicion.

2 Proposed Differential Phase Encoding Algorithm

The technique presented uses Fast Fourier Transform (FFT) and differential encoding for binary data bit insertion and an algorithm which introduces randomness in the selection of starting location of bit stream in each frame.

Firstly, the cover signal is divided into small frames of 10msec each. Each frame is taken into consideration independently. The frequency response of each frame is plotted using Fast Fourier Transform. The figure below shows the frame division of entire cover signal.

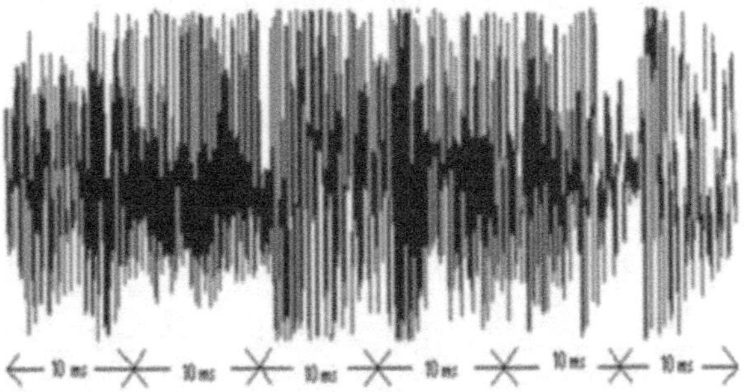

Fig. 1. Frame division of entire cover signal

Taking a single frame into consideration, from its phase response, a random number which is a function of frame number is determined. Now, this number (say y') is approximated to the nearest frequency present to the L.H.S. of y', say 'Y'.

This 'Y' is the starting frequency from which embedding of bits will begin. Let 'n' be the number of data bits to be embedded in a single frame.

2.1 Embedding Data Bits at the Sender

Now, let the phase value (ø) of the starting frequency 'Y' (derived above) be O1. Let us call this Y as Y1.

$$Ø(Y1)= O1$$

Similarly, let the phase value of the next frequency Y2 be O2.

$$Ø(Y2)= O2$$

In this manner, the nth frequency from Y1 (i.e. Yn) along with its phase is represented as:

$$\varnothing(Yn) = On$$

For embedding the first bit of the data block [BIT1= 0/1], the phase value of the next adjacent frequency is adjusted by a minimum extent such that the magnitude of the difference between the rounded lower bound integer phase values of the two frequencies is even/odd respectively.

Thus, if '0' is the first bit to be embedded, then the phase value of the adjacent frequency (O2) is manipulated to obtain an even difference between the rounded lower bound integer phase values $\underline{O1}$ and $\underline{O2}$.

Similarly, if '1' is the first bit to be embedded, then the phase value of the adjacent frequency (O2) is manipulated to obtain an odd difference between the rounded lower bound integer phase values $\underline{O1}$ and $\underline{O2}$.

For:

$$\begin{aligned} \text{BIT1} &= 0 \quad \ldots\ldots \quad |\underline{O1} - \underline{O2}| = \text{EVEN}. \\ &= 1 \quad \ldots\ldots \quad |\underline{O1} - \underline{O2}| = \text{ODD}. \end{aligned}$$

where $\underline{O1}$ = nearest integer lower than O1, $\underline{O2}$ = nearest integer lower than O2.

As differential encoding scheme is used, the next data bit 'BIT2' is embedded using the phase values of frequencies Y+1 and Y+2. Therefore, for:

$$\begin{aligned} \text{BIT2} &= 0 \quad \ldots\ldots \quad |\underline{O2} - \underline{O3}| = \text{EVEN}. \\ &= 1 \quad \ldots\ldots \quad |\underline{O2} - \underline{O3}| = \text{ODD}. \end{aligned}$$

Similarly, the concept can be extended for embedding n bits of information at the frequencies following Y+2. Therefore, for:

$$\begin{aligned} \text{BITn} &= 0 \quad \ldots\ldots \quad |\underline{On} - \underline{On+1}| = \text{EVEN}. \\ &= 1 \quad \ldots\ldots \quad |\underline{On} - \underline{On+1}| = \text{ODD}. \end{aligned}$$

Once the n bits are camouflaged as phase difference between adjacent frequencies in a single frame, the modified spectrum is converted back to time domain using Inverse Fast Fourier Transform (IFFT).

This is followed by considering the frequency spectrum of the next 10 ms frame and working upon it in the manner mentioned above. The stego frames are concatenated back to reconstruct back the entire stego signal. When the reconstructed stego signal is played back, it is very hard to distinguish between the stego signal and the original signal.

2.2 Detecting Bits at the Receiver

Once the stego signal is received at the receiver end, it is again broken down into 10msec frames. For each frame, the same random frequency is determined using the same function which was used at the encoder. Thus, the starting location of bit stream in each frame is obtained. Now using this information, the data bit can be detected using the following method.

If the difference between the rounded lower bound integer phase values on and On+1 is even, then the nth encoded data bit is decoded as '0'.

Similarly, if the difference between the rounded lower bound integer phase values On and On+1 is odd, then the nth encoded data bit is decoded as '1'.

For:

$$| \underline{On} - \underline{On+1} | \;==\; \text{EVEN} \quad \ldots\ldots \quad \text{BITn=0.}$$
$$==\; \text{ODD} \quad \ldots\ldots \quad \text{BITn=1.}$$

3 Experimental Results

The cover signal used for simulation was a 6.8 seconds song which was divided into 680 frames of 10 milliseconds each. The magnitude and phase plot of a single frame taken into consideration is shown in the Figure 2 and 3 respectively. The phase plot of the same frame after embedding block of bits is shown in Figure 4.

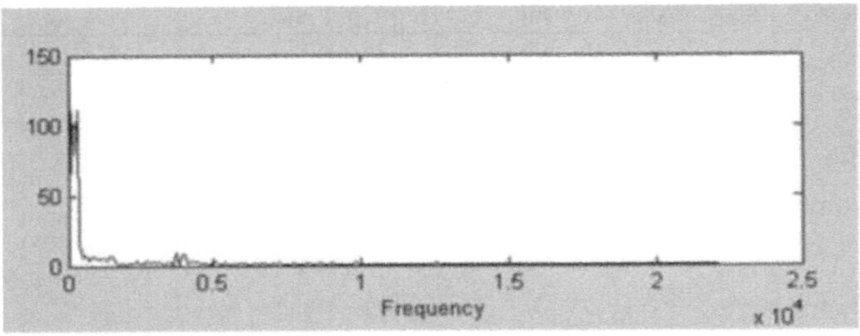

Fig. 2. Magnitude spectrum of a single frame

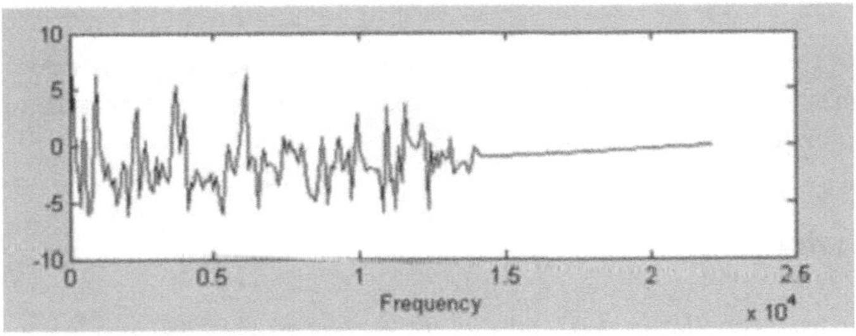

Fig. 3. Phase spectrum of a single frame of original signal

From the phase plot of a single frame of original signal and stego signal, it can be seen that they are almost indistinguishable from each other. The change in the phase plot is almost imperceptible to HAS. In the similar fashion block of n bits are camouflaged in each frame.

From the Table 1, it can be inferred that as the number of bits embedded per frame increases, the similarity between the original and stego cover signal reduces which is

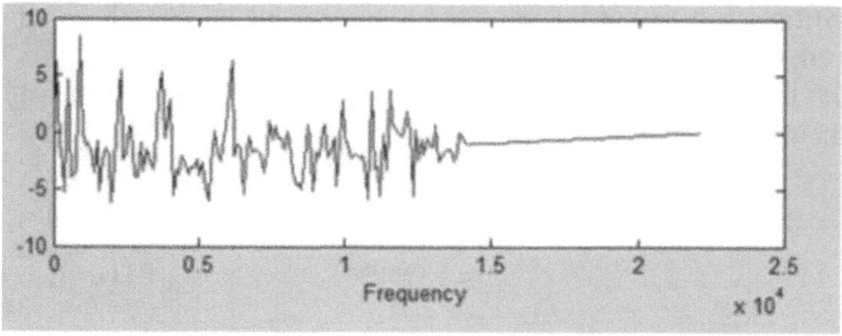

Fig. 4. Phase spectrum of a single frame of stego signal

Table 1. Results of cover signal 1 and cover signal 2

	Number of bits Embedded per Frame	Correlation Coefficient Between Original And Stego Cover Signal	Correlation Coefficient Between Original Data Bits And Decoded Data Bits	Bit Error Rate
Cover Signal 1	25	0.8753	0.9962	0.1895
	10	0.8916	0.9958	0.2106
Cover Signal 2	25	0.9759	0.9974	0.1294
	10	0.9806	0.9994	0.0294

indicated by a decrease in the value of correlation coefficient between cover signals. Also, an increase in bit error rate is observed with an increase in the number of embedded bits per frame.

4 Conclusion

A technique for embedding data bits using differential phase modification has been presented along with its simulated results. From the simulation results it can be concluded that the technique is feasible with almost cent percent data recovery. Also, the proposed technique has following advantages. Firstly, the selection of frequency (Y) is not fixed and varies with every frame of speech signal. This introduces a sort of randomness which makes it difficult for an malicious attacker to analyze the starting location of bit stream even when a stream of frames is analyzed together .Secondly, as the difference between the phase values of the adjacent frequencies is used to embed critical data instead of absolute phase value , the probability of data being altered by noise is reduced by a large extent .Also, the selection of phase as a medium to hide data instead of amplitude increases the noise immunity level of the stego signal.

Thus, it can be concluded that through differential phase encoding, secure and robust steganography can be achieved where critical information can be camouflaged without a hint of suspicion.

References

1. Gopalan, K., Wenndt, S.J., Adams, S.F., Haddad, D.M.: Audio steganography by amplitude or phase modification. In: Proceedings Security and Watermarking of Multimedia Contents, vol. 5020, pp. 67–76.
2. Basu, P.N., Bhowmik, T.: On Embedding of Text in Audio – A case of Steganography. In: International Conference on Recent Trends in Information, Telecommunication and Computing (2010)
3. Gopalan, K., Wenndt, S., Noga, A., Haddad, D., Adams, S.: Cover speech communication via cover speech by tone insertion. In: Proc.2003 IEEE Aerospace Conference, vol. 4, pp. 41647–41653 (2003)
4. Gopalan, K.: Audio Steganography for Covert Data Transmission by Imperceptible Tone Insertion. In: Proc. The IASTED International Conference on Communication Systems And Applications (CSA 2004), Banff, Canada (2004)
5. Bender, W., Gruhl, D., Morimoto, N., Lu, A.: Techniques for data hiding. IBM Systems Journal 35(3/4), 313–336 (1996)
6. Gopalan, K.: Audio steganography using bit modification. In: Proc. IEEE International Conference on Acoustics, Speech, and Signal Processing (ICASSP 2003), vol. 2, pp. 421–424 (2003)
7. Swanson, M.D., Zhu, B., Tewfik, A.H., Boney, L.: Robust audio watermarking using perceptual masking. Signal Processing 66, 337–355 (1998)
8. Wang, Q.S., Sun, S.H.: A novel algorithm for embedding watermarks into digital audio signals. ACTA Acustica 26(5), 464–467 (2001)
9. Gopalan, K., Wenndt, S., Haddad, D.: Steganographic method for covert audio communications. U.S. Patent 7 231 271 (2007)

Audio Steganography Using Spectrum Manipulation

Mark Nathan, Nikhil Parab, and K.T. Talele

Sardar Patel Institute of Technology, Munshi Nagar, Andheri (West), Mumbai-400058, India
{marknathan.147,parabnik17}@gmail.com, kttalele@spit.ac.in

Abstract. This paper describes a method for achieving audio steganography through spectrum modification of the cover signal- the signal into which the useful and valuable information is embedded. The embedded data signal is inserted into the regions of the cover signal spectrum where power of frequency content is low. Also the level of the embedded signal is kept low enough inorder to avoid the crossing of thresholds for audio perception. By performing these steps, the data signal shows no trace of being embedded in the cover signal. This method of audio steganography can be used to robust covert communication.

Keywords: Steganography, Frequency Spectrum.

1 Introduction

Clandestine communication has been gaining steadily growing importance in the field of information technology. Audio steganography mainly deals with camouflaging valuable information into a vast sea of data in a robust, covert and secure manner to achieve its imperceptibility. This method of steganography is very critical for secure government communications and for bank transactions.

The foundations of audio steganography lie in the imperfections of the human auditory system. As stated by K.Gopalan in [3], psychoacoustical, or auditory masking property renders a weak tone imperceptible in the presence of a strong tone in its temporal or spectral neighborhood. This property arises because of the low differential range of the HAS even though the dynamic range covers 80 dB below ambient level. Frequency masking occurs when human ear cannot perceive frequencies at lower power level if these frequencies are present in the vicinity of tone- or noise-like frequencies at higher level. This property of inaudibility of weaker sounds is used in different ways for embedding information.

2 Spectrum Manipulation Algorithm

This method of audio steganography achieves the task of camouflaging valuable information by implementing spectrum manipulation. The two signals involved in steganography are the cover signal and the data signal. The cover signal is the one into which valuable information is embedded. The data signal is the valuable information that is camouflaged and blended in the cover signal.

K. Shah, V.R. Lakshmi Gorty, and A. Phirke (Eds.): ICTSM 2011, CCIS 145, pp. 152–159, 2011.

The spectrum manipulation can be bifurcated into two distinct sets of operations- the operations on the cover signal and the operations on the data signal. We assume that c[n] is the cover signal in time domain and C(jω) is its corresponding frequency domain representation. Similarly, we assume that d[n] is the data signal in time domain and D(jω) is its corresponding frequency domain representation.

The frequency domain representations are calculated using the Fast Fourier Transform to find the Discrete Fourier Transform of the signals.

$$C(j\omega) = \sum_{n=-\infty}^{\infty} c[n]e^{-j\omega n} \Big|_{\omega=\frac{2\pi k}{N}} . \tag{1}$$

where k=0,1,..,N-1 and N: length of the c[n].

$$D(j\omega) = \sum_{n=-\infty}^{\infty} d[n]e^{-j\omega n} \Big|_{\omega=\frac{2\pi k}{M}} . \tag{2}$$

where k=0,1,..,M-1 and M: length of the d[n].

2.1 Operations on the Cover Signal

Firstly, the average energy of the cover signal is determined from its frequency spectrum. Then, from C(jω), a region is selected where the amplitude of the frequency content contained in that region is below the threshold of audibility, thus causing them to be masked by the frequencies of higher amplitude. Within this region, a center frequency is selected. This center frequency will later be used for data signal operations.

Subsequently, the cover signal is passed through a tunable notch filter that has a center frequency equal to that of the selected center frequency and a stop-band width of 3000-4000 Hz. The notch filter thus attenuates the low intensity frequencies in the determined region to a level of infinitesimal energy. The stop-band width is chosen in this manner, as it will be enough to accommodate the major frequency content of the speech data signal.

Also, we have to send the selected center frequency to the receiver in a secure manner. For this purpose, the selected center frequency is used as an argument for a function, along with a randomly generated number and a metric that is unique to the cover signal. This metric can be any feature of the cover signal such as its length in seconds or minutes, size on disk, etc. We have selected the average energy of the cover signal as a metric. The three arguments are given to a function that generates a response. The response, along with the random number is attached along with the final stego signal.

We assume that the transfer function of the notch filter is H(jω). Then the filtered cover signal is:

$$C'(j\omega)=C(j\omega)H(j\omega) . \tag{3}$$

Assuming the average energy to be given by ‘A’, random number is given by ‘R’ and center frequency by ‘f_c’, then the response ‘r’ is calculated as:

$$r = f(A,R,f_c) . \quad (4)$$

2.2 Operations on the Data Signal

The data signal is modulated by the selected center frequency f_c. The modulated data signal is then band-limited by a filter having a pass-band of 3000-4000 Hz. As the data signal we have considered is speech, the band-pass filter thus passes the major frequency content. Thus the modulated data signal is given as:

$$d'[n]=d[n]*\sin(2\pi f_c n) . \quad (5)$$

If we assume that the transfer function of the filter is H'(jω) and the FFT of d'[n] is D'(jω), then the frequency domain representation of t he band-pass modulated data signal:

$$D''(j\omega)=D'(j\omega)H'(j\omega) . \quad (6)$$

Next, scale D''(jω) to levels of low intensity so that the data signal can be effectively masked by the higher energy frequencies. If we assume the scaling factor as 'a'(< 1), then the scaled frequency of D''(jω) is given:

$$D'''(j\omega)= a*D''(j\omega) . \quad (7)$$

While scaling, there is another point to be noted that it should be done in such a way that the average energy content of the resultant stego signal is equal to that of the original cover signal.

2.3 Formation of Resultant Stego Signal

After operations on the data signal, the resultant stego signal is obtained by adding D'''(jω) and C'(jω). The resultant spectrum's Inverse Fast Fourier Transform is calculated to obtain the time-domain representation of the resultant stego signal. If the resultant signal is given by s[n] in time-domain and S(jω) in frequency-domain, then:

$$S(j\omega)= D'''(j\omega)+ C'(j\omega) . \quad (8)$$

And the time-domain signal is calculated as:

$$s[n] = N^{-1} \sum_{k=0}^{N-1} S(j\omega)e^{j\omega n} |_{\omega=2\pi k/N} . \quad (9)$$

where n=0,1…,N-1 and N=length of S(jω).

2.4 Operations to Be Done at the Receiver Side

At the receiver end, the only step of major importance is the detection of the selected center frequency. This is done in the following manner.

After obtaining the stego signal, the receiver will be able to identify regions of low frequency intensity from the FFT of the stego signal. The receiver will also know about the random number 'R' and the response 'r' as these are appended to the resultant stego signal at the transmitter end. Now, the receiver and transmitter will first agree on the feature of the cover signal at the beginning so that only the pair will know what is the third argument for the function. In our case, we assumed that the agreed upon feature is the average energy of the cover signal. As the scaling done at the transmitter end ensures that the average energy content of the stego signal is almost equal to that of the cover signal, the detection is possible. It proceeds as follows. The receiver will input 'R', 'A' and any selected frequency from the region it has identified as low intensity, to the function and obtain a response r'. The receiver will then check if r'=r. If not, then it will proceed to another frequency and the repeat the checking process. As the selected center frequency lies in the region of low frequency intensity, the receiver will ultimately determine that center frequency.

After determining the center frequency, the receiver will then perform the operations described above in a suitable manner to obtain the demodulated data signal from the received stego signal.

3 Experimental Results

Below are the FFT plots of the data signal and the cover signal:

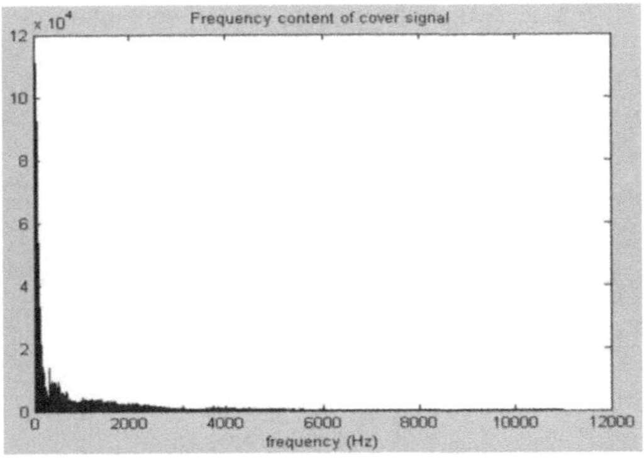

Fig. 1. Frequency spectrum of cover signal

From the plots it can be seen that for the cover signal, majority of the spectral power is concentrated from 0 Hz to 5500 Hz. For the data signal, the spectral power is mainly concentrated from 0 Hz to 5000 Hz.

In most covert communication applications, most data signals consist of speech. Keeping this in mind and knowing that in telephony, voice transmission bandwidth extends from 300 Hz to 3400 Hz, the data signal is bandlimited from 0 to 2000 Hz through spectrum modification. There was no noticeable difference between the bandlimited signal and the original signal in terms of speech content articulation. In order to embed the data signal successfully inside the cover, it is necessary to scale the spectral amplitudes of the frequencies present in the data signal. In the simulation process, this could be achieved by scaling the entire data signal spectrum into the range 0 to 150. This is followed by modulation of the data signal at the carrier frequency. In our simulation, the center frequency was determined as 9000 Hz. Below is the FFT plot of the scaled modulated data signal:

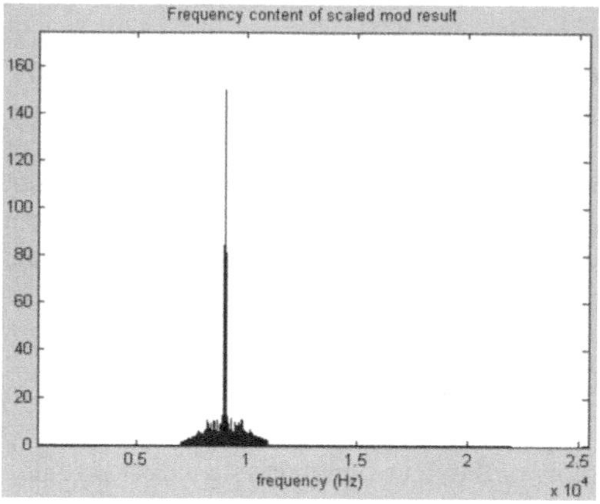

Fig. 2. Frequency spectrum of data signal

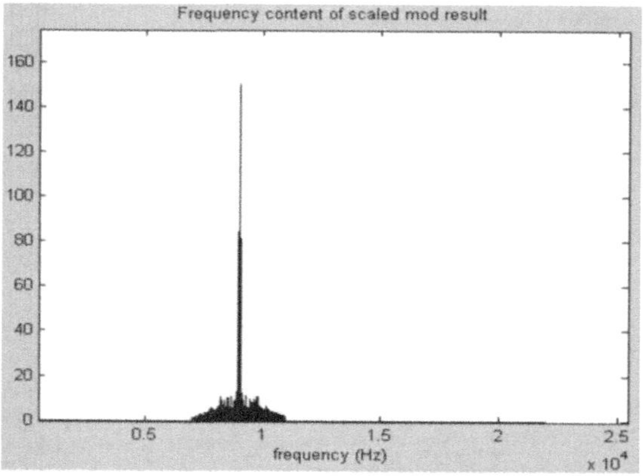

Fig. 3. Frequency spectrum of scaled modulated data signal

Below is the FFT plot of the stego signal containing the data. It can be seen that the FFT plots of both the stego signal and the original cover signal show a high degree of similarity. Another important point to be highlighted in our simulations is that the lengths of the data and cover signals even though they are originally different, should

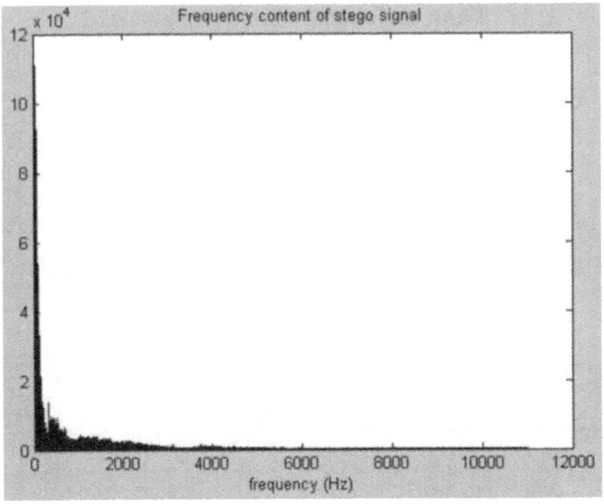

Fig. 4. Frequency spectrum of resultant stego signal

be made equal (usually the greater of the two lengths should be considered). This will ensure that the frequencies present in the frequency vectors of both the cover and data signals are the same.

On converting the stego signal from frequency domain to time domain and then converting the uncompressed .wav stego signal to an .mp3 format causes a glitch to occur. The time domain stego signal when read again indicates a higher number of samples, an increase of roughly 20000. But the size of the stego signal and the original cover signal are the same. This glitch is due to the encoding process used by mp3write( ) function. The glitch can be removed by considering the samples of the stego signal from 1 to length(cover signal). But the tradeoff of this method is that the by truncation, we lose some information and hence the entire spectral content is not obtained. This causes distortion in the demodulated data signal. But the level of distortion during the time frame of the speech signal, that is the data signal, is below levels that can cause discomfort.

To show statistically the similarity between the original cover signal and the resultant stego signal, we had experimented with two different data signals and one common cover signal. Below is a table that lists the correlation coefficients between the original cover and stego signal. In addition to this, we have also tabulated the correlation coefficients between the original data signal and the demodulated data signal obtained from the stego signal. The variable parameter considered for the two data signals is their bandwidth selected for steganography. As per the results, more the bandwidth of data signal taken into consideration, more is the value of the correlation coefficient between the original data signal and the demodulated data signal obtained from the stego signal. At the same time, with increase in the bandwidth of data signal, the correlation coefficient between the original cover and stego signal decreases.

Even though the correlation coefficients between the original data and the demodulated data signal show very low degress of similarity, the content of the data signal when heard is quite articulate. The content can be enhanced through the use of an equalizer.

Table 1. Results of data 1 and data 2

	BANDWIDTH OF DATA SIGNAL (Hz)	CROSS CORRELATION BETWEEN STEGO SIGNAL AND ORIGINAL COVER SIGNAL	CROSS CORRELATION BETWEEN DEMODULATED DATA SIGNAL AND ORIGINAL DATA SIGNAL
DATA 1	2000	1.000	-0.042
	5000	0.9929	0.0136
DATA 2	2000	1.000	0.0054
	4000	0.9929	0.0065

4 Conclusion

A technique for embedding valuable information into a cover signal using spectrum modification has been presented along with the simulated results. The results show that this technique is feasible in terms of implementation and performance. The scope for research is still available for determining the optimum balance between the three entities of data signal bandwidth, cross correlation coefficient between extracted data and original data signal and cross correlation coefficient between the cover and resultant stego signal.

Using spectrum modification, a data signal can have a maximum length equal to the length of the cover signal. Even in the case of the lengths of the cover and data signal being equal, the simulated results achieved were quite desirable. The advantage of this method is that a considerably large amount of valuable data can be camouflaged into the cover signal without leaving a hint of steganography. An additional advantage is that a malicious attacker would not be able to demodulate the embedded data because the algorithm to determine the center frequency is not known to the user or the attacker. Hence, an attempt to find the embedded data would result in a falsely detected signal containing no valuable information. Thus, it can be concluded that through spectrum modification, secure and robust steganography can be achieved where valuable information can be embedded and camouflaged successfully.

References

1. Gopalan, K., Wenndt, S., Noga, A., Haddad, D., Adams, S.: Covert speech communication via cover speech by tone insertion. In: Proc. 2003 IEEE Aerospace Conference, vol. 4, pp. 41647–41653 (2003)
2. Gopalan, K.: Audio steganography using bit modification. In: Proc. IEEE International Conference on Acoustics, Speech, and Signal Processing (ICASSP 2003), vol. 2, pp. 421–424 (2003)

3. Gopalan, K.: Audio Steganography for Covert Data Transmission by Imperceptible Tone Insertion. In: Proc. The IASTED International Conference on Communication Systems And Applications (CSA 2004), Banff, Canada (2004)
4. Al-Najjar, A.J., Alvi, A.K., Idrees, S.U., Al-Manea, A.M.: Hiding encrypted speech using steganography. In: Proceedings of the 7th Conference on 7th WSEAS International Conference on Multimedia, Internet & Video Technologies, vol. 7, pp. 275–281 (2007)
5. Bender, W., Gruhl, D., Morimoto, N., Lu, A.: Techniques for data hiding. IBM Systems Journal 35(3/4), 313–336 (1996)
6. Swanson, M.D., Kobayashi, M., Tewfik, A.H.: Multimedia data embedding and watermarking technologies. Proc. IEEE 86, 1064–1087 (1998)
7. Anderson, R.J., Petitcolas, F.A.P.: On the limits of steganography. IEEE J. Selected Areas in Communications 16(4), 474–481 (1998)
8. Swanson, M.D., Zhu, B., Tewfik, A.H., Boney, L.: Robust audio watermarking using perceptual masking. Signal Processing 66, 337–355 (1998)

The Study on Data Warehouse Modelling and OLAP for Birth Registration System of the Surat City

Desai Pushpal[1] and Desai Apurva[2]

[1] M.Sc. (I.T.) Programme, Veer Narmad South Gujarat University, Surat, Gujarat, India
[2] Department of Computer Science, Veer Narmad South Gujarat University, Surat, Gujarat, India
desaipushpal@yahoo.com, desai_apu@yahoo.com

Abstract. Data Warehousing has been a buzzword in the IT industry, however very little work has been done for e-governance systems in India. The e-governance systems developed by the Surat Municipal Corporation has achieved great success in several years, to service citizens in a more timely, effective, and cost-efficient method. This initiative has resulted in collection of large amount of unexplored and unorganised data. In this paper we proposed Data Warehouse Modelling and Online Analytical Procession (OLAP) for Birth Registration System using Microsoft SQL Server 2008. Our study utilizes data of the Surat city from the year of 2000 to 2009. To query and analyze the data in the data warehouse conveniently and effectively, we designed Data Warehouse using star schema. Our work will help administrators of The Surat Municipal Corporation analyze Birth Registration System data and provide decision-making support for future planning and better service to citizens of the Surat city. Since the research is still in its early stage, the paper mainly focuses on design and implementation of Data Warehouse Modelling and OLAP for Birth Registration System.

Keywords: Data Warehouse, OLAP, and Birth Registration System.

1 Introduction

In the scenario of ever-changing social and business conditions, organization's needs to have access to more and better information. Almost all organizations are now days using computerized systems as the backbone of their operations but the fact is that despite having a large number of powerful computerized systems and a fast and reliable network, access to information that is already available within the organization is very difficult to access. After implementing several computerized systems typically organization's data are scattered across various databases, flat files, physical records store at different geographical locations. All organizations that use computerized systems for their different operations produce large amount of data. Most of the time this data remains in the operational systems, flat files and can't be used by the organization. In this condition only a small portion of this data that is entered, processed and stored is actually available to decision makers. The unavailability of crucial data can cause significant reduction in efficiency of organizations. Many large organizations found themselves with data scattered across multiple platforms and variations of

K. Shah, V.R. Lakshmi Gorty, and A. Phirke (Eds.): ICTSM 2011, CCIS 145, pp. 160–167, 2011.

technology, making it almost impossible for any one individual to use data from multiple sources. It is therefore crucial that organizations have a good source of organization's data that can be used quickly and flexibly by management. Data Warehousing has emerged as a key technology for enterprises that wish to improve their data analysis, decision support activities, and the automatic extraction of knowledge from data.

2 Data Warehouse

Data Warehouse has been defined in a different ways by various authors. Inmon defines Data Warehouse as "a subject-oriented, integrated, time-variant and non-volatile collection of data in support of management's decision" [1]. Here "subject-oriented" means the information in the data warehouse is organized according to subject and provides information for subject-oriented decision making process. The meaning of "integrated" is that the information in the data warehouse is not simply extracted from various operation systems but systematically processed, summarized and reduced, which ensure that the information in data warehouse is consistent and comprehensive for a organization. The "stable" means that, generally speaking, once a piece of data is put in the data warehouse, it will be kept permanently. The "time-varying" means that information in the data warehouse is not only on the present situation or a certain time point's situation, but also on situation of various stages from a former time point to present, on which quantitative analysis and forecasting can be made on development history and trend. Fox provides another perspective to define the data warehouse: "Data Warehouse is the process by which an organization sets up and maintains a central repository for significant portions of its transaction processing or program management data, which can be selective, extracted and organized for analytical applications, user queries and report generation" [2]. Data Warehouses are databases used for storing large amounts of data, collected from multiple data sources. This data is used in knowledge retrieval processes, business intelligence applications, data mining etc. as the organization's primary source of decision making data.

3 The Problem

In recent years governments all over the world have been successful in implementing various E-governance projects. E-governance allows government to service their citizen in a better way. Not only E-governance provide more facilities to citizen but also make government accountable which is crucial in developing nation like India. The E-governance projects implemented in the Surat city have already paid rich dividends for Surat Municipal Corporation and citizens of the city. Surat Municipal Corporation has taken early initiatives in E-governance and hence different manual tasks are computerized and processes are simplified. This has resulted in timely, effective and cost-effective service to citizen of Surat city. Currently many processes like Child Birth registration, Death registration, Vehicle registration, Property registration etc…are fully computerized and data are stored in centralized database system. This automation has resulted in gigabytes of data containing millions of records regarding various aspects of population of the Surat city. All this information is scattered and maintained in different format. Following diagram describes current state of information scattered across various databases, files, reports and websites:

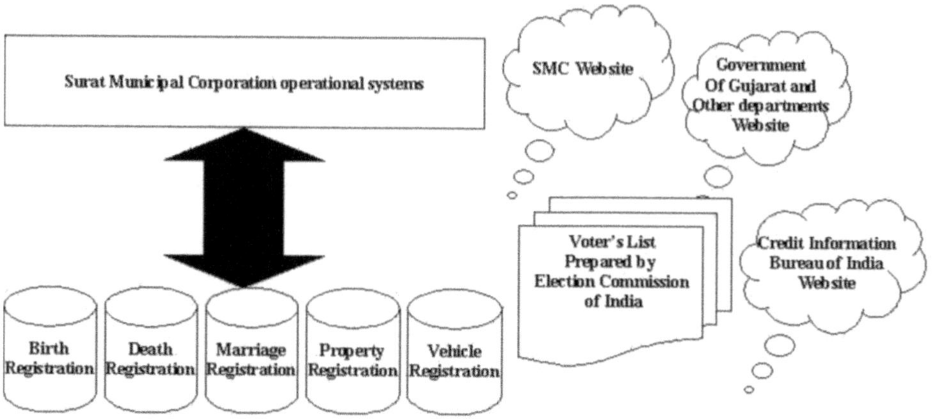

Fig. 1. Current State of Information at Surat Municipal Corporation

Since last few years many government and non-government organizations have invested heavily for computerization of various processes. As various processes are computerized, that has resulted in collection of large amount of data residing in different operational systems. All this data most of the time remains idle and not utilized. By building Data Warehouse on huge amount of data already available, there is good potential of knowledge discovery that can be utilized by different organizations and our society. In this research paper we have focused design and implementation Data Warehouse for Birth Registration data.

4 Research Methodology

We used Microsoft SQL Server 2008 Analysis Service to build Data Warehouse for Birth Registration System. The process of extracting data from Birth Registration Systems and loading into Data Warehouse is commonly known as ETL, which stands for extraction, transformation, and loading. To implement ETL process Microsoft SQL Server 2008 was used and Data Warehouse was established using Microsoft Analysis Services.

4.1 Designing of Conceptual Model

At present, there are two commonly used conceptual models in Data Warehouse: Star Model and Snowflake Model. We have adopted Star model in the Data Warehouse system for reducing scanning time in the fact tables and improving capability of inquiring. Following table contains list of dimensions which connects to fact table:

Table 1. The Dimensions of the Birth Registration System

Subject	Dimension ID
Birth Location	BirthLocation ID
Year	YearID
Religion	ReligionID
Gender	GenderID
Education	EducationID
Zone	ZoneID
Delivery Attention	DeliveryAttentionID
Delivery Method	DeliveryMethodID

4.2 Designing of Logical Model

After analysing various dimensions and subjects for Data Warehouse, we developed Data Warehouse using Star Model. Following diagram shows structure of the star model. The Birth Data fact table is at the center multidimensional Data Warehouse. The fact table contains information about Birth Registration and different keys that connects fact table with various dimension tables. Dimensions are stored in dimension tables that contain dimensional elements and attributes.

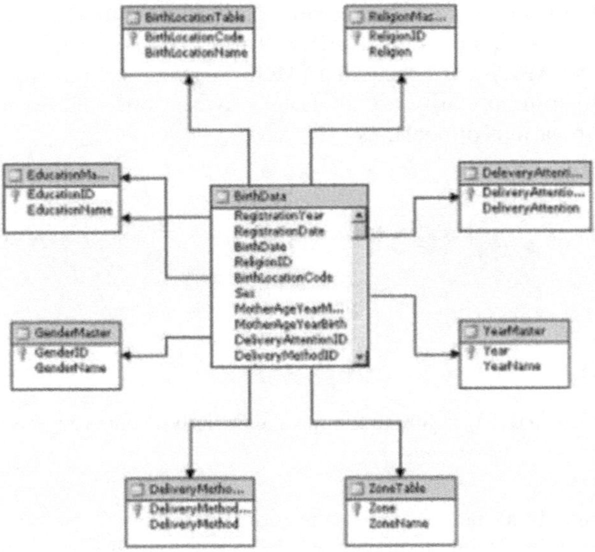

Fig. 2. Star Schema of Birth Registration Data

4.3 Result

Online Analytical Processing is extremely useful for multidimensional data analysis. The objective of using OLAP is to help decision maker utilize data and information effectively. Multidimensional data set is the core mechanism in OLAP for data analysis. Analysis Services of SQL Server 2008, allows us to Slice, Dice, Rotate and Drill that converts data into information. As shown in Figure 2 and Figure 3, OLAP technology is used to construct multidimensional data set for Birth Registration data.

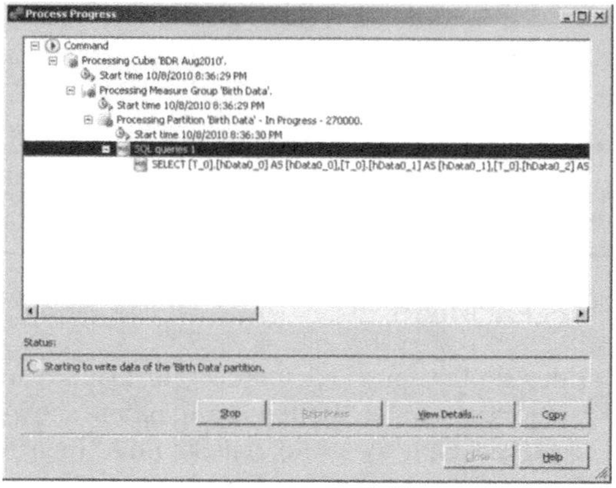

Fig. 3. Processing of multidimensional data

Depending upon the organization requirements fact measures can be selected from the fact table. For example, we have created Birth Registration Cube having Delivery Method, Zone and Religion as three different dimensions. For Cube design, Multidimensional Online Analytical Processing (MOLAP) data storage was chosen and data can be traversed from any arbitrary angle by browse option. Following figure shows browse data with various dimensions.

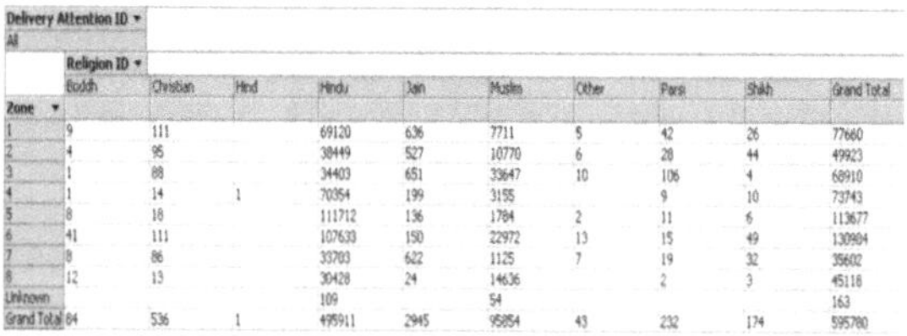

Delivery Attention ID: All; Religion ID

Zone	Boddh	Christian	Hind	Hindu	Jain	Muslim	Other	Parsi	Shikh	Grand Total
1	9	111		69120	636	7711	5	42	26	77660
2	4	95		38449	527	10770	6	28	44	49923
3	1	88		34403	651	33647	10	106	4	68910
4	1	14	1	70354	199	3155		9	10	73743
5	8	18		111712	136	1784	2	11	6	113677
6	41	111		107633	150	22972	13	15	49	130984
7	8	86		33703	622	1125	7	19	32	35602
8	12	13		30428	24	14636		2	3	45118
Unknown				109		54				163
Grand Total	84	536	1	495911	2945	95854	43	232	174	595780

Fig. 4. 3-Dimensional (Delivery Method, Zone and Religion) data browse of Birth Registration Data

Here, it is possible to increase or decrease dimension levels and angles. We can also form combinations of different dimensions from existing dimension tables. In another Cube, we considered four dimensions – Birth Location, Gender, Zone and Year for Birth Registration data. Following figure shows browse data considering four dimensions.

Once we create Cubes in Data Warehouse it is extremely easy for end users to explore data by using different perspectives. In the above-mentioned cube, we can drill data Zone wise, Year Wise, Gender wise and/or Birth location wise. Following figure shows Data Drilling concept where Zone equal to 1, Female & Male Birth Data Count, Year and Birth Location Code.

Birth Location Code				
All				
		Gender ID		
		Female	Male	Grand Total
Zone	Year	Birth Data Count	Birth Data Count	Birth Data Count
⊞ 1		34832	42828	77660
⊞ 2		22856	27067	49923
⊞ 3		32268	36642	68910
⊞ 4		31604	42139	73743
⊞ 5		46751	66926	113677
⊞ 6		58401	72583	130984
⊞ 7		16225	19377	35602
⊞ 8		20833	24285	45118
⊞ Unknown		83	80	163
Grand Total		263853	331927	595780

Fig. 5. 4-Dimensional (Birth Location, Zone, Year and Gender) data browse of Birth Registration Data

Birth Location Code				
(Multiple Items)				
		Gender ID		
		Female	Male	Grand Total
Zone	Year	Birth Data Count	Birth Data Count	Birth Data Count
⊟ 1	2000	3691	4435	8126
	2001	3280	4107	7387
	2002	3382	4449	7831
	2003	4083	5093	9176
	2004	4221	5367	9588
	2005	4281	5508	9789
	2006	3304	3910	7214
	2007	2625	3110	5735
	2008	2784	3182	5966
	2009	3041	3487	6528
	Total	34692	42648	77340

Fig. 6. Data drilling

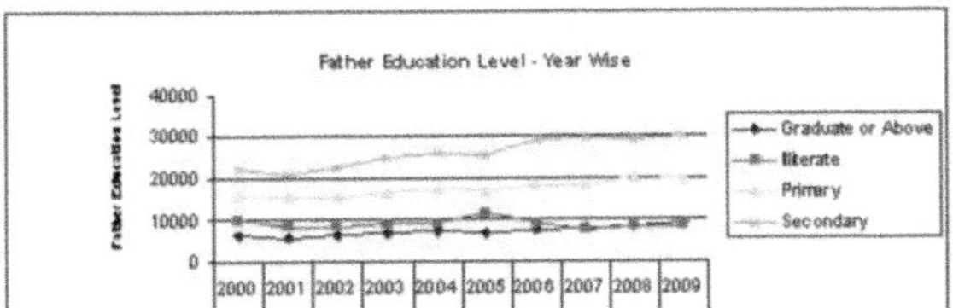

	2000	2001	2002	2003	2004	2005	2006	2007	2008	2009
Graduate or Above	6207	5578	6268	6547	7241	6890	7396	8004	8444	9027
Illiterate	1001	9489	9204	9048	9272	1157	9081	7613	8314	8243
Primary	1566	1528	1552	1669	1746	1691	1817	1806	2021	1992
Secondary	2211	2099	2265	2507	2609	2531	2914	2967	2937	2998

Fig. 7. Chart 1. 2-Dimensional (Father Education and Year) data browse with chart

In another cube, we explored the relationship between Birth Registration and Parents Education Level. We observed that father education information was provided for all records. However, mother education information was provided only in 1,43,604 records out of 5,95,780. For remaining 4,52,176 records mother education information field was blank. The parents education level can be categorised into for levels: Graduate or Above, Illiterate, Primary, and Secondary. Based on these categories and Birth Registration data we developed cube that show relationship between father education level and birth registration year wise.

Similarly, chart was prepared for mother education level and birth registration. To obtain desired result two measures *Father Education Count* and *Mother Education Count* were used in the cube.

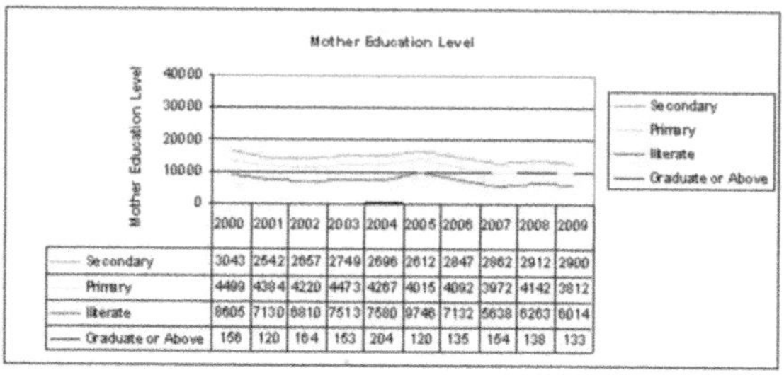

	2000	2001	2002	2003	2004	2005	2006	2007	2008	2009
Secondary	3043	2542	2657	2749	2696	2612	2847	2862	2912	2900
Primary	4499	4384	4220	4473	4267	4015	4092	3972	4142	3812
Illiterate	8605	7130	6810	7513	7580	9746	7132	5638	6263	6014
Graduate or Above	158	120	164	163	204	120	135	164	138	133

Fig. 8. Chart 2. 2-Dimensional (Mother Education and Year) data browse with chart

5 Conclusion

This paper gives an insight on Data Warehouse modelling and OLAP for Birth Registration system, which will provide the effective support to the Municipal Corporation. The traditional system can only collect data but does not provide information of strategic significance to administrators and decision makers of the Municipal Corporation. The main advantage of building Data Warehouse is that administrators of the Municipal Corporation can view data from different perspectives, define various metrics of interest and query data at any level of detail using various methods such as Slice, Dice, Rotate and Drill. Advance visual tools can be used for further analysis or presentation of data.

References

1. Inmon, W.H.: Building the Data Warehouse, 4th edn. John Wiley & Sons, New York (2006)
2. Fox, A.: Data warehousing: avoiding the pitfalls. Behavioral Health Management 20(3), 18 (2000)
3. Tong, L., Yan, C., Po, R.S.: Analysis on Data Warehouse Technology and Its Development Situation. In: IEEE International Symposium on Information Science and Engieering, pp. 486–488 (2008)
4. Fayyad, U., Piatetsky-Shapiro, G., Smyth, P.: The KDD process for extracting useful knowledge from volumes of data. Association for Computing Machinery. Communications of the ACM 39(11), 27–34 (1996)
5. Brabazon, T.: Data mining: A new source of competitive advantage? Accountancy Ireland, 30–31 (1997)
6. Tong, L., Yan, C., et al.: Analysis on Data Warehouse Technology and Its Development Situtation. In: IEEE International Symposium on Information Science and Engineering, pp. 486–488 (2008)

7. Liao, S.h.: Knowledge Management Technologies and applications-Literature review from 1995 to 2002. In: Expert System with Application 25, Pergamon, pp. 155–164 (2003)
8. Neumuth, T., Mansmann, S., et al.: Data Warehousing Technology for Surgical Workflow Analysis. In: 21st IEEE International Symposium on Computer-Based Medical Systems, pp. 230–235 (2008)
9. Kuanfu, W., Xuanzi, H.: Application of Data Warehouse Technology in Data Centre Design. In: IEEE International Conference on Computational Intelligence and Security, pp. 484–488 (2008)
10. Wen, C.P.: Hierarchical Analysis For Discovering Knowledge in Large Databases. In: Information Systems Management, pp. 81–88 (2004)
11. Zhang, X.: A New Modelling Method for the Data Analysis Solution in Business. In: IEEE International Symposium on Electronics Commerce and Security, pp. 175–178 (2008)
12. Xuezhong, B., et al.: Building Clinical Data Warehouse for Traditional Chinese Medical Knowladge Discovery. In: IEEE International Conference on Bio-Medical Engineering and Informatics, pp. 615–620 (2008)
13. Dan-Ping, Z.: A Data Warehouse Based on University Human Resources Management of Performance Evaluation. In: IEEE International Forum on Information Technology and Application, pp. 655–658 (2009)
14. Qian, Z., Qing, X.: The Study on Data Warehouse Modelling and OLAP for Highway Management. In: IEEE International Conference on Measuring Technology and Mechatronics Automation, pp. 416–419 (2009)

Fingerprint Matching by Sectorized Complex Walsh Transform of Row and Column Mean Vectors

H.B. Kekre[1], Tanuja K. Sarode[2], and Rekha Vig[1]

[1] Mukesh Patel School of Technology and Management Engineering, Mumbai
[2] Thadomal Shahani College of Engineering, Mumbai
hbkekre@yahoo.com, tanuja_0123@yahoo.com, rekha.vig@nmims.edu

Abstract. Currently fingerprints are the most popular way of human identification. The area of Automated Fingerprint Identification Systems (AFIS) gives an extended scope for research with increasing database of fingerprints and the requirement of reduced processing time. In this paper, a method which deals with fingerprint identification in the transform domain is considered and the main focus is on the reduction of the processing time. First, the mean of rows (or columns) of the fingerprint image is computed, this converts a two dimensional image signal into one dimension. The one- dimensional Walsh transform of the row (or column) vector is generated and is distributed in a complex plane which is subjected to sectorization to generate the feature vector. The feature vector of a given test image is compared to those present in the database. The scores from row and column transform methods are fused using OR and MAX functions. The results with accuracy of more than 68% (for 8 sectors) and high computational speed show that the method can be used in fingerprint identification in application with requirements of less processing time.

Keywords: Fingerprint identification, Walsh transform, row and column mean vector, sectorization, complex plane.

1 Introduction

Conventionally, verified users have gained access to secure information systems, buildings, or equipment through multiple means: PINs, passwords, smart cards, and so on. However, these security methods have significant vulnerabilities: they can be lost, stolen, or forgotten. In recent years, there has been an increasing use of biometrics, which refers to personal biological or behavioural characteristics used for verification or identification[1]. Biometrics can differentiate between an authorized person as it relies on "something that you are". Automatic fingerprint recognition technology has now rapidly grown beyond forensic applications into civilian applications. In fact, fingerprint-based biometric systems are so popular that they have almost become the synonym for biometric systems [2].

The Automated Fingerprint Identification Systems work in two phases: Enrolment and Identification. In enrolment phase the user (generally one of many) scans and offers his digitized fingerprint for the extraction of its features. These features (or template) are some characteristics of fingerprints which are generally unique to a

K. Shah, V.R. Lakshmi Gorty, and A. Phirke (Eds.): ICTSM 2011, CCIS 145, pp. 168–175, 2011.

fingerprint are of much smaller size than the entire fingerprint and are stored in a database for each user. In the identification phase, the user is granted access to the system if the extracted features of his fingerprint matches to that present in the database.

The AFIS systems are minutiae-based, image-based or textured-based systems [5][6][9][12]. In the first type, the minutiae [11], generally ridge-endings, ridge-bifurcations etc are extracted which form the feature vector (set of features). In these systems the size of feature vector is very small as only the type and location of the minutiae is to be stored. But these types of systems have large requirements of pre-processing of image which include image denoising and enhancement [4]. Image-based representations, constituted by raw pixel intensity information, are prevalent among the recognition systems using optical matching and correlation-based matching. However, the utility of the systems using such representation schemes may be limited due to factors such as brightness variations, image quality variations, scars, and large global distortions present in the fingerprint image. Furthermore, an image-based representation requires a considerable amount of storage.

In this paper, a texture-based fingerprint matching method is explained in which features of a fingerprint are extracted in transform domain[7][8]. The Walsh transform of a fingerprint generates a pattern of sequency distribution described in detail in section 2. The sectorization of complex plane representation of Walsh transform coefficients of the fingerprint image is computed and features are generated for each sector. This method has the advantage over both the above methods; in that it has smaller size of feature vector as well it doesn't need any pre-processing. The added advantage of this method is that the computational time is reduced considerably as processing is done in 1-D instead of 2-D.

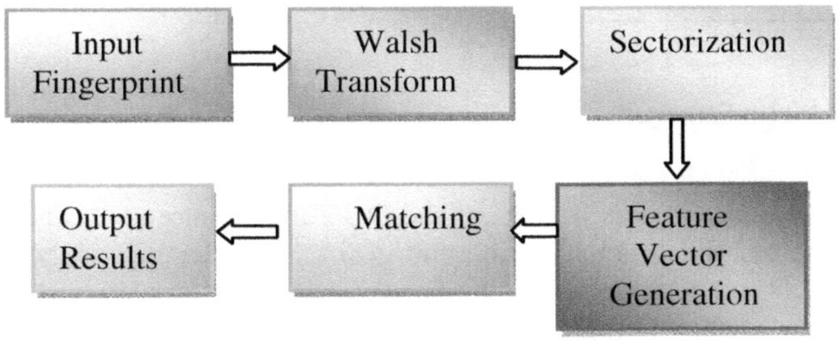

Fig. 1. Flow diagram of Fingerprint Matching Method

Fig. 1 explains the flow of the method used in this paper. The input fingerprint is transformed into sequency domain using Walsh transform, explained in section II. After sectorization feature vectors are generated and they are matched with those of stored fingerprints. Matching gives the output which is the best match.

The rest of the paper is structured as follows. Section 2 describes the Walsh transform technique. Section 3 explains the proposed sectorization method. Section 4 explains how two methods based on row-mean vector and column-mean vector have been fused to obtain better accuracy. The experimental results are given in section 5. Finally, conclusions are given in section 6.

2 Walsh Functions and Transform

In this paper we have used Walsh transform which is a powerful tool of linear system analysis for discrete signals [3]. Images are discrete functions of two dimensions, when acquired by digital acquisition devices. Hence Walsh transform can be aptly used on images to generate coefficients in frequency domain. The Fig. 2 shows the Walsh functions for N=8.

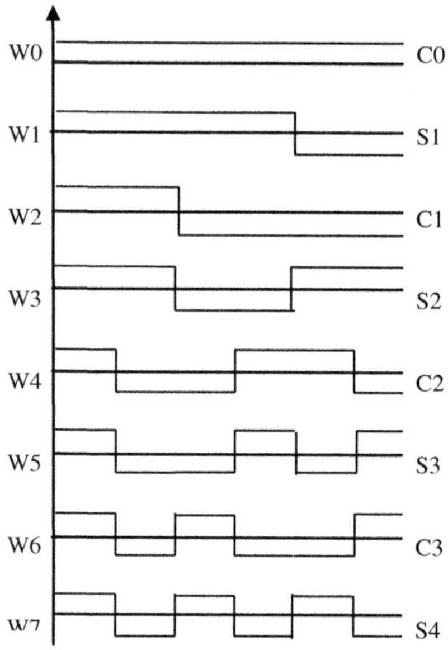

Fig. 2. Walsh Functions

$$W = \begin{pmatrix} 1 & 1 & 1 & 1 & 1 & 1 & 1 & 1 \\ 1 & j & j & j & -j & -j & -j & -j \\ 1 & 1 & -1 & -1 & -1 & -1 & 1 & 1 \\ 1 & j & -j & -j & j & j & -j & -j \\ 1 & -1 & -1 & 1 & 1 & -1 & -1 & 1 \\ 1 & -j & -j & j & -j & j & j & -j \\ 1 & -1 & 1 & -1 & -1 & 1 & -1 & 1 \\ 1 & -j & j & -j & j & -j & j & -j \end{pmatrix} \quad (1)$$

Here C_0 represents the coefficient of the DC component and S_n and C_n represent the sal and the cal (analogous to sine and cosine) coefficients of the nth sequency (analogous to frequency) component. The Walsh functions[10] are generated by Walsh generator as shown in the Fig. 5. First a DC signal C_0 is generated. This function (base function) is then used for the generation of all other Walsh functions. The odd signal S_1 is generated by time compressing C_0 (by half), shifting it (by time equal to that of compressed signal), inverting it and then adding time compressed and shfted signal as shown in Fig. 3. The even signals C_1 is generated by time compressing S_1 (by half), shifting it (by time equal to that of compressed signal), inverting it and then adding time compressed and inverted signal as shown in Fig. 4. The other even and odd functions are generated by the same procedure as mentioned above except that the base signal is its previous even or odd function respectively as shown in Fig. 5.

The Walsh transform can then be represented by a matrix as shown in equation (1) which is generated by sampling the Walsh functions at the middle of the smallest time interval. The samples of the sal functions are multiplied by j to convert them into

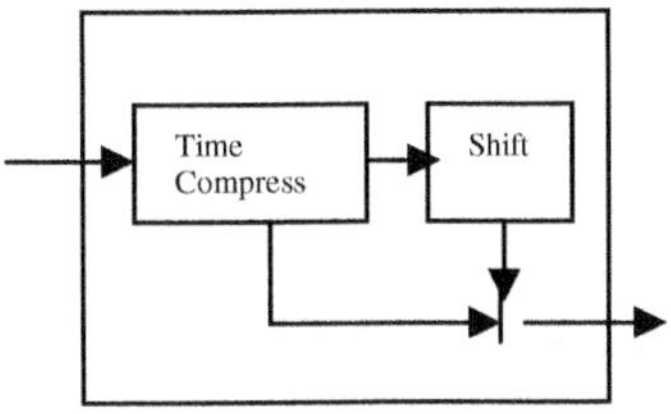

Fig. 3. Time Compress and Shift (TCS) Module

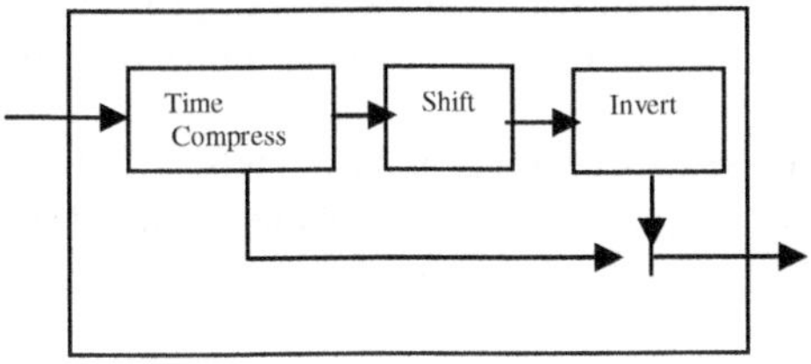

Fig. 4. Time Compress, Shift and Invert (TCSI) Module

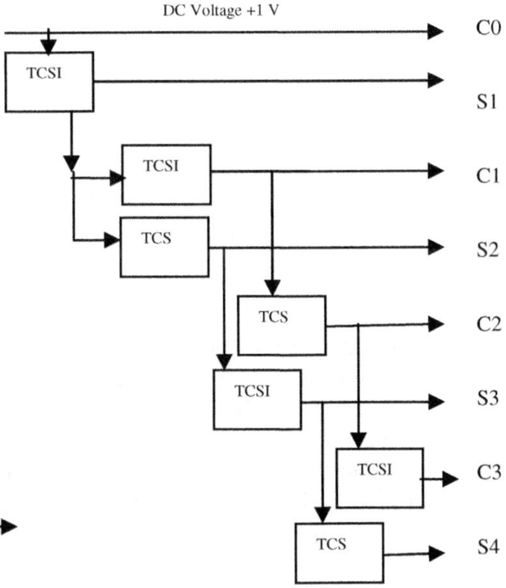

Fig. 5. Walsh Generator

imaginary part of the complex Walsh Transform. The Walsh transform F(u) of a one dimensional discrete signal f(x) is calculated by (2).

$$F(u) = W f' \tag{2}$$

where f′ is the column representation of row vector form of discrete signal f(x). The first value in F(u) represents the DC component, and the next ones the higher sequency components. Now, since images are two dimensional the Walsh transform of an image is generated by two step method. First transform of each row vector is calculated and then the transform of each column vector of the output of first step is calculated to get the final Walsh transformed image.

Alternatively, the 2-D Walsh transform can be calculated by equation (3) where W and W′ (transpose of Walsh matrix) are same, W being symmetric.

$$F(u,v) = (W f(x,y) W') / N \tag{3}$$

This two step method is very computation intensive and hence in this paper we have used a method to reduce the processing time. The Walsh transform is computed on two sets of vector: one is a row vector generated by calculating the mean of each column of the image matrix and the other is the column vector generated by calculating the mean of each row of the image matrix as shown in the Fig. 6.

The Walsh transform of the row-mean vector is calculated by taking its transpose and that of the column-mean vector is calculated directly [13]. This method takes less processing time as row-mean and column-mean vectors are 1-D vectors and Walsh

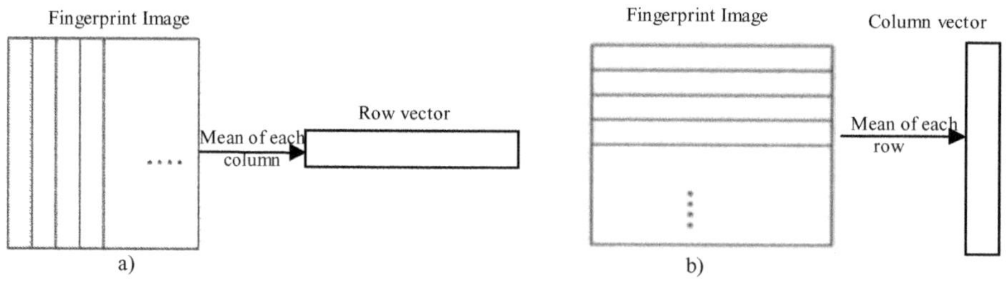

Fig. 6. a) Mean of columns b) Mean of rows of the fingerprint image

transform of 2 1-D vectors each of size N need 2NlogN additions whereas that of one 2-D vector (entire image) will need 2N2logN additions. Hence there is improvement in the processing time by 1/N for this method, though the accuracy is compromised to some extent.

The cal and the sal components of the same sequency are grouped together and are considered to be in the four quadrants of 2-D complex co-ordinate plane as shown in Fig. 7. This complex plane is now sectorized into different numbers of sectors.

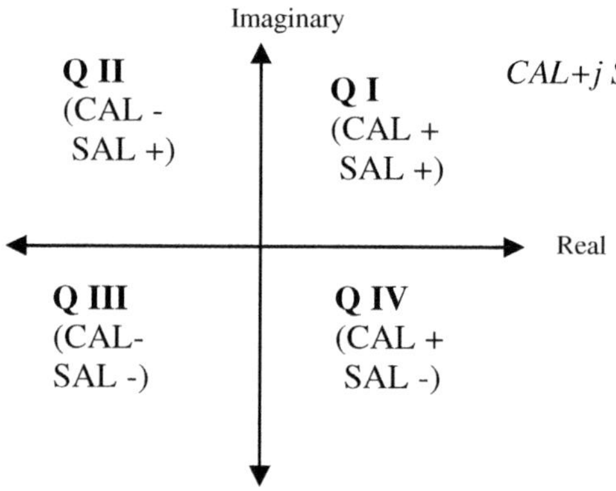

Fig. 7. Complex Walsh Plane

3 Sectorization

The complex plane consisting of same-sequency (sal, cal) components is now sectorized into 4, 8 and 12 radial sectors. A feature vector is generated for each sector, which is the mean value of all the transform coefficients in that sector. This value is unique for each fingerprint as the sequency distribution of each fingerprint is unique in different sectors. As compared to all or those transform coefficients which contain major part of signal energy feature vectors generated using sectorization are much less in number and hence the reduction in processing time and complexity. The division of the complex 2D plane into various sectors is based on the criteria as shown in the tables (Tables 1 and 2). Here, first the complex plane is divided into 4 sectors, the

Table 1. Criteria for Four Sectors

Sectors	Criteria
I, IV, VI and VII	sal < cal
II, III, V and VIII	sal >= cal

Table 2. Criteria for Eight Sectors

Sectors	Criteria
I, VI, VII and XII	sal < cal*tan(π/6)
II, V, VIII and XI	sal >= cal*tan(π/6) and sal < cal*tan(π /3)
III, IV, IX and X	sal >= cal tan(π /3)

division being same as those of the quadrants of a plane as shown in Fig. 7. For more number of sectors i.e. 8 and 12, the 4-sector plane is considered as the basic entity and the criteria shown in the respective tables are applied for division.

The sectors generated are radial and the mean values of the transform coefficients in each sector are calculated as in equation (4), where Mk is the mean and N is the number of coefficients in a sector, which form the features. The DC component, separate means of the sal and cal component and the last sequency component together form the feature vector, and hence the number of features is 2S+2, where S is the number of sectors.

$$M_k = \frac{1}{N} \sum_{i=1}^{N} W_i \tag{4}$$

The sectorization of both row mean transform and column mean transform vectors is performed and feature vectors for both are generated.

4 Fusion

The features obtained from the test image are compared with those obtained from the stored fingerprint in the database, for which the Euclidian distances and absolute distances between the two are calculated and the results matched. The minimum distance gives the best match. The results obtained from both the methods (row mean transform and column mean transform) are then fused together by using OR function to calculate the accuracy in terms of the first position matching and MAX function in terms of total number of matches obtained in the first 7 matches (there are total 8 samples of each fingerprint and one of those is taken as test image). Larger is the number of sectors better is the accuracy obtained in fingerprint identification.

5 Experimental Results

In this experiment, we have used the fingerprint image database containing 168 fingerprint images of 500 dpi and size 256×256 pixels including 8 images per finger from 21 individuals. The set of fingerprints are obtained with rotational (+/- 15o) and shift (horizontal and vertical) variations. The algorithm compared the feature vectors of the test image with those in database and the first 7 matched were recorded. The average number of matches in the first 7 matches gives a good indication of accuracy. The second measure evaluated was whether the first match belonged to the same fingerprint or not. The results shown in Table 3 and 4 are of randomly selected 3 samples of each fingerprint and it has been observed that for 8 sectors the results more accurate than for 4 or 12 sectors as shown in Fig. 8 and 9. We have applied the proposed technique on the database images and results in each category show that our method can satisfactorily identify the fingerprint images with the advantage of reduced computational time. Matching done using absolute distance vis-à-vis Euclidean distance show that the results are almost similar for both, whereas absolute distance method helps in further reduction in computation time as number of multiplications are eliminated. The accuracy rate observed is 66% for 8 sectors with processing time reduced by as much as 1/256 of that of full image methods.

Table 3. Average number of matches

Sectors	Using Euclidean distance	Using absolute distance
4	2.05	1.825
8	2.29	2.05
12	2.16	2.00

Table 4. Accuracy obtained for different sectors

Sectors	Using Euclidean distance	Using absolute distance
4	58.73 %	60.32%
8	66.67 %	66.67%
12	63.50 %	63.50%

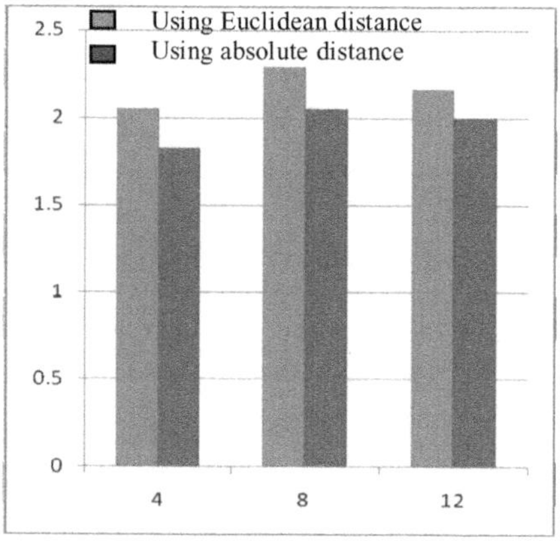

Fig. 8. Average number of matches in the first 8 matches for different sectors

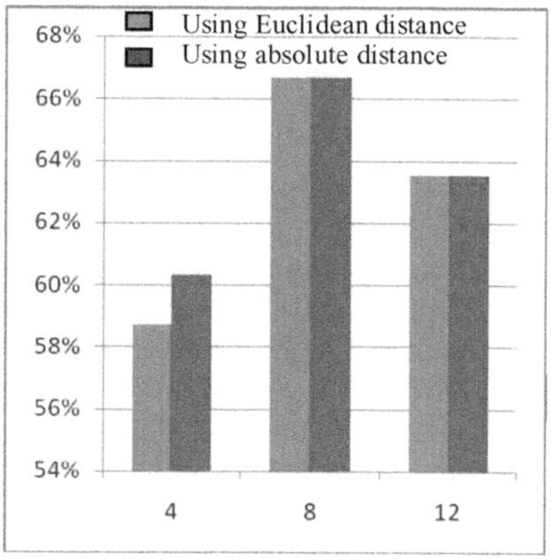

Fig. 9. Accuracy obtained for different sectors

6 Conclusion

In this paper, a fast and unique method for fingerprint matching is described. The technique of sectorization of the Walsh Transform of row and column means of fingerprint images has been used to generate feature vectors and matching is done using the same. This method is computationally very simple and fast as it is based on 1-D transform rather than 2-D transform. It is also considerably independent of shift and rotation of fingerprint images. Experimental results show that our proposed simple yet effective method can be used for fingerprint identification systems where high speed processing is given priority over accuracy.

References

1. Maltoni, D., Maio, D., Jain, A., Prabhakar, S.: Handbook of Fingerprint Recognition. Springer, New York (2003)
2. Jain, L., et al.: Intelligent Biometric Techniques in Fingerprint and Face Recognition. CRC Press, Boca Raton (1999)
3. Gonzalez Rafael, C., Woods Richard, E.: Digital Image Processing, 3rd edn. Prentice-Hall, Englewood Cliffs (2008)
4. Hong, L., Wan, Y., Jain, A.K.: Fingerprint Image Enhancement:Algorithm and Performance Evaluation. IEEE Transaction on Pattern Analysis and Machine Intelligence 20(8) (August 1998)
5. Jang, X., Yau, W.Y.: Fingerprint Minutiae matching based on thelocal and global Structures. In: IEEE 15th Int. Conf. on Pattern Recognition, vol. 2, pp. 1024–1045 (2000)
6. Kekre, H.B., Sarode, T.K., Rawool, V.M.: Fingerprint Identification using Discrete Sine Transform (DST). In: International Conference on Advanced Computing & Communication Technology (ICACCT 2008), Asia Pacific Institute of Information Technology, Panipat, India (2008)
7. Kekre, H.B., Sarode, T.K., Rawool, V.M.: Fingerprint Identification using Principle Component Analysis (PCA). In: International Conference on Computer Networks and Security (ICCNS 2008), Pune (2008)
8. Kekre, H.B., Sarode, T.K., Thepade, S.D.: DCT Applied to Column Mean and Row Mean Vectors of Image for Fingerprint Identification. In: International Conference on Computer Networks and Security (ICCNS 2008), Pune (2008)
9. Jain, A., Ross, A., Prabhakar, S.: Fingerprint matching using minutiae and texture features. In: Int'l conference on Image Processing (ICIP), pp. 282–285 (2001)
10. Walsh, J.L.: A Closed Set of Normal Orthogonal Functions. Amer. J. Math. 45, 5–24 (1923)
11. Berry, J., Stoney, D.A.: The history and development of fingerprinting. In: Lee, H.C., Gaensslen, R.E. (eds.) Advances in Fingerprint Technology, 2nd edn., pp. 1–40. CRC Press, Florida (2001)
12. Ross, A., Jain, A., Reisman, J.: A hybrid fingerprint matcher. In: Int'l conference on Pattern Recognition, ICPR (2002)
13. Kekre, H.B., Mishra, D.: Four Walsh Transform Sectors Feature Vectors for Image Retrieval from Image Databases. International Journal of Computer Science and Information Technology (IJCSIT) 01(02) (2010)

A Decision-Making Methodology for Automated Guided Vehicle Selection Problem Using a Preference Selection Index Method

V.B. Sawant[1], S.S. Mohite[2], and Rajesh Patil[3]

[1] Research Scholar, Mechanical Engineering Department,
Government College of Engineering, Vidyanagar, Karad, Maharashtra - India 415124
[2] Professor, Mechanical Engineering Department,
Government College of Engineering, Vidyanagar, Karad, Maharashtra - India 415124
[3] Associate Professor, Manufacturing Engineering Department,
SVKM's Mukesh Patel School Of Technology, Management & Engineering,
Vileparle Mumbai India
sawantvb@gmail.com, mohitess@yahoo.com, rajesh.patil@nmims.edu,
http://www.geck.ac.in,http://www.nmims.edu

Abstract. Automated guided vehicle selection, a key concern in manufacturing environment is a complex, difficult task and requires extensive technical knowledge with systematic analysis. It is invaluable to justify the selected equipment before actual implementation of the same. This paper presents a logical procedure to select automated guided vehicle in manufacturing environment for a given application. The procedure is based on preference selection index (PSI) method. An automated guided vehicle selection index is proposed that evaluates and ranks automated guided vehicle for the given application. An example is included to illustrate the approach.

Keywords: PSI, TOPSIS, Automated guided vehicle.

1 Introduction

Automated guided vehicles (AGVs) are among the fastest growing classes of equipment in the material handling industry. They are battery-powered, unmanned vehicles with programming capabilities for path selection and positioning. They are capable of responding readily to frequently changing transport patterns and they can be integrated into fully automated intelligent control systems. These features make AGVs a viable alternative to other material handling methods, especially in flexible environments where the variety of products processed results in fluctuating transport requirements. The decision to invest in AGVs and other advanced manufacturing technology has been an issue in the practitioner and academic literature for over two decades. An effective justification process requires the consideration of many quantitative and qualitative attributes. AGV selection attribute is defined as a factor that influences the selection of an automated guided vehicle for a given application. These attributes

K. Shah, V.R. Lakshmi Gorty, and A. Phirke (Eds.): ICTSM 2011, CCIS 145, pp. 176–181, 2011.

include: costs involved, floor space requirements, maximum load capacity, maximum travel speed, maximum lift height, minimum turning radius, travel patterns, programming flexibility, labor requirements, expansion flexibility, ease of operation, maintenance aspects, payback period, reconfiguration time, company policy, etc. In the past very few research had been reported for selection of AGV using multi attribute decision-making methods except [1]. A multi attribute analysis is a popular tool to select best alternative for given applications and the methods are simple additive weighted (SAW) method, weighted product method (WPM),technique for order preference by similarity to ideal solution (TOPSIS), Vlse Kriterijumska Optimizacija Kompromisno Resenje (VIKOR) method, analytical hierarchy process (AHP), graph theory and matrix representation approach (GTMA), etc [2, 3, 4]. However, there is a need for a simple, systematic and logical scientific method or mathematical tool to guide user organizations in taking a proper decision. To the best of our knowledge, no-one has implemented a PSI method for selection of AGV for a given application. The objectives of this paper is to illustrate the PSI method for AGV selection using an example and compare the results with TOPSIS method.

2 PSI Procedure

The steps of PSI procedure can be expressed as follows [5]:

Step 1: Identify the goal; find out all possible alternatives, selection attribute and its measures for the given application.

Step 2: Construct a decision matrix Assume there m alternatives (AGVs) A_i(i = 1,2. . . ,m) to be evaluated against n selection attributes C_j(j = 1,2, . . . ,n). The decision matrix D = x_{ij},i = 1,2. . . ,m; j = 1,2. . . ,n as shown below represents the utility ratings of alternative A_i with respect to selection attribute C_j.

$$\mathbf{D} = \begin{bmatrix} x_{11} & x_{12} & \cdots & x_{1n} \\ x_{21} & x_{22} & \cdots & x_{2n} \\ \cdots & \cdots & \cdots & \cdots \\ x_{m1} & x_{m2} & \cdots & x_{mn} \end{bmatrix} \quad (1)$$

Step 3: The process of transforming attributes value into a range of 0-1 is called normalization and it is required in multi attribute decision making methods to transform performance rating with different data measurement unit in a decision matrix into a compatible unit. The normalized decision matrix is constructed using Eq.(2) and (3). If the expectancy is the larger-the-better (i.e. profit), then the original attribute performance can be normalized as follows:

$$R_{ij} = \frac{x_{ij}}{x_j^{max}} \quad (2)$$

If the expectancy is the-smaller-the-better (i.e. cost), then the original attribute performance can be normalized as follows:

$$R_{ij} = \frac{x_j^{min}}{x_{ij}} \quad (3)$$

where x_{ij} is the attribute measures (i = 1, 2, 3, ... , N and j = 1, 2, 3, . . . , M)

Step 4: Compute preference variation value (PV_j). In this step, preference variation value (PV_j) for each attribute is determined with concept of sample variance analogy using following equation:

$$PV_j = \sum_{i=1}^{n} [R_{ij} - \bar{R_j}]^2 \quad (4)$$

where R_j is the mean of normalized value of attribute j and $R_j = \sum_{i=1}^{n} [R_{ij} - \bar{R_j}]^2$

Step 5: Determine overall preference value (ψ_j).

In this step, the overall preference value (ψ_j) is determined for each attribute. To get the overall preference value, it is required to find deviation (Φ_j) in preference value (PV_j) and the deviation in preference value for each attribute is determined using the following equation:

$$\Phi_j = 1 - PV_j \quad (5)$$

and overall preference value (ψ_j) is determined using following equation:

$$\psi_j = \frac{\Phi_j}{\sum_{j=1}^{m} \Phi_j} \quad (6)$$

The total overall preference value of all the attributes should be one, i.e.$\sum \Phi_j = 1$.

Step 6: Obtain preference selection index (I_i). Now, compute the preference selection index (I_i) for each alternative using following equation:

$$I_i = \sum_{j=1}^{m} (R_{ij} \times \psi_j) \quad (7)$$

Step 7: After calculation of the preference selection index (I_i), alternatives are ranked according to descending or ascending order to facilitate the managerial interpretation of the results, i.e. an alternative is ranked/selected first whose preference selection index (I_i) is highest and an alternative is ranked/selected last whose preference selection index (I_i) is the lowest and so on.

3 Illustrative Example and Results

An industry problem is selected to to demonstrate and validate the PSI method for evaluation of AGV for a given industrial application. The selected company is a medium sized manufacturing enterprise, which is located in Maharashtra, India. The company is an automated manufacturing company dealing with an enormous volume and varieties of products and supplies it to oil refineries. The company wants to purchase a few AGVs to improve on the productivity by reducing its work in process inventory and to replace its old material handling equipment. The decision of which AGV to select is very complex because AGV performance is specified by many parameter for which there are no industry

Table 1. AGV Selection Attribute data

M	L	W	H	MLC	MS	B	P	LH	LS
HK40/O	2.0	0.9	1.5	3628.7	91.4	345	6.3	6.09	45.0
F150	2.6	1.7	2.2	3628.7	67.0	345	9.5	1.8	17.5
P330	2.9	1.4	2.4	3628.7	60.9	560	9.5	1.8	16.5
P325	4.6	1.8	2.5	18143.6	60.9	240	9.5	1.8	11
C530	0.9	1.5	0.4	18143.6	45.7	300	12.7	1.8	12
DT-40	1.2	0.9	1.6	3628.7	60.9	345	25.4	1.8	12
DT-60	1.6	2.4	1.3	3628.7	60.9	345	25.4	1.8	12
RLV/N	2.7	2.0	1.6	6096.2	119.8	300	6.3	3	30
AD100	4.0	3.5	3.2	11339.8	54.8	300	12.7	1.8	12
AD130	3.6	3.04	2.6	11339.8	54.8	300	12.7	1.8	12
T-20	2.9	1.7	1.7	4535.9	41.1	350	6.3	1.8	12
T-40	4.5	2.4	2.4	9071.8	30.4	350	6.3	1.8	12
T-60	5.0	2.7	2.6	13607.7	24.3	350	6.3	1.8	12
T-100	5.8	3.2	3.3	22679.6	18.2	350	6.3	1.8	12
UV-200	3.8	3.4	4.6	9071.8	45.7	345	9.5	1.5	6
UV-600	5.5	4.8	5.8	27215.5	15.2	345	9.5	1.5	2.5

standards. There are more than 114 AGVs from 76 companies worldwide available in market. Above mentioned industry has supplied information about the requirement from material handling equipment which is given below. 1. Load to be carried is greater than 3628 kg. 2. Maximum lift height is 150 mm. 3. Battery capacity has to be more than 200 amp-hr. 4. Budgetary provisions is less than U. S. $20,000. Among nineteen attributes available with system, We have considered for this analysis nine attributes and sixteen feasible AGV models. Among nine attributes, four attributes viz., length(L), Width(W), Height(H) of AGV, maximum load capacity (MLC), maximum travel speed (MS), Battery capacity (B), maximum lift height (LH), Lift speed (LS) attributes are beneficial attributes, i.e. higher values are desired and position accuracy(P)is non-beneficial attributes, i.e. lower values are desired.

The next step is to represent all the information available of attributes in the form of a decision matrix as shown in eq 1. The data given in table 1 are represented as matrix D_{16x9}. But the matrix is not shown here as it is nothing but the repetition of data given in Table 1. Next the procedure given in section 2 is followed to calculate the values of preference selection index (I_i). AGVs are ranked according to descending order to facilitate the managerial interpretation of the results. The results obtained are presented using the PSI and compare with TOPSIS method with entropy weight approach as shown in Table 2. The AGV selection index is calculated for sixteen AGV models, which are used to rank the AGVs. The AGVs are arranged in the descending order of their selection index for TOPSIS method.

Table 2. Result comparision of PSI, TOPSIS with entropy and Average of two methods

Alternatives	PSI	TOPSIS with entropy	Average
HK40/O	0.685	0.523	0.604
F150	0.447	0.743	0.595
P330	0.564	0.752	0.658
P325	0.323	0.604	0.464
C530	0.461	0.646	0.553
DT-40	0.427	0.724	0.575
DT-60	0.475	0.716	0.596
RLV/N	0.549	0.588	0.568
AD100	0.418	0.662	0.540
AD130	0.409	0.683	0.546
T-20	0.405	0.823	0.614
T-40	0.393	0.746	0.569
T-60	0.385	0.677	0.531
T-100	0.363	0.545	0.454
UV-200	0.432	0.701	0.567
UV-600	0.354	0.504	0.429

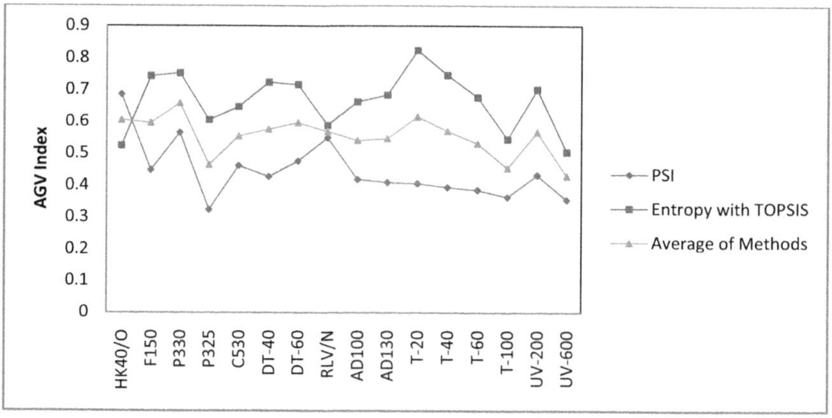

Fig. 1. AGV index using PSI, TOPSIS and Average of methods

From Fig. 1, It can be seen that the match between the PSI, TOPSIS and Average of two methods is good. The top two ranked AGV for PSI is HK40/O and P330, for TOPSIS is T20 and P330, and average of two methods gives P330 as first and T20 second ranked AGV. From the above values it is understood that the AGV designated as P330 and T20 are the right choice for the given industrial application under above methods. We need further scrutiny with respect to there attribute data, so that best AGV can be selected. The proposed approach in the present work ranks the alternatives in a single model.

The proposed AGV selection procedure using a PSI, TOPSIS with entropy and average of two method is a relatively easy and simple approach and can be used for any type of decision making situations. These methodology avoids the approach of relative importance of attributes to rank the alternatives.

4 Conclusions

A PSI methodology based on MADM methods was suggested for the selection of AGV for a given application. These are general methods and are applicable to any type of material handling system. Unlike conventional methods which adopt only one of the assessment attribute, the proposed method considers all of the attribute simultaneously and gives the correct and complete evaluation of the AGV to be selected. The proposed AGV selection index evaluates and ranks AGVs for the given application. The system developed also helps in not just selecting the best AGV, but it can be used for any number of quantitative and qualitative AGV selection attributes simultaneously and offers a more objective and simple AGV selection approach. Further, the PSI results are compared with TOPSIS with entropy and average of two methods are relatively simple and can be used for any type of decision making situations. It may also be worthwhile to incorporate all other multi attribute decision methods to validate the results.

References

1. Sawant, V.B., Mohite, S.S.: Investigations on Benefits Generated by Using Fuzzy Numbers in a TOPSIS Model Developed for Automated Guided Vehicle Selection Problem. In: Sakai, H., et al. (eds.) RSFDGrC 2009. LNCS, vol. 5908, pp. 295–302. Springer, Heidelberg (2009)
2. Hwang, C.L., Yoon, K.P.: Multiple attribute decision-making: methods and applications. Springer, New York (1991)
3. Bodin, L., Gass, S.I.: On teaching the analytic hierarchy process. Computers & Operations Research 30, 1487–1497 (2003)
4. Chakraborthy, S., Banik, D.: Design of a material handling equipment selection model using analytic hierarchy process. International Journal of Advanced Manufacturing Technology 28, 1237–1245 (2006)
5. Maniya, K., Bhatt, M.G.: A selection of material using a novel type decision-making method: Preference selection index method. Journal of Materials and Design 31, 1785–1789 (2010)

Development of a Decision Support System for Fixture Design

Manisha Yadav and Suhas Mohite

Mechanical Engineering Department, Govt. College of Engineering, Karad, PIN - 415124, Dist- Satara, (M. S.)
mohitess@yahoo.com

Abstract. A comprehensive decision support system is developed to design fixtures for machining centers. The CAD interface provided to the system enables it to exchange drawings and data with other commercial CAD software tools. A set of structured queries incorporated in the preprocessor prompts the designer to extract qualitative and quantitative part features. The database, rule base and knowledge base built into the design module assist the designer to select an appropriate fixture body, position on to it various fixture elements, calculate clamping forces, decide number and types of clamps with their locations and orientations, etc. The post processor, finally, generates the bill of material, part drawings and assembly drawing of the designed fixture. The software is implemented in a fixture manufacturing industry. It is observed that, the design lead time for fixture is reduced from a few days to a few hours with fractional efforts and expertise of the designer.

Keywords: fixture design, feature extraction, decision support system.

1 Introduction

Fixture is a device used in machining, inspection, assembly, welding, and other manufacturing operations to locate and hold a work piece firmly in position. Fixture plays an important role in ensuring production quality, shortening production cycle time and reducing production cost. Fixture design, fabrication, and testing consume a substantial portion of the product development time. Traditionally, fixture design is performed manually for the selection of locators, rest-pads, work-supports, fixture-plate and the design of clamps. In a consumer driven market environment, the product mix has large variety and small batches which also means a correspondingly similar need for fixtures. Therefore, there is a need for automating the fixture design activity to ensure efficient and effective fixture design. The computer-based automation of the fixture design activities is commonly referred to as computer aided fixture design (CAFD) [1]. In the next session we will discuss the need of CAFD.

1.1 Need of CAFD

Design of a fixture is one of the important tasks. The design and manufacturing of fixtures requires a few weeks. Of these, the manufacturing lead-time for fixtures has

K. Shah, V.R. Lakshmi Gorty, and A. Phirke (Eds.): ICTSM 2011, CCIS 145, pp. 182–189, 2011.

been considerably reduced due to the availability of standard parts such as standard base plates, support plates, locators, clamps and other accessories, which form the basic building blocks of fixtures. So, once a fixture is conceptualized, manufacturing the fixture is only making the assembly of the modular parts. However, designing fixture requires expertise and knowledge of experienced designer and hence, it is highly human dependent. Availability of sophisticated software such as FIXES, FIXTURENET, etc. is addressing this issue; but these software are not affordable to small scale industries. Also, the needs of these industries are specific and limited to their scope of design and manufacturing activities. This necessitates having simpler and cheaper software tailored for the specific needs of these industries.

In this paper, fixture design software developed with Visual Basic 6.0 as a front-end tool is presented. The database is organized in MS-Access as a back-end tool.A drawing library of fixturing elements drawn in Pro-E is linked to the database [2]. The detailed structure of the software of the developed system is discussed in the next section.

1.2 Structure of the CAFD Software

The CAFD software consists of three modules, viz., preprocessor, processor and post-processor as shown in fig. 1. The information contained in the part drawing is an input to this system. Extracting this is really a challenging task. The most important step in

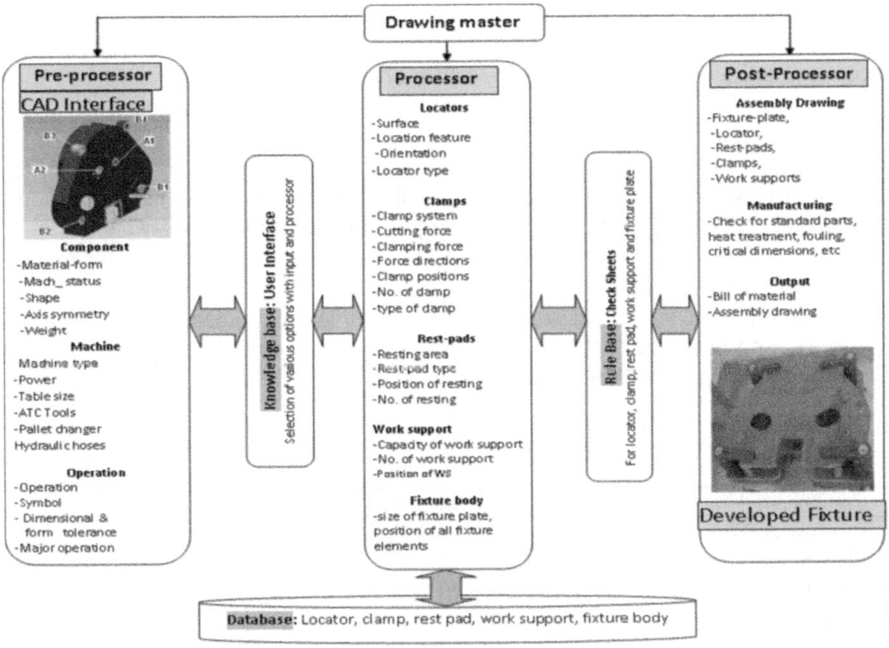

Fig. 1. The structure of CAFD, showing preprocessor, processor and postprocessor stages

fixture design, viz., understanding the part drawing in Pro-E or AutoCAD, is accomplished by a query of the part drawing prompted by questionnaires in the preprocessor. The system is developed in an algorithmic manner in such a way that at every stage it navigates the designer to extract qualitative part features such as shape, material, operations, etc. by selecting from the menu items. Alongside, a database of locators, clamps, work support, etc. integrated with the software provides various options to the design engineer to choose from.

Next, decision making or processing task involves identifying appropriate locating surfaces and choosing suitable locators to constrain maximum number of degrees of freedom of the component. This is followed by computation of the clamping forces and decision on the number and location of clamps, work supports and the base plate. At every stage a series of checkpoints are provided to remove the discrepancies in selection of fixturing elements. The checkpoints are the series of questionnaire provided to foolproof the selection of elements.

In the post processor task, the output is generated as a bill of material (BOM) and drawings of the elements of the fixture. Various case studies have been carried out to assess the effectiveness of the system developed. One such case study is reported in the next section.

2 Case Study

Here, we present a case study on fixture design for a gear-box housing carried out using the CAFD software at Datta Tools Pvt. Ltd. Karad, a tool room solution provider. Figure 2 shows the solid model of the gearbox housing in front and rear view. The following three subsections briefly describe the automated fixture design procedure leading to the bill of material and the assembly drawing.

2.1 Preprocessor Task

In the preprocessor task, details regarding component, machine tool and operation are extracted from 2D CAD drawing interactively as prompted by the software.

2.1.1 Component Details

In this, important physical features of the component are extracted. To enable systematic storage and retrieval of the fixture design case, the drawing is coded for various attributes using group technology guidelines.

Table 1. Component details selected from the menu options provided by the software

Component details	Specifications
Drawing code	02B2Q4V3JG1017
Material	Cast iron
Raw material form	As cast
Shape	2D box
Size	700mm x795mm x136mm
Weight	90 kg
Axis symmetry	Symmetric
Other information	e.g., specific constraints, etc. given by the customer

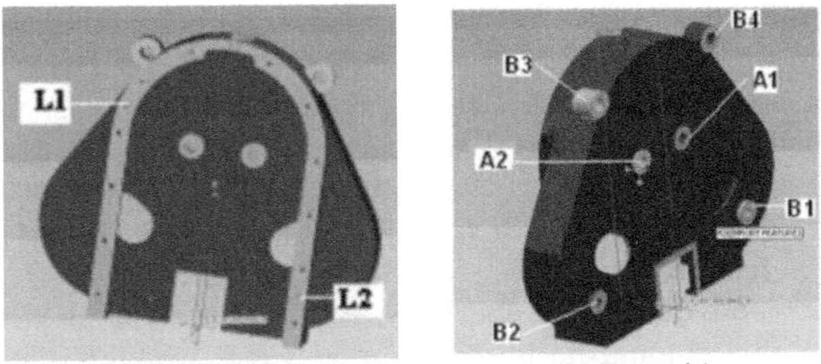

(a) Front side (b) Rear side

Fig. 2. Solid model of the gear-box housing showing location holes

2.1.2 Machine Details

In this, technical specifications of the machine tool on which the component is to be machined are listed. These usually are given by the customer. The fixture must be designed within the constraints and limitations posed by these specifications.

Table 2. Machine specifications fed to the CAFD system as provided by the customer

Machine Parameters	Specifications
Type of machine	Vertical Machining Centre (VMC)
Power	11/15 kW
Table size	1500mm x 810mm
Pallet	Rectangular
Loading capacity	2000kg
Hydraulic hose	Not available
ATC tools	24 No.
Max. tool dia. x length	ø100 mm x 250mm
Max. weight of tool	8 kg

2.1.3 Operation Details

In this, various operations that are already performed and required to be performed on the component are extracted along with the dimensional and the form tolerances [3]. In Table 3, important operations have been listed with dimensions and tolerances.

Thus, all the relevant information regarding the component, machine tool and operations necessary for fixture design are fed to the system. With this, the preprocessor task ends. Next, the design (processor) task is performed for the component with the CAFD software.

Table 3. Operations to be performed on components with dimensions and tolerances

Sr. No.	Operation	Geometric Tolerance	Dimensional Tolerance
1.	Ø 75.0mm Boss face-milling	-	100 μm
2.	Ø 15.8mm Drilling and tapping (5/8 BSW)	-	50 μm

2.2 Preprocessor Task

In the processor, the system assists the designer to take decisions on the selection of the fixture elements with the help of the knowledge and rules incorporated in it. The selection of important fixture elements is described in the next subsection.

2.2.1 Locator Selection

The software helps the designer in locator selection by evaluating errors for different alternatives available among the locating surfaces and selecting the one with minimum error of location [4]. Two location holes L1 and L2, ø6.3H7 diameter are selected for location (see, fig. 2). Figure 3 shows 2D part drawing on a digitized window accessed by the CAFD software. The locators L1 and L2, their types, coordinates, locating surface and orientation are displayed on the screen [5], [6]. Next, we see the clamp selection.

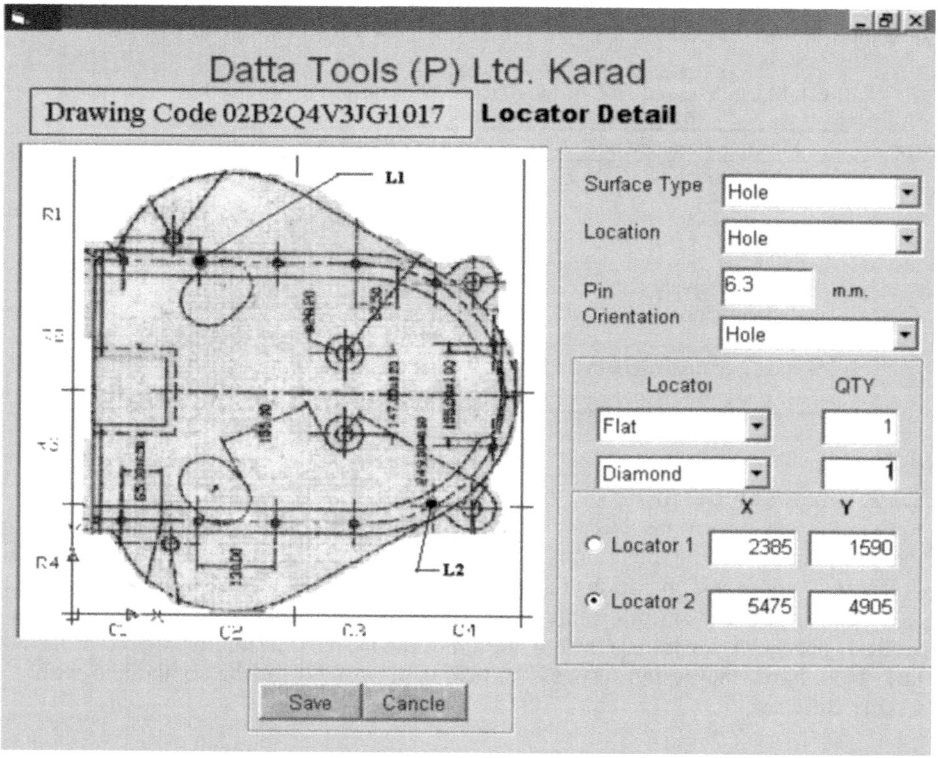

Fig. 3. Locator selection form showing CAD interface and details of location and orientation

2.2.2 Clamp Selection

The clamp selection is made in the software through the links provided to the operation details input. All the values from major operation are selected for the calculation of forces, cutting power and number of clamps. The cutting force is 527 N. Four hydraulic swing clamps provide the clamping force of 4x131.84= 527 N for face milling operation. Therefore, the clamping force is equal to the machining force. Fig. 4 shows 2D part drawing on a digitized window accessed by the CAFD software. The clamps CL1, CL2, CL3 and CL4, their direction, position, type, number of clamp, and machining and clamping force values are displayed on the screen. Next, we see the clamp selection. Next, we see the rest pad and work support selection.

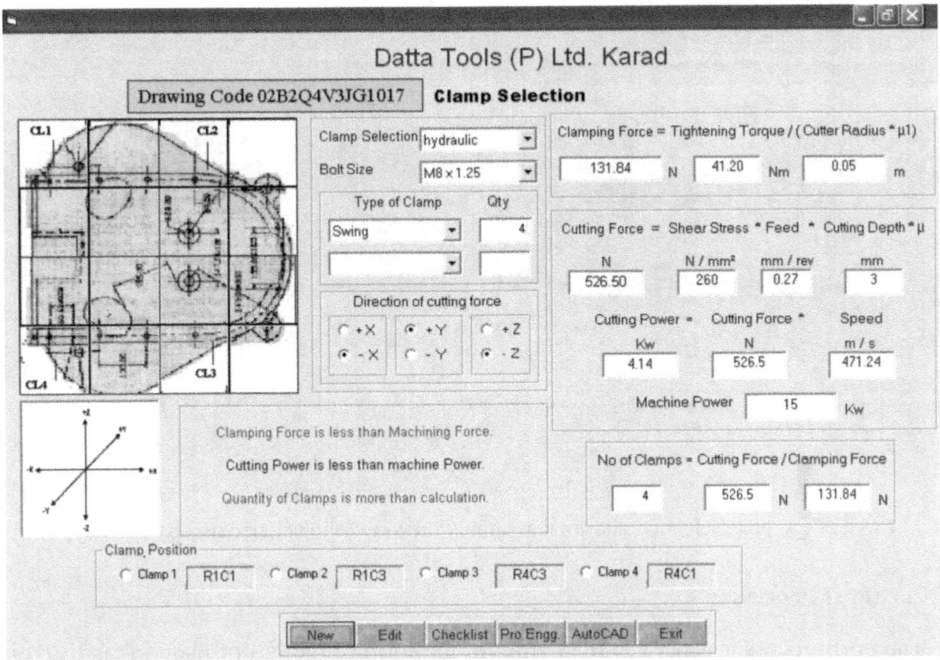

Fig. 4. Clamp selection form showing clamp positions on component drawing (CL1, CL2, CL3, and CL4), type and number of clamps, machining and clamping forces

2.2.3 Rest Pad and Work Support Selection

The software subsequently helps the designer for selection of the rest pads and work supports. The digitized display shows the coordinates of the position of the rest pads and work supports. As major operation of milling is carried on the two bosses of diameter 75 mm, there is only one work support used between them. In the next subsection, the fixture plate selection process is discussed.

2.2.4 Fixture Plate Selection

The software helps to select the fixture plate through the links provided to component details and clamp selection form. The thickness of the fixture plate depends on the weight of the component, its volume and the magnitude of the cutting force. In Fig. 5, the 2D drawing of the component shows the positions of all the fixturing elements on the component along with the selection of the fixture plate. Standard parameterized parts are selected from the database are marked on the base plate with different colours. Next, we discuss the post processor task.

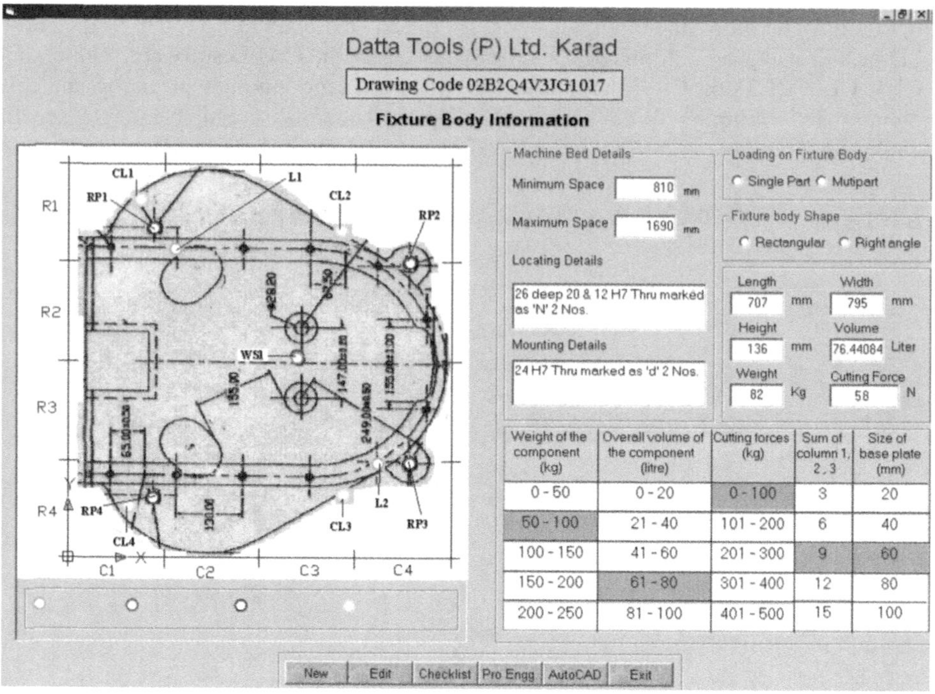

Fig. 5. Base plate selection and final assembly drawing as developed using CAFD system

2.3 Post Processor

The post processor generates the output in the form of the bill of material as listed in Table 4 and the part drawings and the assembly drawing prepared by the designer.

Table 4. The bill of material generated by CAFD software for the designed fixture

Sr. No.	Fixture Elements	Specifications	Qty
1.	Fixture plate	CI / 835x800x60	1
2.	Cylindrical locator	EN-24 / Ø 13.5g_6	1
3.	Diamond locator	EN-24 / Ø 13.5g_6	1
4.	Swing clamps	Model no. 41-5233-21/22	4
5.	Work-support	Model no. 41-0050-02	7
6.	Rest-pads	EN24/Ø66x18.5/HR_C50	4

3 Implementation and Conclusions

The computer assisted fixture design software, CAFD, has proved to be a quick and effective tool for feature extraction and efficient decision making for the selection of fixture elements finally leading to a comprehensive solution in the form of fixture assembly and the bill of material. The CAFD is implemented at Datta Tools Pvt. Ltd.,

Fig. 6. Gear box housing with the component and fixturing elements assembled on fixture plate

Karad, a tool room solution provider. Fig.6 is the final photograph of the manufactured fixture with all elements along with the work piece assembled on to it.

The software assists the designer at every stage of the design by taking into account all input parameters and helps in selecting the major fixturing elements accordingly. The design lead time is reduced from a few days to a few hours. Further, a major part of the repetitive design procedure has become independent of an individual designer thereby relieving the designer to spend more time on creative aspects of the design.

References

1. Boyle, I.M., Rong, K.: CAFixD: A Case-Based Reasoning Fixture Design Method Framework and Indexing Mechanisms. J. of Comp. and Info. Sc. in Engg. 6, 40–48 (2006)
2. Guolin, D., Cai, J., Chen, X.: Application of XML to the graphic exchange tech. of the modular fixture. In: 7th WSEAS Int. Conf. on Robotics, Control & Mfg. Tech., Hangzhou, China, pp. 284–289 (2007)
3. Hunter, R.: A functional tolerance model: an approach to automate the inspection process. J. of Achievements in Mat. and Mfg. Engg. 31(2), 662–670 (2008)
4. Subrahmanyam, S.: Fixturing feature selection in feature based systems. J. Computer Industry 48, 99–108 (2002)
5. King, D.A., Lazaro, A.: Process and Tolerance Considerations in the Automated Design of Fixtures. J. of Mech. Design 116, 480–486 (1994)
6. Quin, G.H., Zang, W.H.: A Machining-Dimension-Based Approach to Locating Scheme Design. J. of Mfg. Science and Engg. 130, 051010-01–051010-08 (2008)
7. Kumar, S., Nee, A.Y.C.: Development of an internet-enabled interactive fixture design system. J. Computer-Aided Design 35, 945–957 (2003)

Capacity Increase for Information Hiding Using Maximum Edged Pixel Value Differencing

H.B. Kekre[1], Pallavi Halarnkar[2], and Karan Dhamejani[3]

[1] Senior Professor, [2] Assistant Professor, [3] B.Tech Student
Computer Engineering Department, SVKM's NMIMS (Deemed-to-be University)
Mumbai, India
hbkekre@yahoo.com, pallavi.halarnkar@gmail.com,
kd.mpstme@gmail.com

Abstract. Steganography is a transmitting secret messages through innocuous cover carriers in such a manner that the existence of the embedded messages is undetectable. In this Paper, the existing Pixel Value Differencing method is modified so as to increase the embedding capacity for a Digital Image. The proposed method involves examining 2X2 pixel group and finding the direction of max slope. This direction is selected for embedding the secret message. In this case since we are selecting the maximum slope the capacity is increased by as much as 6-7% at the cost of slight increase in MSE and AFCPV. However this does not affect the perceptibility and stego image appears to be as good as the original image.

Keywords: Steganography, Information Hiding, PVD, Edged PVD.

1 Introduction

Information hiding techniques have recently become important in a number of application areas. Digital audio, video, and pictures are increasingly furnished with distinguishing but imperceptible marks, which may contain a hidden copyright notice or serial number or even help to prevent unauthorised copying directly[2]. A broad overview of information hiding is available in [1], [2]. An important subdiscipline of information hiding is steganography. Steganography is a transmitting secret messages through innocuous cover carriers in such a manner that the existence of the embedded messages is undetectable. Carriers of such messages may resemble innocent images, audio, video, text, or any other digitally represented code or transmission. The hidden message may be plaintext, ciphertext, or anything that can be represented as a bit stream [3]. In contrast to cryptography, it is not to keep others from knowing the hidden information but it is to keep others from thinking that the information even exists.

The Internet provides an increasingly broad band of communication as a means to distribute information to the masses. Such information includes text, images, and audio to convey ideas for mass communication. Such provide excellent carriers for hidden information and many different techniques have been introduced [8], [10] other carriers for hidden information include storage devices [9] and TCP/IP packets [11].

K. Shah, V.R. Lakshmi Gorty, and A. Phirke (Eds.): ICTSM 2011, CCIS 145, pp. 190–194, 2011.

Many different methods of hiding information in images exist. These methods range from Least Significant Bit (LSB) or noise insertions, manipulation of image and compression algorithms, and modification of image properties such as luminance. Other methods include palette-based image steganographic method using Colour Quantisation[12], image steganographic method using : RLC & Modular Arithmetic[13], Variable Embedding Length[14] and Pixel-Value Differencing and LSB replacement methods among others. An introduction to steganography and its application to digital images is available in [4]. Various algorithms on Video steganography are given in [5], [6], [7].

2 Pixel-Value Differencing Steganography

Embedding Stage

In the embedding phase a difference value d is computed from every non-overlapping block of two consecutive pixels, say p_i and p_{i+1} of a given cover image. The way of partitioning the cover image into two-pixel blocks runs through all the rows of each image in a zigzag manner. Assume that the gray values of p_i and p_{i+1} are g_i and g_{i+1}, then d is computed as $g_{i+1} - g_i$ which may be in the range from -255 to 255. A block with d close to 0 is considered to be an extremely smooth block, whereas a block with d close to -255 or 255 is considered as a sharply edged block. The method only considers the absolute values of d (0 through 255) and classifies them into a number of contiguous ranges, such as R_k where $k=1,2,\ldots,q$. These ranges are assigned indices 1 through n. The lower and upper bound values of R_k are denoted by l_k and u_k, respectively. The width of R_k is u_k - l_k+1.In PVD method, the width of each range is taken to be a power of 2.

Every bit in the bit stream should be embedded into the two-pixel blocks of the cover image. Given a two-pixel block B with gray value difference d belonging to k^{th} range, then the number of bits, say n, which can be embedded in this block, is calculated by $n=\log_2(u_k-l_k+1)$ which is an integer. A sub-stream S with n bits is selected from the secret message for embedding in B. A new difference d'then is computed with equation 1.

$$d' = \begin{cases} l_k + b & d \geq 0 \\ -(l_k + b) & d < 0 \end{cases} \tag{1}$$

where b is the value of the sub-stream S . Because the value b is in the range $[0, u_k-l_k]$, the value of d' is in the range from l_k to u_k . If we replace d with d', the resulting changes are presumably unnoticeable to the observer. Then b can be embedded by performing an inverse calculation from d' to yield the new gray values (g_i', g_{i+1}') for the pixels in the corresponding two-pixel block (p_i, p_{i+1}) of the stego-image. The inverse calculation for computing (g_i', g_{i+1}') from the original gray values (g_i, g_{i+1}) of the pixel pair is based on a function given in equation 2.

$$(g_i', g_{i+1}') = \begin{cases} (g_i - \lceil m/2 \rceil, g_{i+1} + \lfloor m/2 \rfloor \text{if d is even} \\ (g_i - \lfloor m/2 \rfloor, g_{i+1} + \lceil m/2 \rceil \text{if d is odd} \end{cases} \tag{2}$$

where m is d'-d .the embedding is only done for pixels which their new values would fall in the range of [0,255].

Retrieving Stage

In the extracting phase, the original range table is necessary. It is used to partition the stego-image by the same method used for the cover image. Calculate the difference value $d'(p_i, p_{i+1})$ for each block of two consecutive pixels Then, find the optimum R_i of the d' same as in the hiding phase. Subtract l_i from d' (p_i, p_{i+1}) and b_0 is obtained. The b_0 value represents the secret data in decimal number. Transform b_0 into binary with t bits, where $t = [\log_2 (u_k'-l_k'+1)]$, where u_k' and l_k' are the upper and lower bounds obtained in the two pixel block of the stego image. The t bits can stand for the original secret data of hiding.

3 Maximum Edged Pixel Value Differencing Method

In the original PVD method a difference value d is computed from every non-overlapping block of two consecutive pixels, say p_i and p_i+1 of a given cover image. Partitioning the cover image into two-pixel blocks runs through all the rows of each image in a zigzag manner. However the direction of the edge is not considered while taking the difference of the two-pixel block, Here in this method rather than a two pixel block, we consider a four pixel block. In this block , the direction of the edge is found and max difference is considered as higher the edge difference , higher is the embedding capacity. The embedding and retrieval of secret message is same as the above original PVD method.

4 Experimental Results

For Experimental purpose we have used four cover images, Lena, Baboon, Pepper, Mahalakshmi and Ganpati. All the cover images used were 256X256 gray scale. The

Fig. 1. (a) Original Image

Fig. 1. (b) Stego Image

Fig. 2. (a) Message Image

Fig. 2. (b) Retrieved Image

message images used is also a gray scale image of an ATM card. Value of RMSE, PSNR, percentage of bytes changed and AFCPV, for embedding a message image of an ATM Card using PVD and Edged PVD are given in Table No 1. The embedding capacity obtained in both the methods is given in Table No 2.

Table 1. Value of RMSE, PSNR, percentage of bytes changed and AFCPV, for embedding a message image of an ATM Card using PVD and Edged PVD

	Pixel Value Differencing(PVD)			
	Lena	Baboon	Mahalakshmi	Pepper
MSE	2.63	21.90	19.45	12.05
RMSE	1.62	4.68	4.41	3.47
PSNR	43.92	34.72	35.24	37.32
%Bytes Changed	46.80	38.63	39.21	42.48
AFCPV	0.009347	0.021569	0.019585	0.017278
	Edged Pixel Value Differencing Method			
MSE	2.85	27.06	23.49	13.24
RMSE	1.69	5.20	4.84	3.63
PSNR	43.56	33.80	34.42	36.91
%Bytes Changed	46.31	36.90	38.12	41.86
AFCPV	0.009561	0.023768	0.020985	0.017698

Table 2. Embedding Capacity Obtained in Bytes

Cover Image	**PVD**	**Edged PVD**
Lena	12,830	13,003
Baboon	17,608	18,616
Mahalakshmi	17145	18,569
Pepper	15730	16,394

5 Conclusion

In this paper we have suggested a variation of existing PVD method called as Edged PVD to increase the embedding capacity of the stego image. The method suggested not only improves the embedding capacity by 6-7% as given in Table 2. Original and Stego images obtained in EdgedPVD method are shown in Figure 1(a) and 1(b). The proposed method involves examining 2X2 pixel group and finding the direction of max slope. This direction is selected for embedding the secret message. In this case since we are selecting the maximum slope the capacity is increased by as much as 6-7% at the cost of slight increase in MSE and AFCPV. However this does not affect the perceptibility and stego image appears to be as good as the original image.

References

1. Moulin, P., O'Sullivan, J.A.: Information – Theoretic Analysis of Information Hiding. In: IEEE International Symposium on Information Theory, Boston, MA (October 1999)
2. Peticolas, F.A.P., Anderson, R.J., Kuhn, M.G.: Information Hiding – A Survey. In: Proceeding of IEEE (July 1999)
3. Johnson, N.F., Jajodia, S.: Staganalysis: The Investigation of Hiding Information. IEEE, Los Alamitos (1998)
4. Johnson, N.F., Jajodia, S.: Exploring Steganography: Seeing the Unseen. IEEE Computer 31(2) (February 1998)
5. Pan, F., Xiang, L., Yang, X.-Y., Guo, Y.: Video steganography using motion vector and linear block codes. In: 2010 IEEE International Conference, Beijing (July 2010)
6. Liu, B., Liu, F., Yang, C., Sun, Y.: Secure Steganography in Compressed Video Bitstreams. In: Third International Conference on Availability, Reliability and Security, ARES 2008 (2008)
7. Mozo, A.J., Obien, M.E., Rigor, C.J., Rayel, D.F., Chua, K., Tangonan, G.: Video steganography using Flash Video (FLV). In: Instrumentation and Measurement Technology Conference, I2MTC 2009. IEEE, Los Alamitos (2009)
8. Anderson, R. (ed.): IH 1996. LNCS, vol. 1174. Springer, Heidelberg (1996)
9. Anderson, R., Needham, R., Shamir, A.: The Steganographic File System. In: Proc. Information Hiding Workshop, Portland, Oregon, USA (April 1998) (to be published)
10. Bender, W., Gruhl, D., Morimoto, N., Lu, A.: Techniques for Data Hiding. IBM Systems Journal 35(3&4), 313–336 (1996)
11. Handel, T.G., Stanford III., M.T.: Hiding Data in the OSI Network Model, pp. 23–38 (1996)
12. Wang, X., Yao, T., Li, C.-T.: A palette-based image steganographic method using colour quantisation. In: ICIP 2005 IEEE International Conference on Image Processing (2005)
13. Naimi, H.M., Ramazannia, B.: New Image Steganographic Method Using RLC & Modular Arithmetic. In: IEEE International Conference on Signal Processing and Communications, ICSPC 2007 (2007)
14. Kim, K.-J., Jung, K.-H., Yoo, K.-Y.: Image Steganographic Method with Variable Embedding Length. In: International Symposium on Ubiquitous Multimedia Computing, UMC 2008 (2008)

A Review of Handoff Optimization Techniques in Data Networks

Kushal Adhvaryu and Vijay Raisinghani

Department of Computer Engineering, Department of Information Technology
Mukesh Patel School of Technology Management and Engineering,
NMIMS (Deemed-to-be) University
kushal.adhvaryu@hotmail.com, vijay.raisinghani@nmims.edu

Abstract. On a mobile device, uninterrupted reception of multimedia streams is essential for a good user experience. However, uninterrupted reception is not possible when the mobile device is moving. When a user changes her location, the Mobile Node (MN) may change its current cell. A considerable amount of time is consumed in handoff procedure. Handoff latency causes an interruption in data transfer to the MN. There are many ways of improving the handoff procedure so that delay sensitive applications do not suffer due to handoff latency. Some of the mechanisms are early transfer of context to the new Base Station (BS), early channel scanning, efficient inter-BS communication, introducing extra buffers between BS and MN, using a proxy for seamless connectivity, etc. In this paper we review some of these techniques. Most of the applications are applicable on Wi-Fi domain.

Keywords: Packet Loss, Buffer, Delay, Handoff, Resource Allocation, Wireless Network, Handoff Latency, Channel Scanning, Context.

1 Introduction

Using a mobile, it is possible to stay connected to the Internet as a user moves across different location using different mobile devices. However, mobile Internet access has many challenges. In this paper term Base Station (BS) is used for entities like BS in a mobile network or access routers and access points in a data network etc. As the user moves, the Mobile Node (MN) may change its cell and therefore there may be one or more handoffs. The MN receives all the data packets from its connected BS. During the handoff process the MN gets disconnected from the current BS and gets connected to a new BS. There is a period of time during handoff process, when the MN is not connected to any BS. It is not possible to send data packets to the MN during that period of handoff which can eventually lead to packet loss. Longer the handoff procedure and handoff delay, longer may be the interruption at the MN. Also the packets in flight may be lost during handoff. After handoff, the MN connects itself to a new BS and disconnects from the old BS. The packet transmission to the MN runs through the new BS.

The problem of handoff latency can be overcome by improvements at the BS [1] [2] [4] [5] [6] [9], improvements at the MN [7] or introducing an intermediate node

K. Shah, V.R. Lakshmi Gorty, and A. Phirke (Eds.): ICTSM 2011, CCIS 145, pp. 195–200, 2011.

[3] [8] between the BS and the MN. It is observed that most of the techniques suggest improvements at the BS only. Very few consider the MN for handoff performance improvement. Many authors have suggested introducing an intermediate node for better handoff performance. Many schemes propose that if the initial phase of the handoff procedure, like channel scanning, is completely done by the BS then it is possible to reduce handoff latency [1] [5]. Some techniques suggest transferring context of a MN and BS, like all basic information of MN, free channels etc, to new BS well ahead of actual handoff time [2] [6]. Reserving resources of a cell especially for the MNs performing handoff is also a solution [4]. Intermediate nodes acting as proxies and buffers can be used to improve handoff performance [3] [8]. In this paper the solutions that we discuss are based on BS improvement or introduction of an intermediate node. Most of the techniques are applicable to Wi-Fi.

The rest of the paper is organized as follows: In section 2 different techniques to overcome handoff latency and packet loss problems are explained and analyzed. In section 3, conclusion and future work is given.

2 Handoff Optimization Techniques

In this section we review schemes which propose solutions to reduce handoff latency and packet loss. Most of the solutions work on Layer 2; and some of the solutions work on Layer 2 and Layer 3 together. Some solutions also consider application layer improvements. These schemes for handoff improvement can be divided as follows:

A) *Handoff delay reduction schemes.*

1. In references [1] [5], authors have suggested handoff delay can be reduced by forcing BS to perform some part of the initial phase of the handoff procedure.
2. Handoff procedure can also be improved by transferring context of the network to the other BSs [2].

B) *Using buffer to store data packets.*

1. Some of the techniques in references [3] [6] [8] suggest that extra buffers can be created at BSs, MNs and on intermediate nodes.

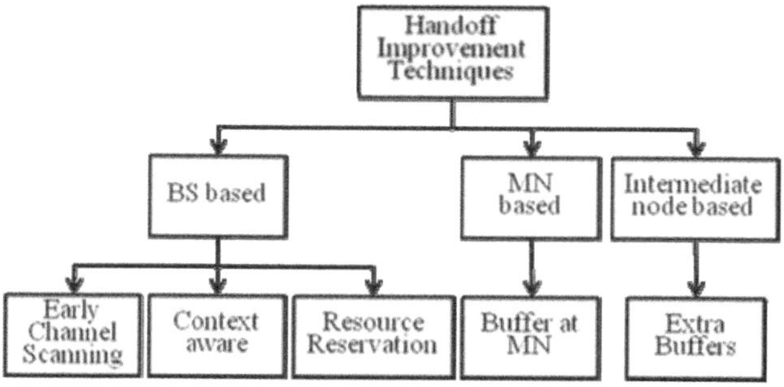

Fig. 1. Classification of techniques for Handoff improvement

2.1 Solutions Based on Base Stations

Various techniques have been proposed that suggest changes at the BS for handoff optimization. We review these schemes in following section.

2.1.1 Early Channel Scanning

Handoff begins with channel scanning phase, which forms a long part handoff [5]. The MN wastes a significant amount of time, in channel scanning which it can use to send and receive data from the currently connected BS. This makes handoff procedure longer. Wu et. al. have proposed a scheme of performing the channel scanning phase of the handoff procedure at the BS [5]. In this approach, the BS actively scans for available free channels of neighbouring BSs simultaneously when MNs and other neighbouring BSs are communicating with it. These channels can be used by MNs when the MNs want to perform handoff. The authors have shown that this method can reduce handoff time by 10 seconds. . A similar strategy has been proposed in reference [1] which reduces time taken by *inquiry phase* in Bluetooth. The authors have shown that time and power needed for completing handoff procedure can be reduced.

Analysis: This approach can increase the load of the BS unnecessarily. For a significant time period MNs may not be able to communicate with the BS. This scheme may eventually decrease the speed of the actual data communication of the BS. If multiple MNs want to perform the handoff together then it would be difficult for the BS to satisfy the requirements of each MN.

2.1.2 Improvement of Downlink Queue

The BS in wireless networks contains several queues like *Input-Queue, Inter-BS Queue* and *Downlink Queue* which store packets in them, before forwarding them finally to the MN or to the next BS [9]. When handoff occurs, it becomes necessary that packets stored in the Downlink Queue are retrieved to forward them to the new BS. The current architecture of the BS does not allow retrieving packets that are stored in the *Downlink Queue* [9], which causes packet loss. Nguyen and Sasase have proposed architecture for the BS using Mobile IP (Internet Protocol). In this approach multiple different virtual queues at IP layer for Input and Inter-BS communication are created. And at Radio layer multiple different Downlink Queues exist. In this architecture packets are transferred from the Input Queues to the Downlink Queues when packets are to be forwarded to the MN. If the MN performs handoff, packets are transferred back from the Downlink Queue to its corresponding buffer at the IP layer.

Analysis: Implementing this procedure may be complicated as the architecture of the BS needs to be changed. Changes may have to be made at each BS of the network which may consume time. Fetching packets back from the radio layer is not an easy task because there might be some packets on MAC layer which are to be transmitted in a few moments. The size of the buffers at IP layer should be dynamic and large enough to avoid any kind of packet loss and it should take care that amount packet forwarding during handoff process is less. If the packets are put on wireless medium just before the MN initiates handoff procedure, this kind of packets may not be recovered.

2.1.3 Improving Architecture of Mobile IP

Hsieh et. al. in [7] have proposed an architecture for improvement of Mobile IP which is divided in two parts 1) Network registration time and 2) Address resolution process. According to the current location of the MN, the BS to which the MN will handoff is decided [7]. In this scheme, when handoff procedure is initiated by the MN, the current stream of packets is simultaneously broadcast to the current BS to which the MN is attached, and to the most probable BS to which the MN will connect itself. This helps in reducing the latency involved in transferring the packet stream buffered at the old BS to the new BS. Simulcast ensures that the packet stream is already available at the new BS. Thus probability of packet loss is reduced [7].

Analysis: Changing the technology at each site may be difficult. If the current location of the MN is not available, it may not be able to predict the best BS for the MN after handoff. Buffers at the BSs should be large enough to store data packets for all the MNs performing handoff. Even after storing packets at the new BS, if the MN does not handoff to that BS, the BS will have to drop data packets. This may cause wastage of resources.

Few other techniques based on BS are discussed here. Mishra et. al. in [2] have also proposed a way of transferring context of a MN, like clock, device address, copy of the current data stream, etc., to the new BS, before the actual handoff process starts. This would reduce the handoff latency. However, if the MN does not perform handoff after initiating the handoff procedure, the new BS will have to drop data packets which have context information of the MN and the data stream. This may lead to wastage of resources Ramanathan et. al. have suggested reservation of resources in the new cell for MNs performing handoff in reference [4]. However, this can leave fewer resources for upcoming MNs.

In this section we discussed some techniques to improve handoff management by proposing solutions at a BS. Now in next section we will focus on handoff management process using intermediate node.

2.2 Buffer Management Using Intermediate Node

Dutta et. al. in reference [3] have proposed the use of extra buffers to store all data packets during handoff which are to be sent to the MN. This buffer can be implemented at any intermediate node, or it can be the BS too, in the network. When the MN completes the handoff procedure these buffered packets are forwarded to the MN. During handoff, the MN also needs to transfer packets to the BS. These packets also can be stored in buffer at these intermediate nodes.

A similar strategy is proposed by Bellavista et. al. in reference [8]. According to their solution, proxies containing buffers at intermediate nodes can be implemented between the BS and the MNs. Packets can be stored at these proxies during communication

Analysis: The buffers would increase delay of packets. This delay may not be acceptable for delay sensitive traffic. Cost of setting up network components may be high, as these approaches suggest use of extra buffers. If too many nodes are connected to the same intermediate node, then it may not be possible for that intermediate node to address needs of all MNs. Some multimedia applications running on the MN, may require a large amount of buffer which would reduce the buffer available for the MNs.

In this section, we discussed the solutions to overcome problems of handoff like latency and packet loss, by introduction of buffers at intermediate. Solutions which propose improvement at MN are not considered here, as most of the solutions propose use of extra buffers at the MN. Implementing extra buffers at the MN does not efficiently solve the problem of handoff latency and packet loss significantly. In the next section we present the conclusion and future work direction.

3 Conclusion and Future Work

In this paper we reviewed techniques to improve handoff management by reducing handoff latency or reducing packet loss during handoff. Some of the techniques suggested are BS scan for channels instead of MN [1] [5], changing structure of Mobile IP [7], reserving resources for MNs in a cell [4], buffering packets at intermediate nodes during handoff [3] [8] etc. Early channel scan by the BS may load the BS heavily. Reserving resources for some MNs in BS may lead to scarcity of resources for other MNs. Changing structure of Mobile IP would be difficult to implement in the existing network. Introducing intermediate nodes with buffers at those nodes would change network topology and communication procedure may also change. Interested readers can refer [6], [10], [11], [12] and [13] for more schemes related to handoff process improvement.

The techniques that we have seen in previous sections propose many solutions to optimize handoff process. The early channel scan for free channels of other BSs is done on data channels which may interrupt data communication. Research is needed to investigate the amount of interruption on the data channel and to ensure that this interruption is minimized. The problem of scalability of buffers is unanswered. It is not known how buffers will be able to accommodate large number of MNs. Further, the size of buffers should be large enough handle multimedia traffic of high data rates. This needs further research.

References

1. Chung, S., Yoon, H., Cho, J.: A Fast Handoff Scheme for IP Over Bluetooth. In: IEEE International Conference on Parallel Processing Workshops, pp. 51–55 (2002)
2. Mishra, A., Shin, M., Arbaugh, W.: Context Caching Using Neighbor Graphs for Fast Handoffs in a Wireless Network. In: IEEE International Conference on Computer Communication, pp. 351–361 (2004)
3. Dutta, A., Van den Berg, E., Famolari, D., Fajardo, V., Ohba, Y., Taniuchi, K., Kodama, T., Schulzrinne, H.: Dynamic Buffering Control Scheme for Mobile Handoff. In: IEEE International Symposium on Personal, Indoor Mobile Radio Communication, pp. 1–11 (2006)
4. Ramanathan, P., Sivalingam, K.M., Agrawal, P., Kishore, S.: Dynamic Resource Allocation Schemes During Handoff for Mobile Multimedia Wireless Networks. IEEE J. Select. Areas Commun. 17(7), 1270–1283 (1999)
5. Wu, H., Tan, K., Zhang, Y., Zhang, Q.: Proactive Scan: Fast Handoff with Smart Triggers for 802.11 Wireless LAN. In: IEEE International Conference on Computer Communication, pp. 749–757 (2007)

6. Bellavista, P., Cinque, M., Cotroneo, D., Foschini, L.: Integrated Support for Handoff Management and Context Awareness in Heterogeneous Wireless Networks. In: International Workshop on Middleware Pervasive Ad-Hoc Computing, pp. 145–152 (2005)
7. Hsieh, R., Zhou, Z.G., Seneviratne, A.: SMIP: A Seamless Handoff Architecture for Mobile IP. In: IEEE International Conference on Computer Communication, pp. 1774–1784 (2003)
8. Bellavista, P., Corradi, A., Foschini, L.: Proactive Management of Distributed Buffers for Streaming Continuity in Wired-Wireless Integrated Networks. In: 10th IEEE/IFIP Network Operations Management Symposium, pp. 351–360 (2006)
9. Nguyen, H.N., Sasase, I.: Downlink Queuing Model and Packet Scheduling for Providing Lossless Handoff and QoS in 4G Mobile Networks. IEEE Trans. Mobile Comput. 5(5), 452–462 (2006)
10. Gazis, V., Alonistioti, N., Merakos, L.: Toward a Generic "Always Best Connected" Capability in Integrated WLAN/UMTS Cellular Mobile Networks (and Beyond). IEEE Wireless Communication 12(3), 20–29 (2005)
11. Nakajima, N., Dutta, A., Das, S., Schulzrinne, H.: Handoff Delay Analysis and Measurement for SIP Based Mobility in IPv6. In: IEEE International Conference Communication, pp. 1085–1089 (2003)
12. Kounavis, M.E., Campbell, A.T., Ito, G., Bianchi, G.: Design, Implementation, and Evaluation of Programmable Handoff in Mobile Networks. ACM/Kluwer Mobile Networks Applications 6(5), 443–461 (2001)
13. Guo, C., Guo, Z., Zhang, Q., Zhu, W.: A Seamless and Proactive End-To-End Mobility Solution for Roaming Across Heterogeneous Wireless Networks. IEEE J. Select. Areas Commun. 22(5), 834–848 (2004)

A Review of Congestion Control Mechanisms for Wireless Sensor Networks

Shivangi Borasia and Vijay Raisinghani

Department of Information Technology
Mukesh Patel School of Technology Management and Engineering
NMIMS (deemed to be university)
shivangi.borasia@nmims.edu,
vijay.raisinghani@nmims.edu

Abstract. Wireless sensor network (WSN) plays an important role in many application areas like in military surveillance, health care etc. A WSN is deployed with a large number of sensor nodes in a wide geographical area. These nodes collect information depending on type of the application and transmit the data towards the sink node. When a large number of sensor nodes are engaged in transmitting data, there is a possibility of congestion in the network. Congestion is one of the critical issues in WSNs because it has direct impact on energy efficiency of sensor nodes, and the application's throughput. Congestion degrades overall channel capacity and increases packet loss rate. In order to handle these problems, an efficient congestion control mechanism required. a number of congestion control mechanism have been proposed in literatures. Any congestion control mechanism consists of congestion detection, congestion notification and rate adjustment mechanisms. Some of the mechanisms reviewed in this paper are CODA, PCCP, FACC, Fusion, and Siphon. We discuss pros and cons of each of these mechanisms.

Keywords: wireless sensor network, congestion control, energy efficient.

1 Introduction

Wireless sensor networks [1] are composed of a large numbers of tiny radio-equipped sensor nodes. A wireless sensor network has distinct characteristics like unique network topology, peculiar traffic characteristics-like periodic data observations or event driven bursty traffic, resource constraint for battery life and bandwidth availability and small message size. There are a wide range of applications of wireless sensor networks, such as health care, military surveillance etc. Many wireless sensor network applications require that the readings or observations collected by sensors be stored at some central location. Congestion can occur while collecting the data and sending it towards the central location over the wireless sensor network.

Our focus in this paper is on various congestion control protocol designs and their analysis. Congestion occurs in a part of a network when the network traffic increases to an extent such that it results in decreased network throughput. One of the reasons for congestion is when packet arrival rate exceeds packet service rate. Another reason

K. Shah, V.R. Lakshmi Gorty, and A. Phirke (Eds.): ICTSM 2011, CCIS 145, pp. 201–206, 2011.

can be link level conditions such as contention, interference and bit errors. As congestion occurs in the network, packet loss increases. Hence, the numbers of retransmissions also increase. These retransmissions impact the life of nodes, as they have limited battery power. Whenever new sensor node is added to a sensor network, it introduces the possibility of more interference, and hence congestion. Further, sensor networks' traffic pattern also contributes to congestion, the traffic is event driven and would be bursty. In wireless sensor networks congestion can occur at sink node or at the source nodes. Congestion causes several problems in WSNs like buffer overflow at receiver, longer queuing time for packets and packet loss. There are three important components any congestion control mechanism should have: *congestion detection, congestion notification and rate adjustment.*

Congestion detection: Congestion detection is a mechanism by which irregularities in normal traffic pattern are detected. Congestion detection mechanisms can be classified into local congestion detection and global congestion detection [2]. Local congestion detection takes place at intermediate nodes where congestion detection is carried out by local indicators of congestion such as queue occupancy and channel condition. On the other hand global congestion detection is carried out at the sink where end-to-end attributes such as inter packet delay and frequency of packet losses can be used to infer congestion.

Congestion notification: Once congestion has been detected, congestion information should be propagated from the congested node to upstream traffic nodes and source node. Congestion notification can be implicit or explicit. In explicit congestion notification, the source node sends special control messages to notify the sensor nodes about congestion. On the other hand, in implicit congestion notification, there is no need of sending special control messages; congestion is notified by setting a congestion notification (CN) bit in data packet header.

Rate adjustment: After receiving congestion notification, the node adjusts its packet sending rate accordingly. Rate adjustment can be end-to-end or hop-by-hop. If congestion is notified by setting single CN bit then in order to adjust the rate, AIMD (additive increase multiplicative decrease) scheme or its variants are used .Apart from that, if additional information about congestion could be piggybacked with the data packet then exact rate adjustment can be done.

This paper reviews the existing congestion control protocols for WSNs. The remainder of the paper is organized as follows. Section 2 presents existing congestion control schemes and section 3 presents summary and analysis of the existing congestion control protocols.

2 Existing Congestion Control Mechanism for WSNs

Congestion control mechanisms differ in congestion detection, congestion notification, or rate-adjustment mechanisms as shown in Figure 1. We review various congestion control approaches in the section below.

2.1 Congestion Avoidance and Detection (CODA)

CODA [3] is energy efficient congestion control mechanism designed for WSNs. In CODA, congestion is detected by a receiver node using present and past channel loading condition (time period for which the channel is busy) as well as current buffer occupancy. To notify about congestion, receiver node broadcasts suppression message to upstream nodes. Suppression messages are propagated as backpressure signal towards upstream nodes, in a hop-by-hop manner. An upstream node decides whether to further propagate the message or not, depending on its own local network condition. Upon receiving the backpressure message node adjusts its sending rate based on the local congestion control policy like AIMD (additive increase, multiplicative decrease) or packet drop.

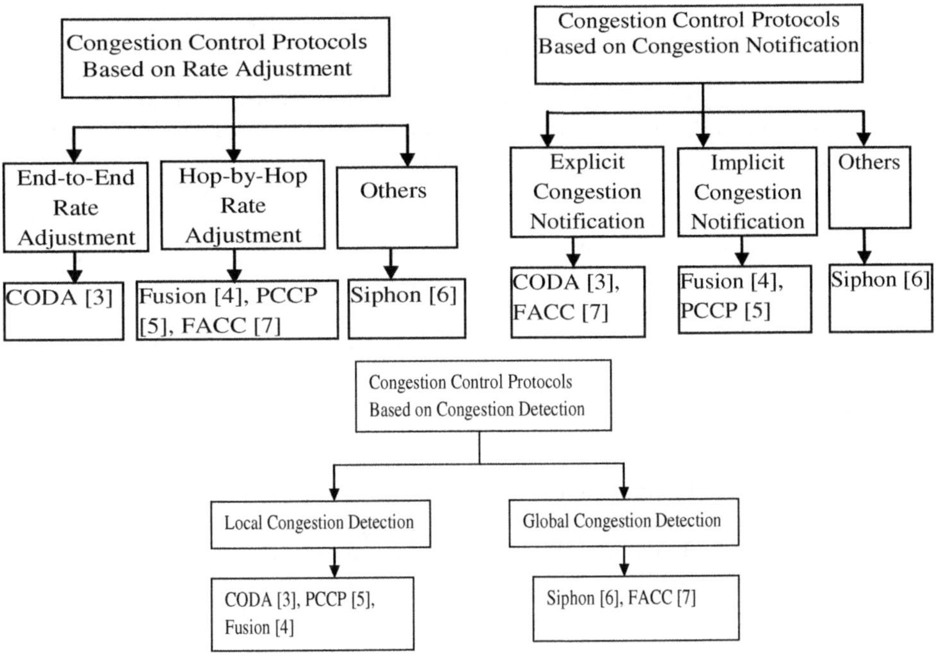

Fig. 1. Classification of existing congestion control mechanisms

2.2 Fusion

Fusion [4] addresses congestion mitigation problem as well as fairness guarantees among sources. In Fusion hop by hop flow control mechanism is used for congestion detection as well as congestion mitigation. Congestion is detected through queue occupancy and channel sampling technique at each intermediate node. If these parameters exceed predefined threshold then the node sets the CN bit in every outgoing packet this is called implicit congestion notification. The sources limit their transmission rate based on the token bucket Rate limitation is done according to token bucket approach. Each sensor node is allowed to transmit until it has at least one token and each transmission costs one token. Further it maintains fairness among the sensor nodes by giving each node a chance to transmit.

2.3 Priority Based Congestion Control Protocol (PCCP)

Priority based congestion control protocol [5] is a congestion control mechanism based on node priority index that is introduced to reflect the importance of each sensor node. Nodes are assigned a priority based on the function they perform and its location. Nodes near the sink have a higher priority. The congestion is detected based on the ratio of sending rate to the packet arrival rate. If the sending rate is lower, it implies that congestion has occurred. The congestion information is piggybacked in data packet header along with the priority index. Nodes adjust their sending rate depending on the congestion at the node itself.

2.4 Fairness Aware Congestion Control (FACC)

FACC [7] is a congestion control mechanism, which controls the congestion and achieves fair bandwidth allocation for each flow of data. FACC detects the congestion based on packet drop rate at the sink node. In FACC nodes are divided in to two categories near sink node and near source node based on their location in WSNs. When a packet is lost, then the near sink nodes send a Warning Message (WM) to the near source node. After receiving WM the near source nodes send a Control Message (CM) to the source node. The source nodes adjust their sending rate based on the current traffic on the channel and the current sending rate. After receiving CM, flow rate would be adjusted based on newly calculated sending rate.

2.5 Siphon

Siphon [6] aims at controlling congestion as well as handling funneling effect. Funneling effect is where events generated under various work load moves quickly towards one or more sink nodes, which increases traffic at sink which leads to packet loss. Virtual sinks are randomly distributed across the sensor network which takes the traffic load off the already loaded sensor node. In siphon initially VS discovery is done. Virtual sink discovery is initiated by the physical sink by as explained in [6]. Node initiated congestion detection is based on past and present channel condition and buffer occupancy as in CODA [3]. After congestion detection traffic is redirected from overloaded physical sink to virtual sinks. It is done by setting redirection bit in network layer header.

3 Summary and Analysis of Congestion Control Mechanisms

In this paper we have reviewed congestion control mechanisms such as CODA [3], Fusion [4], PCCP [5], Siphon [6], and FACC [7]. CODA is an energy efficient approach because it uses hop by hop backpressure congestion detection that deal with local congestion. It consumes less battery power of sensor nodes to detect congestion locally. It can handle persistent congestion using feedback based mechanism called Closed-loop, multi-source regulation. In CODA upstream congestion control mechanism

Table 1. Summary of congestion control protocol for WSNs

Features	Protocols				
	CODA[3]	**Fusion[4]**	**PCCP[5]**	**Siphon[6]**	**FACC[7]**
Congestion detection	Queue length And channel Status	Queue length	Packet Interarrival time and Packet service time	Queue length and application fidelity	Packet Drop At the Sink node
Congestion notification	Explicit	Implicit	Implicit	-	Explicit
Rate Adjustment	AIMD-like end-to-end rate adjustment	Stop-and-start hop-by -hop rate ad ment	Exact hop-by-hop rate adjustment	Traffic redirection	Hop-by-Hop rate adjustment
Provision for Energy efficiency	Yes	No	Yes	No	No
Provision for Fairness	No	Yes	Yes	No	Yes

is used so only sensor nodes can notify the sink node about congestion. The sink node cannot notify about congestion to sensor nodes, hence it is a unidirectional approach. Further, suppression messages that are used for congestion notification can be lost due to link break. In this situation it is difficult to notify upstream nodes about congestion. Further, in CODA one of the rate adjustment policies is packet drop, which might lead to loss of data packets that contain critical data or observations. Fusion provides fairness among sources as it uses token bucket approach so each sensor nodes gets opportunity to send the data by capturing a token. If the token is lost, each node strives for capturing token in order to send data. Priority based scheme PCCP [5] uses implicit congestion notification which avoids need of additional control messages hence overhead is reduced. However, if the data packet in which contains CN bit is lost, piggybacked congestion information like priority index and congestion degree also would be lost. Thus proper rate adjustment would not be done. Another approach that provides fairness is FACC [7]. In order to notify congestion and adjusting the sending rate CM (control message) and WM (warning message) are used. These messages add an overhead in the wireless sensor network. If any of these messages are lost because of path break that would leads to a problem in congestion notification as well as rate adjustment. Last approach we discussed is based on traffic redirection mechanism called Siphon [6]. In siphon if the physical sink node redirects message to one of the nearby virtual sink only, then that VS will be overloaded with traffic. In Table 1 we have summarized the various congestion control protocols.

4 Conclusion and Future Direction

All the congestion control mechanisms we discussed here have common objective of mitigating congestion in wireless sensor networks. First, we discussed the important components of congestion control: congestion detection, congestion notification and rate adjustment. Overall congestion control mechanisms that we discussed are divided into categories based on congestion notification, rate adjustment and congestion detection mechanism. We highlighted the working of congestion control mechanisms and critically analyze them. Although these congestion control techniques are promising there are still there are many challenges to solve in wireless sensor network to handle congestion control efficiently. In protocols reviewed in this paper, we observed that none of the protocol provide a mechanism for packet recovery. This is essential for critical applications where minimal packet loss is required. None of the protocols discuss the scenario of loss of congestion control information. If this information is lost the congestion control mechanism could fail. These two shortcomings need to be investigated further.

References

1. Akyildiz, I.F., Weilian, S., Sankarasubramaniam, Y., Cayirci, E.: A Survey on Sensor Network. IEEE Communication Mag. 40(8), 102–114 (2002)
2. Misra, S., Woungang, I., Misra, S.C.: Guide to Wireless Sensor Networks. Springer Publication, London (2009)
3. Wan, C.Y., Eisenman, S.B., Campbell, A.T.: CODA: Congestion Detection and Avoidance in Sensor Networks. In: 1st International Conference on Embedded networked Sensor Systems, Los Angeles, pp. 56–67 (2000)
4. Hull, B., Jamieson, K., Balakrishnan, H.: Mitigating congestion in wireless sensor networks. In: 2nd International Conference on Embedded Networked Sensor Systems. Maryland (2004)
5. Wang, C., Sohraby, K., Lawrence, V., Li, B.: Priority Based Congestion Control in Wireless Sensor Networks. In: IEEE International Conference on Sensor Networks, Ubiquitous and Trustworthy Computing, Taiwan, pp. 22–31 (2006)
6. Wan, C.Y., Eisenman, S.B., Campbell, A.T., Crowcroft, J.: Siphon: Overload Traffic Management Using Multi-radio Virtual Sinks in Sensor Networks. In: Proceedings of the 3rd International Conference on Embedded Networked Sensor Systems, San Diego, pp. 116–129 (2005)
7. Xiaoyan, Y., Xingshe, Z., Zhigang, L., Shining, L.: A Novel Congestion Control Scheme in Wireless Sensor Networks. In: 5th International Conference on Mobile Ad-hoc and Sensor Networks, Fujian, pp. 381–387 (2009)

Image Retrieval Using Texture Features Extracted as Vector Quantization Codebooks Generated Using LBG and Kekre Error Vector Rotation Algorithm

H.B. Kekre[1], Tanuja K. Sarode[2], Sudeep D. Thepade[3], and Srikant Sanas[4]

[1] Sr. Professor, [2] Assistant Professor, [3] Associate Professor, [4] M.Tech. Student
[1,3,4] MPSTME, SVKM's NMIMS (Deemed-to-be University), Mumbai, India
[2] TSEC, Bandra (w), Mumbai
hbkekre@yahoo.com, tanuja.sarode@gmail.com,
sudeepthepade@gmail.com, shri.sanas@gmail.com

Abstract. In this paper novel methods for image retrieval based on texture feature extraction using Vector Quantization (VQ) is proposed using Linde-Buzo-Gray (LBG) and newly introduced Kekre Error Vector Rotation (KEVR) algorithms for texture feature extraction and their results are compared. The image is first splitted into blocks of size 2x2 pixels (each pixel with red, green and blue component). A training vector of dimensions 12 is created using this block. Collection of all such training vectors is a training set. To generate the texture feature vector of the image, popular LBG and KEVR algorithms are applied on the initial training set to obtain codebooks of size 16, 32, 64 128, 256 and 512. These codebooks formed feature vectors for CBIR. The proposed image retrieval techniques are tested on generic image database having 1000 images. From the results it is observed that KEVR based CBIR give slight improvement over LBG based CBIR. Overall in all codebook sizes KEVR gives best results with higher precision.

Keywords: CBIR, Texture, Vector Quantization, LBG, KEVR.

1 Introduction

Advances in internet technology have motivated people to communicate and express by sharing images, video, and other forms of online media [1]. The problems of acquiring, storing and transmitting the images are well addressed, capabilities to manipulate, index, sort, filter, summarize, or search through image database lack maturity. Modern image search engines [2] retrieve the images based on their visual contents, commonly referred to as Content Based Image Retrieval (CBIR) systems [3]. There are various applications of CBIR systems like fabric and fashion design, interior design as panoramic views [4,5,6], art galleries [4], museums, architecture/engineering design [4], weather forecast, geographical information systems, remote sensing and management of earth resources [7,8], scientific database management, medical imaging, trademark and copyright database management, the military, law enforcement and criminal investigations, intellectual property, picture archiving and communication systems, retailing

K. Shah, V.R. Lakshmi Gorty, and A. Phirke (Eds.): ICTSM 2011, CCIS 145, pp. 207–213, 2011.

and image search on the Internet. Typical CBIR systems can organize and retrieve images automatically by extracting some features such as color, texture, shape from images and looking for similar images which have similar feature [2]. Generally CBIR systems have two phases. First phase is feature extraction (FE), a set of features, called feature vector, is generated to accurately represent the content of each image in the database. A feature vector is much smaller in dimension as that of the original image [9, 10]. The second phase is matching phase which requires similarity measurement (SM) between the query image and each image in the database using their features computed in first phase so that the most similar images can be retrieved [11, 12]. A variety of feature extraction techniques are available in literature like color based feature extraction techniques include color histogram, color coherence vector, color moments, circular ring histogram [13], BTC extensions [10,12]. Texture based feature extraction techniques such as co-occurance matrix [14], Fractals [15], Gabor filters [15], variations of wavelet transform [1], Kekre transform [16,17]. Effort has been made to extend image retrieval methodologies using combination of color and texture as the case in [17] where Walshlet Pyramids are introduced. The combination of color and texture feature extraction methods for CBIR outperforms the CBIR methods that use just color and texture features [2, 7].

In section 2 texture feature extraction using VQ based methods viz. LBG and newly proposed KEVR are discussed. In section 3, technique for image retrieval using vector quantization is proposed. Results and discussion are given in section 4 and conclusions are presented in section 5.

2 VQ Based Methods

Vector Quantization (VQ) [19-24] is an efficient technique for data compression. VQ has been very popular in variety of research fields such as video-based event detection [25], speech data compression, image segmentation, CBIR [2,7,11], face recognition, iris recognition, data hiding etc. VQ can be defined as the mapping function that maps k-dimensional vector space to the finite set $\mathbf{CB = \{ C_1, C_2, C_3, \ldots, C_N\}}$. The set CB is called codebook consisting of N number of codevectors and each codevector $C_i = \{c_{i1}, c_{i2}, c_{i3}, \ldots\ldots, c_{ik}\}$ is of dimension k. The codebook is the feature vector of the entire image and can be generated by using clustering techniques. The method most commonly used to generate codebook is the Linde-Buzo-Gray (LBG) algorithm [7]. The drawback of LBG algorithm is that the cluster elongation is 135^0 with horizontal axis in two dimensional cases which results in inefficient clustering. The disadvantage of LBG is overcome in Kekre's Error Vector Rotation (KEVR) [26] algorithm. To generate the codebook, the image is first divided into fixed size blocks, each forming a training vector $\mathbf{X_i = (x_{i1}, x_{i2}, \ldots\ldots, x_{ik})}$.The set of training vectors is a training set. This training set is initial cluster. The clustering algorithms like LBG and KEVR are then applied on this initial cluster to generate the codebook of desired size. LBG [20] is standard VQ codebook generation algorithm. The KEVR algorithms for codebook generation are discussed. In Kekre Error Vector Rotation algorithm (KEVR) algorithm two vectors v_1 & v_2 are generated by adding error vector to the codevector. Euclidean distances of all the training vectors are computed with vectors v_1 & v_2 and two clusters are formed based on closest of v_1 or v_2. The codevectors of the two clusters are computed and then both clusters are splitted by adding and subtracting error vector rotated in k-dimensional space at different angle to both the codevector. This modus operandi is repeated for every cluster and every time to split the clusters error ei is added and subtracted from the codevector and two vectors v_1 and v_2 is generated. Error vector e_i is the ith row of the error matrix of dimension k. The error vectors matrix E is given in equation 1.

$$E = \begin{bmatrix} e_1 \\ e_2 \\ e_3 \\ e_4 \\ e_5 \\ . \\ . \\ . \\ . \\ e_k \end{bmatrix} = \begin{bmatrix} 1 & 1 & 1 & 1 & \cdots & 1 & 1 & 1 \\ 1 & 1 & 1 & 1 & \cdots & 1 & 1 & -1 \\ 1 & 1 & 1 & 1 & \cdots & 1 & -1 & 1 \\ 1 & 1 & 1 & 1 & \cdots & 1 & -1 & -1 \\ 1 & 1 & 1 & 1 & \cdots & -1 & 1 & 1 \\ \cdots & \cdots & \cdots & \cdots & \cdots & \cdots & & . \\ \cdots & \cdots & \cdots & \cdots & \cdots & \cdots & & . \\ . & & & & & & & \\ . & & & & & & & \\ \cdots & \cdots & \cdots & \cdots & \cdots & \cdots & & . \end{bmatrix} \quad (1)$$

Note that these error vector sequences have been obtained by taking binary representation of numbers starting from 0 to k-1 and replacing zeros by ones and ones by minus ones. Algorithm for KEVR Codebook generation can be explained by following steps

Step 1: Divide the image into non overlapping blocks and convert each block to vectors thus forming a training vector set.
Step 2: initialize i=1;
Step 3: Compute the centroid (codevector) of this training vector set.
Step 4: Add and subtract error vector e_i from the codevector and generate two vector v_1 and v_2.
Step 5: Compute Euclidean distance between all the training vectors belonging to this cluster and the vectors v_1 and v_2 and split the cluster into two.
Step 6: Compute the centroid (codevector) for clusters obtained in the above step 5.
Step 7: increment i by one and repeat step 4 to step 6 for each codevector.
Step 8: Repeat the Step 3 to Step 7 till codebook of desired size is obtained.

3 Image Retrieval Using VQ Based Techniques

Image retrieval based on content requires extraction of features of the image, matching these features with the features of the images in the database and retrieving the images with the most similar features. Here, paper discusses the feature extraction technique based on vector quantization.

A. Proposed Feature Extraction Technique

i. Divide the image into blocks of size 2x2 (Each pixel having red, blue and green component, thus resulting in a vector of 12components per block)
ii. Form the training set/ initial cluster from these vectors.
iii. Compute the initial centroid of the cluster.
iv. Obtain the codebook of desired size using LBG/KEVR algorithm. This codebook represents the feature vector/signature of the image.
v. Repeat steps 2-6 for each image in the image database.
vi. Store the feature vector obtained in step 5 in the feature vector database.

B. Query Execution

For a given query image compute the feature vector using the proposed feature extraction technique. To retrieve the most similar images, compare the query feature vector with the feature vectors in database. This is done by computing the distance between the query feature vectors with those in feature vector database. Euclidian distance as given in equation (6) and correlation coefficient are most commonly used as similarity measure in CBIR. Here Euclidian distance is used as a similarity measure. The proposed KEVR based codebook generation proves to be better than LBG based codebook generation in CBIR.

4 Results and Discussions

The proposed CBIR techniques are implemented in Matlab 7.0 on Intel Core 2 Duo Processor T8100, 2.1 GHz, 2 GB RAM machine to obtain results. The results are obtained on the general database consisting of 1000 images from 11 different categories (some of these are taken from [17]). To test the proposed method, from every class five query images are selected randomly. So in all 55 query images are used. To check the performance of proposed technique we have used precision and recall. The standard definitions of these two measures are given by following equations.

$$\text{Precision} = \frac{\text{Number\_of\_relevant\_images\_retrieved}}{\text{Total\_number\_of\_images\_retrieved}} \tag{2}$$

$$\text{Recall} = \frac{\text{Number\_of\_relevant\_images\_retrieved}}{\text{Total\_number\_of\_relevent\_images\_in\_database}} \tag{3}$$

The crossover point of precision and recall acts as performance measure of CBIR technique. Higher value of precision-recall at crossover point indicates better performance of image retrieval method. Results are obtained using LBG and KEVR for the codebook of sizes 16x12, 32x12, 64x12, 128x12, 256x12 and 512x12.

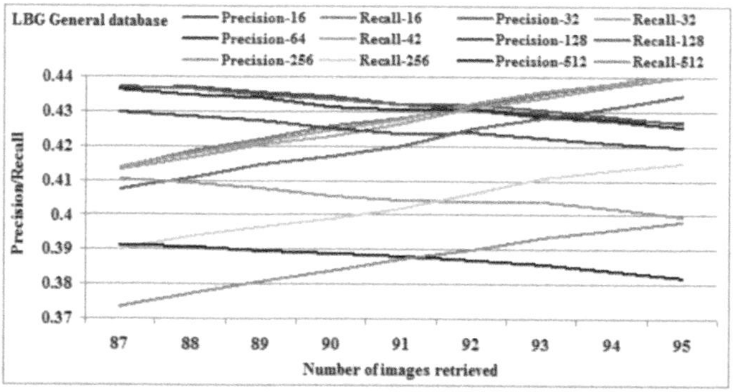

Fig. 1. Cross-over points of average precision and recall using LBG-CBIR for the different codebook sizes varying from 16x12 to 512x12

Figure 1 shows the average precision/ recall values for LBG-CBIR techniques with various codebook sizes, showing the best performance given by 32x12 codebook size. Figure 2 shows the precision/recall values of different codebook sizes for KEVR-CBIR methods, resulting 128x12 to be best codebook size for image retrieval. Figure 3 shows the comparative analysis of LBG-CBIR and KEVR-CBIR for all codebook sizes from 16x12 to 512x12. Everywhere the KVER-CBIR outperforms LBG-CBIR.

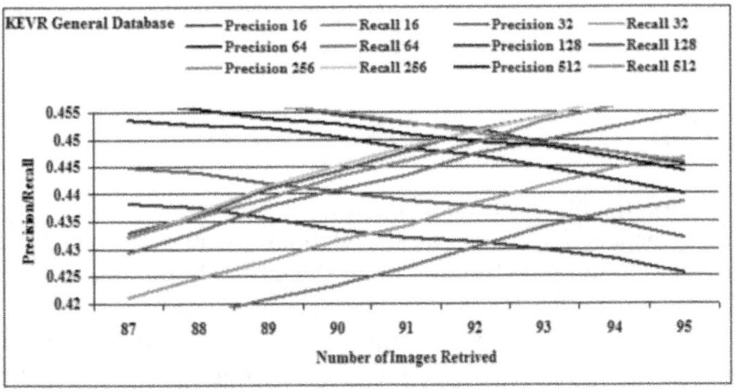

Fig. 2. Cross-over points of average precision and recall using KEVR-CBIR for the different codebook sizes varying from 16x12 to 512x12

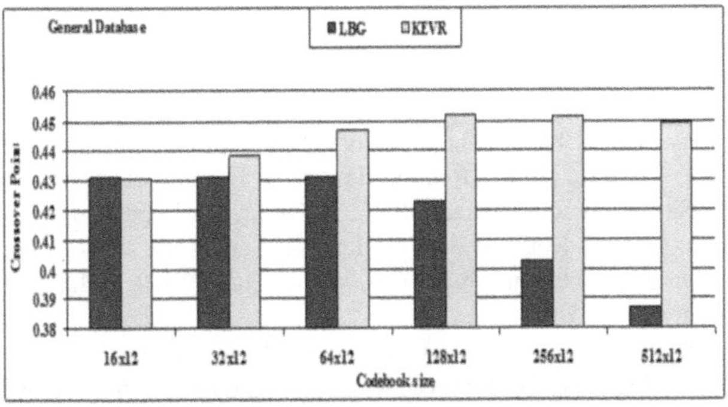

Fig. 3. Crossover points of Average Precision and Average Recall plotted against the size of the codebook varying from 16x12 to 512x12 for the proposed CBIR methods

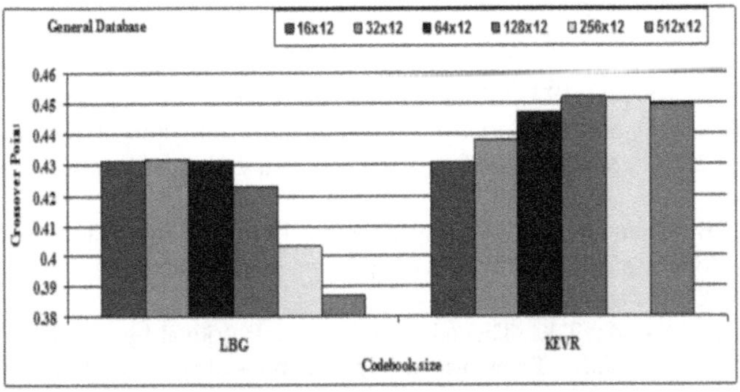

Fig. 4. Crossover points of Average Precision & Recall plotted against the codebook generation techniques for codebook sizes (16x12 to 512x12) for the proposed CBIR method

Figure 4 compares the individual performances of LBG-CBIR and KEVR-CBIR for various codebook sizes. In LBG-CBIR the codebook size 32 gives best performance and then if codebook size is increased further the performance of CBIR deteriorates. In case of KEVR-CBIR the increasing size of codebook improves the performance of image retrieval from codebook size 16x12 to 128x12 and then the performance decreases due to voids being created in codebooks.

5 Conclusion

The use of vector quantization codebooks as feature vectors for image retrieval is proposed in the paper. The paper used codebook generation techniques such as of Linde-Buzo-Gray (LBG) and newly introduced Kekre's Error Vector Rotation (KEVR) algorithms for texture feature extraction. These codebooks extracted with sizes 16, 32, 64, 128, 256 and 512 are used in proposed CBIR techniques. Thus the two codebook generation algorithms and six different codebook sizes per algorithm result in 12 proposed image retrieval techniques. Results of the proposed CBIR techniques are compared. The newly proposed KEVR based CBIR outperforms LBG based CBIR.

References

1. Saha, S.K., Das, A.K., Chanda, B.: CBIR using Perception based Texture and Color Measures. In: 17 Int. Conf. on Pattern Recognition(ICPR 2004), vol. 2 (August 2004)
2. Kekre, H.B., Sarode, T.K., Thepade, S.D., Vaishali, S.: Improved Texture Feature Based Image Retrieval using Kekre's Fast Codebook Generation Algorithm. In: Springer-Int. Conf. on Contours of Computing Tech. (Thinkquest-2010), March 13-14, pp. 13–14. BGIT, Mumbai (2010)
3. Kekre, H.B., Thepade, S.D., Athawale, A., Shah, A., Verlekar, P., Shirke, S.: Walsh Transform over Row Mean and Column Mean using Image Fragmentation and Energy Compaction for Image Retrieval. Int. Journal on Comp. Science and Engg. (IJCSE) 2S(1) (January 2010)
4. Kekre, H.B., Thepade, S.D.: Creating the Color Panoramic View using Medley of Grayscale and Color Partial Images. WASET Int. Journal of Electrical, Computer and System Engineering (IJECSE) 2(3) (Summer 2008), `http://www.waset.org/ijecse/v2/v2-3-26.pdf`
5. Kekre, H.B., Thepade, S.D.: Rotation Invariant Fusion of Partial Images in Vista Creation. WASET International Journal of Electrical, Computer and System Engineering (IJECSE) 2(2) (Spring 2008), `http://www.waset.org/ijecse/v2/v2-2-13.pdf`
6. Kekre, H.B., Thepade, S.D.: Scaling Invariant Fusion of Image Pieces in Panorama Making and Novel Image Blending Technique. Int. Journal on Imaging (IJI) 1(A08) (Autumn 2008), `http://www.ceser.res.in/iji.html`
7. Kekre, H.B., Sarode, T.K., Thepade, S.D.: Image Retrieval by Kekre's Transform Applied on Each Row of Walsh Transformed VQ Codebook. Invited at ACM-Int. Conf. & Workshop on Emerging Trends in Tech. (ICWET), TCET, Mumbai, February 26-27 (2010); uploaded on ACM Portal

8. Kekre, H.B., Thepade, S.D., Athawale, A., Shah, A., Verlekar, P., Shirke, S.: Energy Compaction and Image Splitting for Image Retrieval using Kekre Transform over Row and Column Feature Vector. Int. Journal of Comp. Science &Network Security 10(1) (January 2010), http://www.IJCSNS.org
9. Kekre, H.B., Thepade, S.D.: Image Retrieval using Color-Texture Features Extracted from Walshlet Pyramid. ICGST Int. Journal on Graphics, Vision & Image Processing (GVIP) 10(1), 9–18 (2010)
10. Kekre, H.B., Thepade, S.D.: Using YUV Color Space to Hoist the Performance of Block Truncation Coding for Image Retrieval. In: Proc. of IEEE Int. Advanced Computing Conference 2009 (IACC 2009), Thapar University, Patiala, India, March 6-7 (2009)
11. Kekre, H.B., Sarode, T.K., Thepade, S.D.: Image Retrieval using Color-Texture Features from DCT on VQ Codevectors obtained by Kekre's Fast Codebook Generation. ICGST-Int. Journal GVIP 9(5), 1–8 (2009)
12. Kekre, H.B., Thepade, S.D.: Color Based Image Retrieval using Amendment Block Truncation Coding with YCbCr Color Space. Int. Journal on Imaging (IJI) 2(A09), 2–14 (Autumn 2009), http://www.ceser.res.in/iji.html
13. Xiaoling, W.: A Novel Cicular Ring Histogram for Content-based Image Retrieval. In: First International Workshop on Education Technology and Computer Science (2009)
14. Zhang, J., Li, G.-l., He, S.-w.: Texture-Based Image Retrieval By Edge Detection Matching GLCM. In: 10th Int. conf. on High Perf. Computing and Comm. (September 2008)
15. Song, X., Li, Y., Chen, W.: A Textural Feature Based Image Retrieval Algorithm. In: Proc. of 4th Int. Conf. on Natural Computation (October 2008)
16. Kekre, H.B., Thepade, S.D.: Image Retrieval using Non-Involutional Orthogonal Kekre's Transform. Int. Journal of Multidisciplinary Research & Advances in Engg. (IJMRAE) 1(I) (2009), http://www.ascent-journals.com
17. http://wang.ist.psu.edu/docs/related/Image.orig (last referred on September 23, 2008)
18. Saha, S., Das, A., Chanda, B.: CBIR using Perception based Texture and Color Measures. In: 17th Int. Conf. on Pattern Recognition (ICPR 2004), vol. 2 (August 2004)
19. Gray, R.M.: Vector quantization. IEEE ASSP Mag., 4–29 (April 1984)
20. Linde, Y., Buzo, A., Gray, R.M.: An algorithm for vector quantizer design. IEEE Trans. Commun. COM-28(1), 84–95 (1980)
21. Kekre, H.B., Sarode, T.K.: An Efficient Fast Algorithm to Generate Codebook for Vector Quantization. In: 1st Int. Conf. on Emerging Trends in Engg. and Technology, ICETET-2008, Raisoni COE, Nagpur, India, July 16-18, pp. 62–67 (2008)
22. Kekre, H.B., Sarode, T.: Fast Codebook Generation Algorithm for Color Images using Vector Quantization. Int. Journal of Comp. Sci. & IT 1(1) (January 2009)
23. Kekre, H.B., Sarode, T.K.: Fast Codevector Search Algorithm for 3-D Vector Quantized Codebook. WASET Int. Journal of cal Comp.Info. Science & Engg (IJCISE) 2(4), 235–239 (Fall 2008), http://www.waset.org/ijcise
24. Kekre, H.B., Sarode, T.K.: Fast Codebook Search Algorithm for Vector Quantization using Sorting Technique. In: ACM Int. Conf. on Advances in Computing, Comm. and Control (ICAC3-2009), FCRCE, Mumbai, January 23-24, pp. 317–325 (2009)
25. Liao, H., Chen, D., Su, C., Tyan, H.: Real-time event detection and its applications to surveillance systems. In: IEEE Int. Symp. Circuits & Systems, Kos, Greece, pp. 509–512 (May 2006)
26. Kekre, H.B., Sarode, T.K.: New Clustering Algorithm for Vector Quantization using Rotation of Error Vector. Int. Journal of Computer Science & Information Security 7(03) (2010)

Palm Print Identification Using Fractional Coefficients of Sine/Walsh/Slant Transformed Palm Print Images

H.B. Kekre[1], Sudeep D. Thepade[2], Arvind Viswanathan[3], Ashish Varun[3], Pratik Dhwoj[3], and Nikhil Kamat[3]

[1] Senior Professor, [2] Associate Professor, [3] B.Tech IT Students
Mukesh Patel School of Technology Management and Engineering,
SVKM's NMIMS (Deemed to be University), Mumbai-56
hbkekre@yahoo.com, sudeepthepade@gmail.com

Abstract. The paper presents performance comparison of palm print identification techniques based on fractional coefficients of transformed palm print image using three different transforms like Sine, Walsh and Slant. In transform domain, the energy of image gets accumulated towards high frequency region; this characteristic of image transforms is exploited here to reduce the feature vector size of palm print images by selecting these high frequency coefficients in transformed palm print images. Three image transforms and 12 ways of taking fractional coefficients result into total 36 palm print identification methods. A database of 2350 palm print images is used as a test bed for performance comparison of the proposed palm print identification methods with help of false acceptance rate (FAR) and genuine acceptance rate (GAR). The experimental results in Sine and Walsh transform have shown performance improvement in palm print identification using fractional coefficients of transformed images. In all Sine transform at fractional coefficients of 0.78% gives best performance as indicated by higher GAR value. Thus the task of speeding up palm print identification with better performance is achieved to make it more suitable for real time applications.

Keywords: Palm Print, Biometric, Sine Transform, Walsh Transform, Slant Transform.

1 Introduction

In palm print recognition many of the same matching characteristics observed in fingerprint recognition can be found to make it one of the most well-known and best publicized biometric authentication technique. The information presented in a friction ridge impression is used in both palm and finger biometrics to generate feature set or signature. This information is combination of ridge characteristics and ridge structure of the raised portion of the epidermis. The data represented by these friction ridge impressions allows a determination that corresponding areas of friction ridge impressions [1]. Fingerprints and palms have been used for over a century as a trusted form of identification because of their uniqueness and permanence. Due to some difficulties in computing capabilities and live-scan technologies, the automation in palm recognition

K. Shah, V.R. Lakshmi Gorty, and A. Phirke (Eds.): ICTSM 2011, CCIS 145, pp. 214–220, 2011.

has been slower [11]. Entire palm is scanned by some palm recognition systems, while some systems need the palms to be segmented into smaller areas. The searching of smaller data sets can help in improvising the reliability within either a fingerprint or palm print system. The three main categories of palm matching techniques are minutiae-based matching, correlation-based matching, and ridge-based matching. The most widely used technique is minutiae-based matching, relies on the location, direction, and orientation of each of the minutiae points. Simply lining up the palm images and subtracting them to determine if the ridges in the two palm images correspond is the basic principle of correlation-based matching. Ridge pattern landmark features such as sweat pores, spatial attributes, and geometric characteristics of the ridges, local texture analysis are used in ridge-based matching. Comparatively ridge-based matching is faster and overcomes some of the difficulties associated with extracting minutiae from poor quality images [12]. The advantages and disadvantages of each approach vary based on the algorithm used and the sensor implemented. Minutiae-based matching typically attains higher recognition accuracy, although it performs poorly with low quality images and does not take advantage of textural or visual features of the palm [13]. Processing using minutiae-based techniques may also be time consuming because of the time associated with minutiae extraction. Correlation-based matching is often quicker to process but is less tolerant to elastic, rotational, and translational variances and noise within the image. Some ridge-based matching characteristics are unstable or require a high-resolution sensor to obtain quality images. The distinctiveness of the ridge-based characteristics is significantly lower than the minutiae characteristics. In all the existing methods of palm print identification computational complexity is major drawback making it inappropriate for real time applications. The paper extends the concept of using fractional coefficients of transformed images as feature vectors [7][8][9][10] to palm print identification using three transforms namely Sine, Walsh and Slant. The performance of biometric identification techniques is generally measured using genuine acceptance rate and false acceptance rate [3] which are elaborated in sections 1.1 and 1.2.

1.1 Genuine Acceptance Rate (GAR)

Depending on the choice of threshold, the legitimate users' palm prints that are genuinely accepted by the system can be from none to all images. A legitimate user is an individual that is verified against the given database. The threshold of the genuinely accepted data divided by the number of all legitimate user data is called Genuine Acceptance Rate (GAR). Its value is one, if all data is accurate, and zero if no legitimate user is accepted.

1.2 False Acceptance Rate (FAR)

Depending on the choice of threshold, the impostor's palm prints that are falsely accepted by the system can be from none to all images. Impostor is an individual that is verified against a different stored database. The threshold of the falsely accepted data divided by the number of all impostor data is called False Acceptance Rate (FAR). Its value is one, if all impostor data are falsely accepted, and zero if none of the impostor data is accepted.[4]

2 Proposed Palm Print Identification Techniques

As shown in figure-1 the higher frequency coefficients of transformed image can be selected to form twelve feature vector sets as 100%, 50%, 25%, 12.5%, 6.25%, 3.125%, 1.5625%, 0.7813%, 0.39%, 0.195%, 0.097% and 0.048% of total number of coefficients in transformed image. Each of these feature vector selections when done with a specific image transform do form a different palm print identification method. The three transforms used here are Sine Transform, Walsh Transform and Slant Transform. Three image transforms and twelve feature vector selection methods result into total 36 palm print identification techniques discussed and analyzed in the paper.

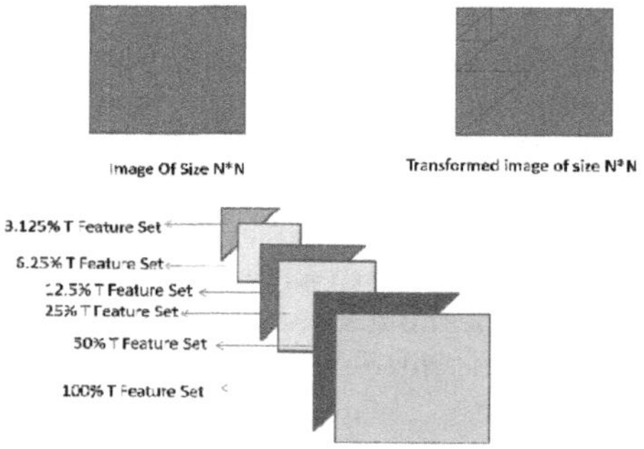

Fig. 1. The Technique Of Fractional Coefficients [7][8][9][10]

3 Implementation

The proposed palm print identification methods have been implemented in MATLAB 7.0 using a computer with Intel Core 2 Duo Processor T8100(2.1GHz) and 2 GB RAM. The techniques are tested on palm print image database [14] of 2350 images of 235 individuals (1175 left palm prints and 1175 right palm prints). Per individual, database has ten palm print images, five samples of each left and right palm prints. To test the performance of proposed palm print identification techniques, in all 1175 queries are fired on each the left and the right palm print databases to compute FAR and GAR values per query. Finally the average FAR and average GAR values of all queries are computed and these results taking into consideration the individual, (left or right) palm prints are compared with consideration of both left and right together as a query. The similarity measure used for matching of query with database images used here is Mean Square Error (MSE) [11].

4 Results and Discussions

Figure-2 shows the average genuine acceptance rate values (GAR) of proposed palm print identification techniques implemented using Sine transform. The GAR values of considering only left palm print (DST L), only right palm print (DST R) and both (DST L_R) are shown in figure 2 with average (AVG) of all these 3 GARs. The figure shows improvement in GAR values with reducing amounts of fractional coefficients up to 0.78%, indicating that fractional coefficients give better performance than all transformed coefficient data in Sine transform. Further it can be observed that considering both left and right palm prints together increases the ruggedness of the system in all fractional coefficients as shown by higher GARs. In all 0.78% fractional coefficients of Sine transformed palm print recognition proves to be better. Even the speed of palm print recognition improves with decreasing amount of fractional coefficients considered as feature vectors because of less number of comparisons needed in matching the query features with database image features resulting into faster palm print identification. The average genuine acceptance rate values (GAR) of proposed palm print identification techniques implemented using Walsh transform is plotted in figure-3. The figure shows the GAR values of considering only left palm print (WALSH L), only right palm print (WALSH R) and both (WALSH L_R) with average (AVG) of all these 3 GARs. The figure shows improvement in GAR values with reducing amounts of fractional coefficients up to 0.78%, indicating that fractional coefficients give better performance than all transformed coefficient data in Walsh transform. Further it can be observed that considering both left and right palm prints together increases the ruggedness of the system in all fractional coefficients as shown by higher GARs. In all 0.78% fractional coefficients of Walsh transformed palm print recognition proves to be better.

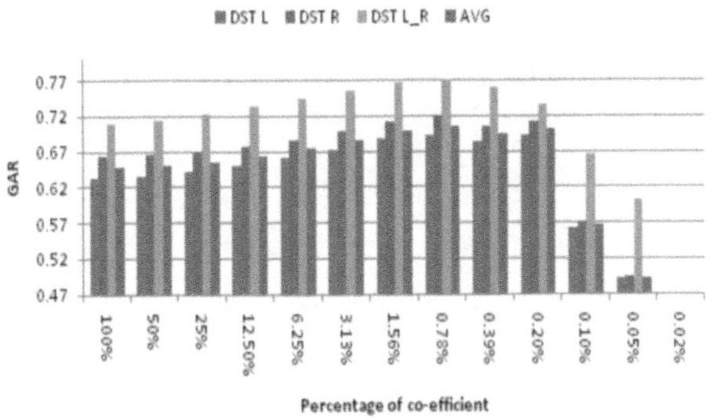

Fig. 2. Average Genuine Acceptance Rate of proposed palm print identification methods using Sine transform

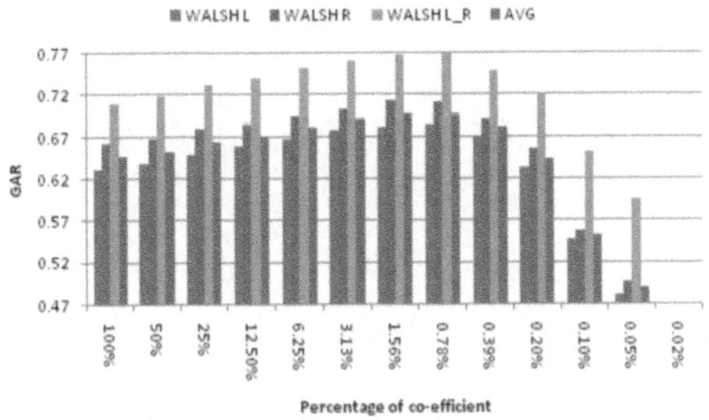

Fig. 3. Average Genuine Acceptance Rate of proposed palm print identification methods using Walsh transform

Even the speed of palm print recognition improves with decreasing amount of fractional coefficients considered as feature vectors because of less number of comparisons needed in matching the query features with database image features resulting into faster palm print identification. The average genuine acceptance rate values (GAR) of proposed palm print identification techniques implemented using Slant transform is plotted in figure-4. The figure shows the GAR values of considering only left palm print (SLANT L), only right palm print (SLANT R) and both (SLANT L_R) with average (AVG) of all these 3 GARs. The figure shows a gradual degradation in GAR values with reducing amounts of fractional coefficients. Further it can be observed that considering both left and right palm prints together increases the ruggedness of the system in all fractional coefficients as shown by higher GARs. 50% of fractional coefficients of Slant transformed palm print recognition proves to be the best in terms of performance [6].

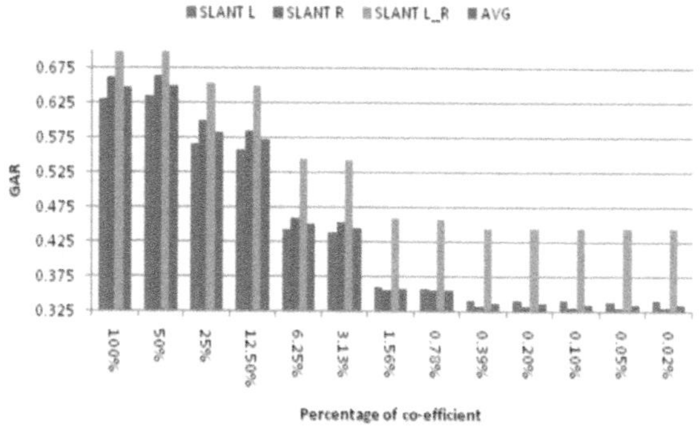

Fig. 4. Average Genuine Acceptance Rate of proposed palm print identification methods using Slant transform

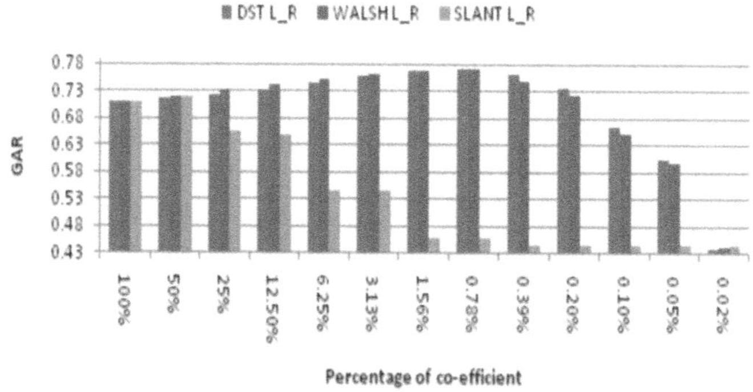

Fig. 5. Average Genuine Acceptance Rate of proposed palm print identification methods using Sine, Walsh and Slant transform

The comparison of the average genuine acceptance rates(GAR) of proposed palm print identification techniques implemented using Sine, Walsh and Slant transforms is plotted in figure-5. The figure shows the GAR values of considering only both left and right palm prints(DST L_R,WALSH L_R,SLANT L_R). The figure shows improvement in GAR values of Sine and Walsh transforms and degradation in GAR values of Slant transform. Further it can be observed that considering both left and right palm prints together increases the ruggedness of the system in all fractional coefficients as shown by higher GARs. In all 0.78% fractional coefficients of Sine transformed palm print recognition proves to be better. Even the speed of palm print recognition improves with decreasing amount of fractional coefficients considered as feature vectors because of less number of comparisons needed in matching the query features with database image features resulting into faster palm print identification.[13].

5 Conclusion

Palm print identification techniques using fractional coefficients of transformed palm print images are discussed in paper. The concept helps in time complexity reduction for palm print recognition making it more suitable to real time applications. Three transforms namely Walsh, Slant and Sine are considered for experimental analysis. All the proposed palm print identification techniques are tested on database having 2350 palm print images. Experimental results have proved the worth of fractional coefficients based techniques using Walsh and Sine transforms. In Walsh and Sine transforms the best performance is observed at 0.78 % fractional coefficients (with size 16x16 for 128x128 image size), indicating tremendous savings in time for palm print identification. Further it is observed that use of both left as well as right palm prints together increases the sturdiness of system than using either of left/right for identification.

References

1. http://www.biometricscatalog.org/NSTCSubcommittee/Documents/Palm%20Print%20%20Recognition.pdf (last referred on November 29, 2010)
2. Bong, D.B.L., Tingang, R.N., Joseph, A.: Palm Print Verification System. In: Proc. of the World Congress on Engineering 2010, WCE 2010, London, June 30 - July 2, vol. I (2010)
3. http://www.ccert.edu.cn/education/cissp/hism/039-041.html (last referred on November 29, 2010)
4. Zhang, D., Kong, W., You, J., Wong, M.: Online Palmprint Identification. IEEE Transactions on Pattern Analysis and Machine Intelligence 25, 1041–1050 (2003)
5. http://www.digital-systems-lab.net/doc/shk_hartleytransform.pdf (last referred on November 29, 2010)
6. http://ralph.cs.cf.ac.uk/papers/Geometry/TruncationSlant.pdf (last referred on November 29, 2010)
7. Kekre, H.B., Thepade, S.D., Maloo, A.: Image Retrieval using Fractional Coefficients of Transformed Image using DCT and Walsh Transform. International Journal of Engineering Science and Technology (IJEST) 2(4), 362–371 (2010), http://www.ijest.info/docs/IJEST10-02-04-05.pdf

8. Kekre, H.B., Thepade, S.D., Maloo, A.: Dr.H.B.Kekre, Sudeep D.Thepade, Akshay Maloo,: Performance Comparison of Image Retrieval Using Fractional Coefficients of Transformed Image Using DCT, Walsh, Haar and Kekre's Transform. CSC Int. Journal of Image Processing (IJIP), Computer Science Journals 4(2), 142–157 (2010), http://www.cscjournals.org
9. Kekre, H.B., Thepade, S.D.: Improving the Performance of Image Retrieval using Partial Coefficients of Transformed Image. International Journal of Information Retrieval, Serials Publications 2(1), 72–79 (2009)
10. Kekre, H.B., Thepade, S.D., Sarode, T.K.: DCT Applied to Column mean and Row Mean Vectors of Image for Fingerprint Identification. In: Int. Conf. on Computer Networks and Security, ICCNS-2008, September 27-28. VIT, Pune (2008)
11. Kumar, A., Wong, D.C., Shen, H.C., Jain, A.K.: Personal Verification Using Palm print and Hand Geometry Biometric. In: Kittler, J., Nixon, M.S. (eds.) AVBPA 2003. LNCS, vol. 2688, Springer, Heidelberg (2003)
12. Saad, E.S.M., Eladawy, M.I.: Eng. Rasha Fathy Aly: Person Identification Using Palm prints. In: NRSC (2008)
13. Parashar, S., Vardhan, A., Patvardhan, C., Kalra, P.K.: Design and Implementation of a Robust Palm Biometrics Recognition and Verification System. In: 6th Indian Conf. on CV, Graphics & IP,
14. IIT Delhi Touchless Palmprint Database (Version 1.0), http://web.iitd.ac.in/~ajaykr/Database_Palm.htm

Image Compression Using Halftoning and Huffman Coding

H. B. Kekre, Sanjay R. Sange, Gauri S. Sawant, and Ankit A. Lahoty

MPSTME, NMIMS University, Mumbai – 400056

Abstract. Halftoning is the printing technology in which each pixel in halftone image is represented by single bit. Hence halftoning gives 87.5% compression ratio. Modified Huffman encoding technique is used on halftone image for further compression of image data. This algorithm achieves a high compression ratio that ensures optimum utilization of network resources and storage. In our earlier work a small operator of size 3x3 is used, which effectively takes only one tap operation. Floyd-Steinberg operator which takes 5 tap operations has been used. Thus factors, like computational complexity, memory space and image quality, have been considered. The proposed algorithm has been implemented on MATLAB platform and has been tested on various images of size 256x256. The image quality measurement has been done using Mean Square Error and Structural Similarity Index parameters. The proposed technique can be used for storage of images in this hybrid compressed form, and low bit rate data transmission for video conferencing.

Keywords: Halftone, Huffman Coding, Symbol, Symbol length, Compression Ratio.

1 Introduction

With technology advancements a kind of data explosion has taken place. Data is being created every moment. In such a situation it becomes essential to optimize the use of already constrained resources. The algorithm proposed in this paper is a hybrid of two well known compression techniques, Halftoning and Huffman Coding.

Halftoning technique is being vastly employed in printing industry. It generates halftone image through the use of dots varying in size, in shape or in spacing. Various halftoning operators are explained in [1] with each one having its own significance. Standard halftoning techniques like Floyd-Steinberg provides better results than random or ordered halftone operators are presented in [2]-[5]. Low computation error diffusing halftoning operator and quantization process have been used in this proposed algorithm [6]-[7].

Modified Huffman Encoding is a lossless compression method that assigns codes to the distinct data values based on the probability of their occurrence in the data. Huffman coding and decoding algorithms are expressed in [8]-[10]. Combination of Huffman coding with other algorithms has been explored in [11].

In section 2 of the paper we talk about halftoning method, section 3 explains Huffman Encoding, section 4 explains about Huffman Decoding, in section 5 the

K. Shah, V.R. Lakshmi Gorty, and A. Phirke (Eds.): ICTSM 2011, CCIS 145, pp. 221–226, 2011.

proposed algorithm is explained and the results obtained the proposed algorithm are discussed in section 6 and section 7. Section 8 states the references.

2 Halftoning Method

To convert continuous image into halftone image operator is used. The operator is convolved with continuous image. After convolution of halftoning operator on image, quantization process is used to convert it into binary value. Both the operators shown in Fig.1 and Fig.2 are 3X3 in size. Floyd-Steinberg half toning operator requires 5 tap effectively 4 tap operation. The small operator requires 3 tap effectively 1 tap operation. Hence the overall computations and processing of entire image is reduced drastically using small operator. The halftoning process gives a compression ratio of 8:1. The pixel value in each plane of color image is represented by 8-bit in comparison to a halftone image where the same is represented by 1-bit. The halftoning operators shown in Fig. 1 and Fig. 2 have been individually applied upon a continuous color image to obtain a halftone image.

Once the halftoning operator is applied to the image, quantization is performed which introduces quantization error called blue noise that degrades the image quality [7].

0	0	0
0	X	1
0	1	3

Fig. 1. Small Operator

0	0	0
0	X	7
3	5	1

Fig. 2. Floyd-Steinberg Operator

3 Modified Huffman Encoding

Huffman coding procedure is based on two observations [8].

1. More frequently occurred symbols will have shorter code words than symbol that occur less frequently.
2. The two symbols that occur less frequently will have the same length.

The Huffman code is designed by merging the lowest probable symbols and the process is repeated until probabilities of two symbols (simple or compounded) are left and thus a code tree is generated and Huffman codes are obtained from labeling of code tree [8]. We call a symbol as simple if it's one of the distinct data values of the data. A symbol is called compounded if it formed as a result of sum of two symbols. The compounded symbols don't form a part of the image data, they are only required by the algorithm to assign codes.

Now the codes are assigned, the upper arm symbol is given code 1 and the lower arm symbol is given code 0. If the symbol is compounded, then the current code is added as a prefix to the symbols from which it is derived. For better understanding, the coding procedure can be visualized as a tree as shown in Fig. 3.

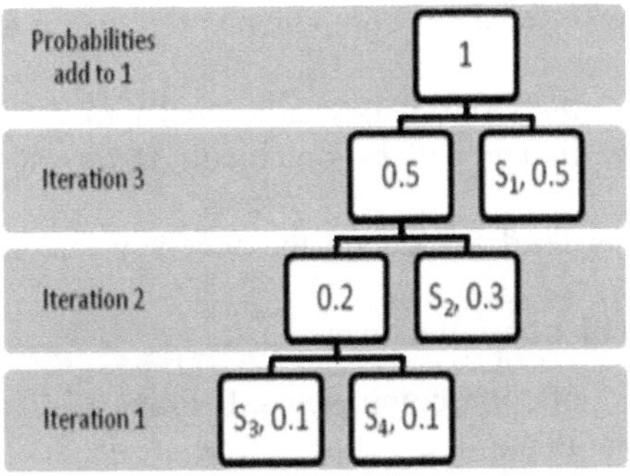

Fig. 3. Huffman Coding Tree Visualization

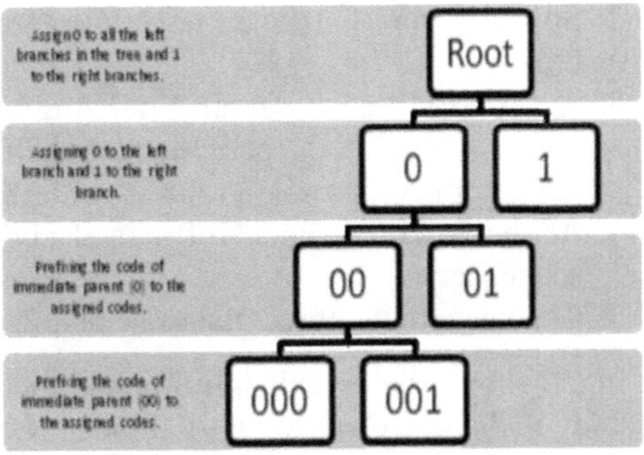

Fig. 4. Huffman Code Assignment

While assigning the codes, we start from the root of the tree, assign 0 to the left branch and 1 to the right branch. This procedure is followed for sub-branches, except that the code of the immediate parent is prefixed to the code assigned as shown in Fig. 4. Now the codes assigned are as follows:

In our algorithm, the codes are used to prepare a dictionary that consists of the symbols and their corresponding codes. Scanning through the image, the symbols are replaced by the corresponding codes. The coded sequence of the halftone image data has been considered in group of 16 bits in size to minimize the network resources and storage required. Each of the 16 bits groups is converted into their decimal equivalent. This is the Huffman Encoded halftone image.

4 Huffman Decoding

The dictionary prepared at the time of modified Huffman encoding is used to decode the received coded sequence. As evident from the example in Table 1, no code assigned to a symbol repeats as a part of another code. Each code is unique and this helps in decoding the coded sequence easily. The decoded sequence is then arranged in a matrix of the size of the original image and thus the Huffman decoded image is obtained, which is identical to that of halftone image.

Table 1. Assigned Codes for Symbols

Symbol	Code
S_1	1
S_2	01
S_3	000
S_4	001

5 Proposed Algorithm

Listed below are the steps for the proposed algorithm:

1. Apply halftoning operator on each of the split image and do the concatenation to obtain color halftone image.

2. Split the color halftone image into three primary color components; red, green and blue.
3. Apply Modified Huffman Encoding on all the three planes individually to get the coded sequence. Hence encoded halftone image will be obtained. At this stage maximum compression ratio is achieved.
4. At the receiving end Huffman decoder is used to decode the received encoded sequence.
5. Concatenate the three planes so as to get color halftone image back.

Fig. 5 shows the overall splitting and merging of color components. R-Plane, G-Plane and B-Plane refer to Red Plane, Green Plane and Blue Plane respectively. H-Encoding and H-Decoding refer to Modified Huffman Encoding and Huffman Decoding respectively.

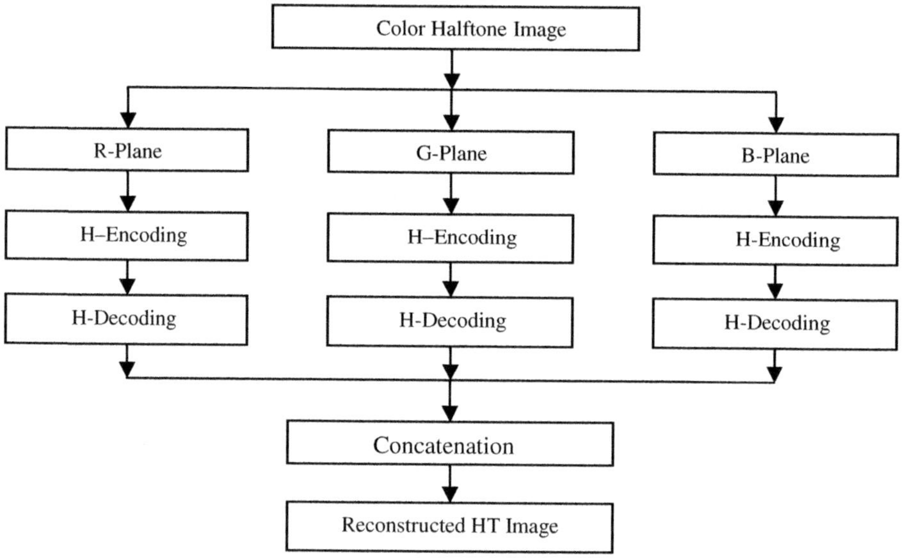

Fig. 5. Block Diagram for the Proposed Algorithm

6 Results

Mean Square Error (MSE), Structural Similarity (SSIM) index and Compression Ratio (CR) are the performance measurement parameters for various images. Table 3 shows the results for different images. All the images used for testing are of size 256x256. In Table 2, PSNR refers to Peak Signal-to-Noise ratio, S refers to small operator and FS refers to Floyd-Steinberg operator.

Sample image Rohitsmile.bmp and results of various operations performed on it by proposed algorithm are shown in Fig. 6 to Fig. 10.

Fig. 6. Original Image

Fig. 7. Halftone Image using Small Operator

Fig. 8. Halftone Image using Floyd-Steinberg Operator

Fig. 9. Huffman Decoded Image for Fig. 7

Fig. 10. Huffman Decoded Image for Fig. 8

7 Discussions

From images in Fig. 9 and Fig. 10 it is clear that result obtained through Floyd-Steinberg operator is better than the result obtained through the proposed Small operator, though the computation complexity is increased.

Contribution towards Modified Huffman Encoding is that, we have trunked the halftone image into 16 bits and considered as a symbol. The major advantage of this consideration is that, we get fixed number of symbol irrespective of the variations in image. We have obtained constant Compression Ratio of 2.99 between halftone image and Huffman Encoded image. The proposed technique can be used in Video Conferencing and live Cricket match, where camera moves from individual person towards crowd. The variations in individual person image are less as compared to the variations in crowd image. Using proposed technique the overall processing time and compression ratio for both the type of images will be same.

References

1. Sange, S.: A Survey on: Black and White and Color Half toning Techniques. SVKM's NMIMS University, MPSTME, Journal of science, Engineering & Technology Management 1(2), 7–17 (2009)
2. Floyd, R.W., Steinberg, L.: An adaptive algorithm for spatial grayscale. Proc. SID 17/2, 75–77 (1976)
3. Wong, P.: Inverse half toning and Kernel estimation for error diffusion. IEEE Trans. Image Processing 4, 486–498 (1995)
4. Hein, S., Zakhor, A.: Halftone to continuous–tone conversion of Error-diffusion coded images. IEEE Trans. Images Processing 4, 208–216 (1995)
5. Kite, T.D., Evans, B.L., Bovik, A.C.: Modeling and Quality Assessment of Half toning by Error Diffusion. IEEE Transaction on Image Processing 9(5) (May 2000)
6. Sange, S.R.: Image data compression using new Halftoning operators and Run Length Encoding. Springer Explorer and Springer CS Digital Library, pp. 224–230
7. Kekre, H.B., Sange, S.: Restoration of Color Halftone image by using Fast Inverse Half toning Algorithm. In: International Conference on Advances in Recent Technologies in Communication and Computing. IEEE, Los Alamitos (2009), ISBN: 978-0-7695-3845-7/09
8. Pujar, J.H., Kadlaskar, L.M.: A New Lossless Method of Image Compression and Decompression Using Huffman Coding Techniques. Journal of Theoretical and Applied Information Technology
9. Saravanan, C., Ponalagusamy, R.: Lossless Grey-scale Image Compression using Source Symbols Reduction and Huffman Coding. International Journal of Image Processing (IJIP) 3(5)
10. Aggarwal, M., Narayan, A.: Efficient Huffman Decoding. In: 2000 International Conference on Image Processing, Proceedings,
11. Tehranipour, M.H.: Nourani: Mixed RL-Huffman encoding for power reduction and data compression in scan test. In: 2010 Third International Symposium on Intelligent Information Technology and Security Informatics (IITSI), April 2-4 (2010)

Dual Band H-shape Microstrip Antennas

A.A. Deshmukh[1], K.P. Ray[2], P. Thakkar[1], S. Lakhani[1], and M. Joshi[1]

[1] DJSCOE, Vile – Parle (W), Mumbai – 400 056, India
draadeshmukh@djscoe.org
[2] SAMEER, I.I.T. Campus, Powai, Mumbai – 400 076, India

Abstract. The resonance frequency formulation for hexagonal microstrip antenna using its equivalence to circular microstrip antenna is discussed and it gives closer agreement with simulated results for fundamental as well as higher order mode. The compact variations of H-shaped microstrip antennas are proposed. The dual band U-slot or pair of rectangular slots cut H-shaped microstrip antennas is proposed. The dual band response with broadside radiation pattern is realized. To understand the slot mode, the modal analysis over wide frequency range using the surface current distribution generated using the IE3D software is studied. It was observed that the dual frequency response is due to the coupling between the fundamental and higher order mode of H-shaped microstrip antenna.

Keywords: Circular microstrip antenna, H-shaped microstrip antenna, Compact H-shaped microstrip antenna.

1 Introduction

The multi-band microstrip antenna (MSA) is realized either by placing an open circuit nearly quarter wavelength or short circuit nearly half wavelength stub on the edges of the patch [1]. The stub offers capacitive and inductive impedance around the resonance frequency of MSA and realizes dual band response. However since the stub is placed on the edges of the patch it increases the overall patch area and the radiation from the stub affects the radiation pattern of the MSA. The dual band response is also realized by cutting the slot at an appropriate position inside the patch [2]. Since the slot is cut inside the patch it neither increases the patch size nor it largely affect the radiation pattern of the patch and hence it is more frequently used technique to realize dual band MSA. In this paper, a modified variation of the rectangular microstrip antenna (RMSA), a Hexagonal microstrip antenna (HMSA) is discussed. For the given patch dimension, the HMSA has higher resonance frequency as compared to RMSA [1]. The direct resonance frequency formulation for the HMSA is not available. The frequency formulas for HMSA have been calculated by using its equivalence with the rectangular MSA (RMSA) [1]. In this paper, the modal distributions of regular HMSA (wherein all side lengths of HMSA are equal) are studied and it was observed that HMSA distribution is similar to the modal distributions of circular MSA (CMSA). Thus by equating

K. Shah, V.R. Lakshmi Gorty, and A. Phirke (Eds.): ICTSM 2011, CCIS 145, pp. 227–232, 2011.

the area of HMSA with that of the equivalent CMSA, the resonance frequency formulation for HMSA is proposed. The formulations obtained using this method agrees closely with the simulated results for different substrates as well as at different frequencies. The compact variations of HMSA by placing the shorting posts along zero field line are proposed. Also various dual band configurations of HMSA by cutting the pair of rectangular slot or U-slot inside the patch are proposed. All these configurations give broadside radiation pattern at the dual frequencies. The modal distributions for pair of rectangular slots cut HMSA for different slot lengths are studied. It was observed that the pair of slots alters the resonance frequency of next higher order mode and along with fundamental mode realizes dual band response. All these MSAs were first analyzed using the IE3D software [3] followed by experimental verification in dual band HMSAs. The glass epoxy substrate (ε_r = 4.3, h = 1.6 mm, tan δ = 0.02) is used for the simulations as well as the measurements. The HMSAs are fed using the SMA connector of inner wire diameter of 0.12 cm.

2 H-shaped MSA

The regular HMSA is shown in Fig. 1(a). For the dimensions shown in Fig. 1(a), the first order resonance frequency of the HMSA is 915 MHz as shown in its surface current distributions in Fig. 1(b). This current distribution is similar to the TM_{11} mode current distributions in CMSA. Due to this similarity between HMSA and CMSA, the resonance frequency formulation for HMSA is derived using frequency equation for CMSA as given below. The resonance frequencies obtained using the IE3D simulation for S = 2 to 6 cm and that obtained using equivalent CMSA whose radius is calculated by using above equations, are shown in Fig. 2(a, b). The % error between the HMSA frequency and equivalent CMSA frequency is calculated by using equation (5). For the entire range a close agreement between the two results is obtained. The current distribution at next frequency for HMSA is shown in Fig. 1(c).

$$a_H = 2.598S^2 \tag{1}$$

$$a_C = \pi r_c^2 \tag{2}$$

Equating the two areas gives,

$$r_c = S\sqrt{2.598/\pi} \tag{3}$$

$$f_r = \frac{K_{mn} c}{2 r_c \pi \sqrt{\varepsilon_r}} \tag{4}$$

$$\% \text{ error} = \frac{f_{cmsa} - f_{hmsa}}{f_{hmsa}} \tag{5}$$

where,

a_H = area of HMSA, a_C = area of CMSA, r_C = equivalent radius of CMSA

K_{mn} = 1.84118 (TM_{11} mode)

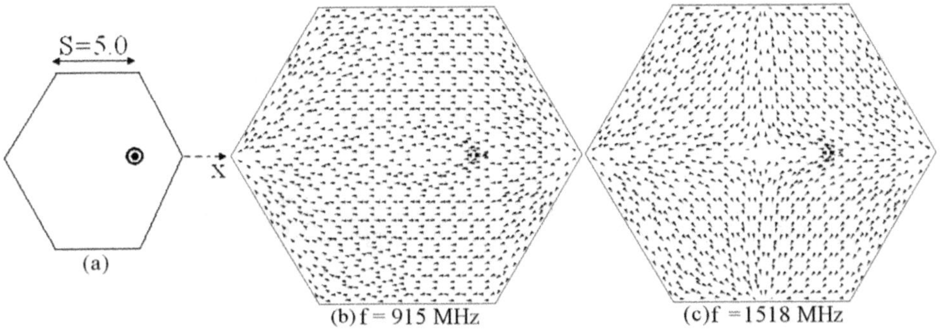

Fig. 1. (a) HMSA and its (b, c) Surface current distribution at first two frequencies

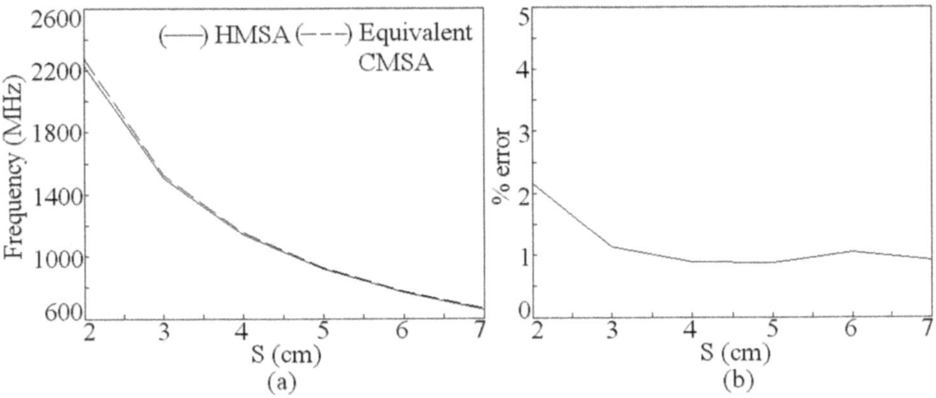

Fig. 2. (a) Resonance frequency and (b) % error plots for HMSA for different S

3 Compact HMSAs

As the HMSA shows similar modal variations to that of the CMSA, the compact HMSAs are realized by placing the shorting posts along the zero field line of the first mode of HMSA. The resonance frequency of HMSA is 915 MHz with a BW of 12 MHz. Since the field is symmetrical across the feed point axis, by removing the bottom half of HMSA, a compact half HMSA is realized as shown in Fig. 3(b). This HMSA operates at 928 MHz with a BW of 11 MHz. Since the area of the MSA is reduced its resonance frequency has increased. By placing the shorting posts along the zero field line of HMSA, shorted half HMSA is realized as shown in Fig. 3(c). The resonance frequency of this shorted HMSA is 926 MHz. By using the symmetry of the shorted half HMSA along the feed point axis a super compact half shorted half HMSA is obtained as shown in Fig. 3(d). This configuration operates at 936 MHz. The results for all these compact variations of HMSA are summarized in Table 1. The area reduction by a factor of 75% is realized super compact half shorted half HMSA. The radiation pattern for HMSA is in the broadside direction. The E and H-planes are aligned along

Table 1. Comparison between various compact HMSA configurations

HMSA shown in	f_r (MHz)	BW (MHz)	Area (cm^2)	Feed point (cm)
Fig. 3(a)	915	10	65	2.2
Fig. 3(b)	928	9	32.5	1.37
Fig. 3(c)	926	15	32.5	1.36
Fig. 3(d)	936	8	16.25	0.76

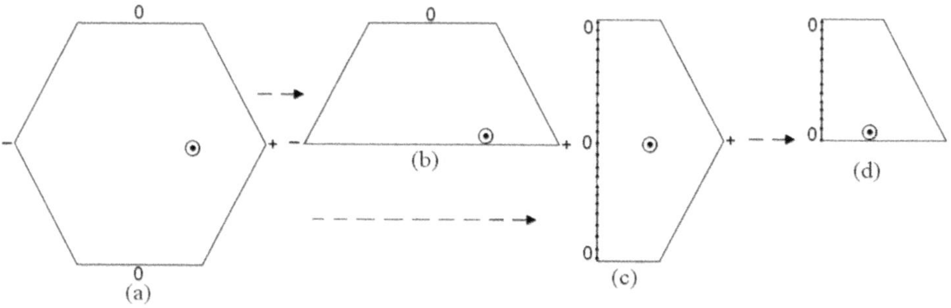

Fig. 3. (a) HMSA, (b) half HMSA, (c) shorted half HMSA and (d) half shorted half HMSA

$\Phi = 0^0$ and 90^0, respectively. In shorted MSAs, since the field distributions along the periphery of the shorted HMSA is changed the pattern shows higher cross polarization levels in the E and H-planes.

4 Dual Band H-shaped MSAs

The dual band pair of slots cut HMSA is shown in Fig. 4(a). The pair of slots is cut on one of the edges of the patch. The slots are taken to be nearly quarter wave in length. Since the shape of the HMSA resembles to the letter 'E', this MSA is called as E-shaped HMSA. The feed point is located on the other side of the slots. To optimize for the dual frequencies, the feed point position, slot dimensions and the separation between them is changed. The optimized return loss (S_{11}) plots are shown in Fig. 4(b). The simulated dual frequencies and BW's are 898 and 1186 MHz and 15 and 12 MHz, respectively. The response is experimentally verified and the measured frequencies and BW's are 890 and 1172 MHz and 14 and 13 MHz, respectively. The radiation pattern at the dual frequencies is in the broadside direction with cross- polarization levels less than 15 dB as compared to that of the co-polar levels. Similarly dual frequency response is realized by cutting the pair of slots inside the HMSA as shown in Fig. 4(c). The slot width is taken to be 0.2 cm. In this dual band HMSA the slot are taken to be nearly half wave in length. By optimizing the slot length the dual band response is realized and the simulated return loss plots are shown in Fig. 4(b). The simulated dual frequencies and BW's are 915 and 1009 MHz and 18 and 13 MHz, respectively. The measured frequencies and BW's are 901 and 1018 MHz and

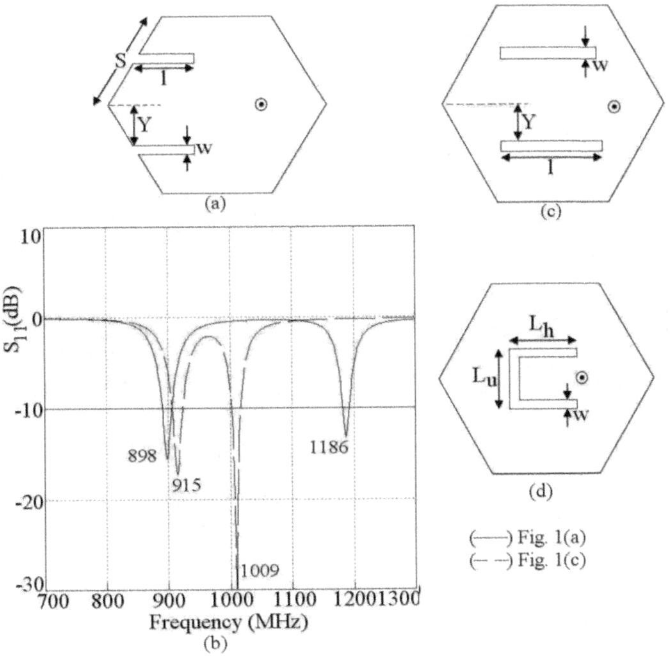

Fig.4. (a) E-shaped HMSA and its (b) simulated S_{11} plots, (c) Pair of slots cut HMSA and (d) U-slot cut HMSA

15 and 11 MHz, respectively. The radiation pattern at the dual frequencies is in the broadside direction with cross-polarization levels less than 20 dB as compared to that of the co-polar levels. Since the slots are cut symmetrical to the feed point axis this dual band HMSA shows lower cross polarization levels as compared to the dual band HMSA as shown in Fig. 4(a). The dual band HMSA can also be realized by cutting the U-slot in the centre of the HMSA as shown in Fig. 4(d). The inner U- slot length nearly equals half the wavelength.

5 Modal Analysis of Dual Band HMSAs

In dual band slot cut MSAs, the slot length at desired frequency is nearly taken to be quarter wave or half wave in length. However this simpler approximation does not give closer results. Therefore an in-depth analysis of slot cut MSAs is needed. The surface current distributions for the first two modes of the HMSA (Fig. 1(b, c)) were compared with the surface current distributions of E-shaped HMSA for slot length of 1.5 cm as shown in Fig. 5(a, b). It is observed that the pair of slots reduces the second order TM_{21} mode resonance frequency of the patch from 1518 to 1246 MHz and the second frequency is governed by modified TM_{21} mode. The first frequency (TM_{11}) is also slightly reduced from 915 to 900 MHz due to pair of slots.

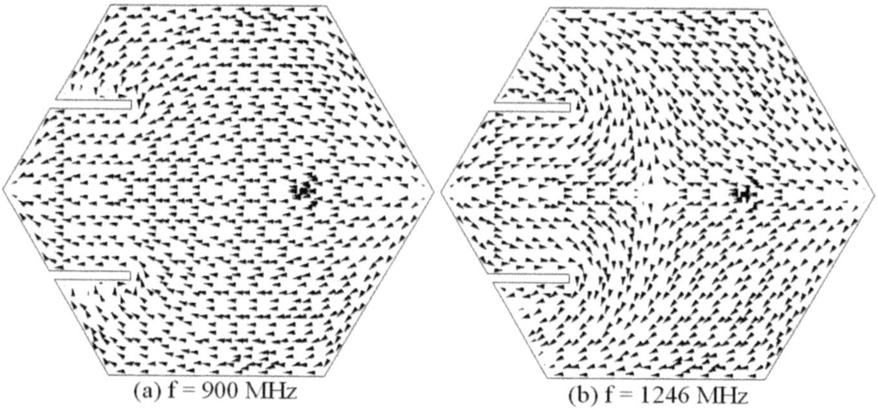

Fig. 5 (a, b) Surface current distributions at dual frequencies for E-shaped HMSA

6 Conclusions

The resonance frequency formulation for HMSA using its equivalence to CMSA is proposed. The frequencies calculated using the proposed formulation closely agrees with HMSA frequencies. The compact variations of shorted HMSAs are proposed. They give reduction in patch area with nearly the same resonance frequency. The dual band configurations of HMSAs by cutting the pair of rectangular slots or U-slot are proposed. The modal analysis for dual band E-shaped HMSA is carried out. It was observed that the slots do not introduce any mode but reduces the second order resonance frequency of HMSA and along with the fundamental mode realizes dual band response.

References

1. Kumar, G., Ray, K.P.: Broadband Microstrip Antennas. Artech House, USA (2003)
2. Wong, K.L.: Compact and Broadband Microstrip Antennas. John Wiley & sons,nc., New York (2002)
3. IE3D 12.1, Zeland Software, Freemont, USA

Higher Accuracy of Hindi Speech Recognition Due to Online Speaker Adaptation

Ganesh Sivaraman*, Swapnil Mehta**, Neeraj Nabar***, and K. Samudravijaya

School of Technology & Computer Science,
Tata Institute of Fundamental Research,
Homi Bhabha Road, Mumbai, India

Abstract. Speaker Adaptation is a technique which is used to improve the recognition accuracy of Automatic Speech Recognition (ASR) systems. Here, we report a study of the impact of online speaker adaptation on the performance of a speaker independent, continuous speech recognition system for Hindi language. The speaker adaptation is performed using the Maximum Likelihood Linear Regression (MLLR) transformation approach. The ASR system was trained using narrowband speech. The efficacy of the speaker adaptation is studied by using an unrelated speech database. The MLLR transform based speaker adaptation technique is found to significantly improve the accuracy of the Hindi ASR system by 3%.

Keywords: Automatic Speech Recognition (ASR), online speaker adaptation, Maximum Likelihood Linear Regression (MLLR), Hindi Speech recognition.

1 Introduction

Automatic Speech Recognition (ASR) systems provide a user-friendly interface to computers. ASR systems are being increasingly used in a variety of special tasks such as voice dialing in mobile phones, voice operated aids to handicapped, spoken document summarization, information retrieval etc. Yet, the accuracy of ASR systems is not high enough to be used widely by the general public as a replacement to a keyboard.

The foremost challenge facing ASR systems is the mismatch between the training and the testıng (actual use) conditions. For example, a system trained using speech data recorded in a quiet environment will have poor recognition accuracy when tested with speech data in presence of noise. Another challenge is caused by the variation of pronunciation of words by different people. Yet another variation in speech signal is caused by different voice characteristics of different people due to anatomical differences. Quality of telephone handsets and limited bandwidth of telephone channel aggravate the problem. In order to handle such varied changes in speech signal, an ASR system has to be trained well taking into consideration all possible pronunciations

* Dept. of Electrical & Electronics, BITS Pilani K.K. Birla Goa Campus.
** Dept. of Information Technology, National Institute of Technology – Durgapur.
*** Dept. of Electronics Engineering PVPP College of Engineering, Mumbai.

K. Shah, V.R. Lakshmi Gorty, and A. Phirke (Eds.): ICTSM 2011, CCIS 145, pp. 233–238, 2011.

for every word spoken by a large number of people. Thus, a huge amount of training data is essential for training a good speech recognition system.

1.1 Types of ASR

Based on the usage, ASR systems can be classified as Speaker Dependent (SD) and Speaker Independent (SI) systems. SD systems are trained to recognize the speech of only one particular speaker at a time. On the other hand, SI systems can recognize speech from anyone. Only one acoustic model is trained using training data from many speakers; this single model is used for recognition of speech by anyone whose voice it might not have 'seen'. SD systems have higher recognition accuracies than SI systems. Thus, it is preferable to use SD systems. However, in order to train SD systems, a large amount of **speaker specific** training data would be required which is not possible to have in case of a multi-user application such as telephonic enquiry system.

Speaker adaptation is a technique that uses a new speaker's voice sample to retrain a SI system to recognize the new speaker's speech better. However, it is not practical to demand the user to provide speech data for adapting the ASR system to his voice, in case of a voice interface, to be used by the general public (for example, information retrieval over telephone/mobile channel). A solution would be to use the spoken query (text spoken by the caller) to adapt acoustic models. However, as the enquiry of the user is short (just a few seconds), model adaptation has to be carried out with tiny amount of adaptation data. Since the transcription (text) of the query is not known, such an adaptation is called unsupervised speaker adaptation. Since the adaptation is carried out, on the spot, just using the caller's query speech, the approach is called online speaker adaptation.

While the advantage of online speaker adaptation has been studied for western languages [2,3,4], we are not aware of such a study for Indian language speech. This paper reports a study of improvement in the accuracy of a speaker independent, Hindi speech recognition system with the addition of an unsupervised online speaker adaptation module. The rest of the paper is organized as follows. A brief theory on speech recognition and speaker adaptation is given in section 2. Section 3 provides the experimental details. Section 4 deals with the experimental results and discussions. The conclusions are presented in section 5.

2 Theory of Speech Recognition and Speaker Adaptation

An overview of the basic concepts of Automatic Speech Recognition and speaker adaptation of acoustic models is provided in this section.

2.1 Automatic Speech Recognition

Speech Recognition is a specific case of pattern recognition. In pattern recognition, a set of reference patterns are stored and the test patterns are compared for matching with the reference patterns for recognition. The speech recognition system implemented here employs Hidden Markov Models (HMM) [1] for representing speech sounds. A HMM is a stochastic model; it does not store a set of reference patterns. A HMM consists of a number of states, each of which is associated with a probability density function. The parameters of a HMM comprises of the parameters of the set of probability density functions, and a transition matrix that contains the probability of transition between states. A lot of training data consisting of well-chosen application specific sentences are recorded from various people. Speech signals are analyzed to extract features useful for recognizing different speech sounds. These features and the associated transcriptions are used to estimate the parameters of HMMs. This process is called ASR system training. In case of Continuous Speech Recognition, the goal is to determine that sequence of words whose likelihood of matching the test speech is the highest. The training procedure involves the use of forward – backward algorithm. The recognition is done using Viterbi decoding.

2.2 Speaker Adaptation

Speaker adaptation is a technique which reduces the difference between training and testing conditions by transforming the acoustic models using a small set of speaker specific speech data. It is also necessary to have the correct word transcriptions for the adaptation data for robust adaptation. There are two types of speaker adaptation. In supervised adaptation, the text of the adaptation speech is known to the system. Supervised adaptation provides good improvement in the recognition rates as it is same as re-training the system. However, this process is not practical in a system designed for a telephonic enquiry that is used by practically everyone, and the user does not have the patience to provide enough adaptation data. Unsupervised adaptation is the approach adopted in this paper. The system automatically adapts itself to the present speaker at the same time as he keeps using the system. The system uses the first sentence spoken by a new speaker as the adaptation data. The (possibly incorrect) output of the SI system is assumed as the correct transcription. Using this transcription and the adaptation data, the system transforms its acoustic models in order to recognize the speech of the current user better. Then, it re-recognizes the unknown utterance with adapted models, hopefully resulting in better recognition accuracy. Since the speaker adaptation is carried out as and when a speaker is using the system, this approach is called online speaker adaptation.

A popular speaker adaptation method that needs small amount of data is Maximum Likelihood Linear Regression (MLLR). The MLLR method assumes that the SI acoustic models can be transformed into speaker adapted models by a simple linear transformation.

$$\mu_{new} = A_n \, \mu_{old} + b_n. \tag{1}$$

As shown in the above formula, the mean vectors of the SI models (μ_{old}) can be transformed linearly into the mean vectors of the adapted models (μ_{new}). The sound clusters are shifted and rotated to better represent the new speaker's voice. The matrix A_n and the vector b_n are the parameters to be found by maximizing the likelihood of the adaptation data [2,3]. Thus, MLLR method puts the linear regression into the Baum-Welch estimation framework. The MLLR transformation allows speaker adaptation of all the sounds even if there are only a few of them present in the adaptation data [4]. In this paper, we have used such a global MLLR transform because of scarcity of adaptation data.

3 Experimental Details

The experiment was carried out using a Hindi speech recognition system developed using the Sphinx speech recognition toolkit [5,6]. A tutorial on implementing a Hindi ASR system can be found at [7]. The basic acoustic units were context dependent phonemes (triphones) represented by 5-state, semi-continuous HMMs. The output probability distributions of states were represented by Gaussian mixture densities; 256 global Gaussian density functions were used to generate the Gaussian mixtures. The language model used in the system was trigram grammar.

3.1 Speech Databases

The experiment on speaker adaptation was carried out on a SI Hindi speech recognition system. The SI system was trained using the database containing 924 random sentences in Hindi. These sentences were spoken by 92 different speakers, of which 56 were males and 46 females. The total number of words in the dictionary was 2731. The speech data was recorded over telephone channel, and hence band limited to 4 kHz.

In order to test the efficacy of online speaker adaptation, we used another Hindi speech database, called the "rlwRes" database. This multi-speaker, continuous speech database consists of spoken queries related to railway reservation availability. The database consists of 1581 sentences spoken by 92 speakers, both male and female. The speech data was recorded using a desktop microphone system in a quiet environment. This wideband speech was down sampled to 8 kHz, and then used as test data.

3.2 Speaker Adaptation and Evaluation Methodologies

The online speaker adaptation experiment was carried out as follows: An utterance from the test set ("rlwRes" database) was recognized using the SI system. Then, the system was adapted to the test speaker's voice. From the test speech and the output of the SI system, the MLLR transformation algorithm generated the transformation matrices A_n and b_n. These matrices were used to transform the acoustic models on the fly, and thus adapt to the new speaker's voice. The same test utterance was re-recognized using the speaker adapted models. This process was repeated, one by one, for all the sentences of the "rlwRes" database. At the end, the recognition accuracies, before and after adaptation, were computed. The accuracy was calculated for percentage correct word matches and complete sentence matches.

The formula for calculation of percentage correct and accuracy are given as follows.

$$Percentage\ correct = (N - D - S)/N \times 100\,\% \tag{2}$$

$$Accuracy = (N - D - S - I)/N \times 100\,\% \tag{3}$$

where

- N = Number of words or sentences correctly recognized.
- D = Number of unrecognized/missed words (Deletion errors).

– S = Number of times a word was misrecognized as another word (Substitution errors).
– I = Number of extra words inserted between correctly recognized words. (Insertion errors).

4 Results and Discussion

The speech recognition accuracies before and after online speaker adaptation are presented in Table 1. The percentage of words and sentences correctly recognized before (i.e., without) and after online speaker adaptation are listed in the table. We observe that the MLLR transform based speaker adaptation gives an improvement of 3% in the percentage of correctly recognized words as well as sentences. It may be noted that the word recognition error has reduced from 15.5% to 12.6%, a reduction by a factor of 0.19. The corresponding relative error reduction for sentences is 0.06.

Table 1. Speech recognition accuracies with and without online speaker adaptation

Performance Measure	Without adaptation	After speaker adaptation	Relative error reduction
Words correct	84.5%	87.4%	0.19
Words accuracy	83.8%	86.8%	0.19
Sentences correct	50.6%	53.4%	0.06

The "rlwRes" database is unrelated to the database used for training the acoustic models. There was little overlap between speakers of the two databases. Yet, the online adaptation of acoustic models to the test speaker just using the test utterance (about 2-3 seconds duration) reduces the word error rate by a factor of 0.19. This is the benefit of online speaker adaptation. However, one should note that MLLR based speaker adaptation of acoustic models is a highly compute-intensive process. So, the approach described in this paper is practical only when speech recognition is performed on a powerful computer. In other words, this approach is not suitable for ASR in embedded systems.

Although this method of unsupervised adaptation yielded a 3% increase in word accuracy, the efficiency of this adaptation method crucially depends upon the accuracy of the base SI system. If the accuracy of the SI system is not good, the initial transcript provided by the SI system will be erroneous. Consequently, the system will adapt to the new speaker improperly, and there might not be much (or any) improvement in the recognition accuracy. In fact, online speaker adaptation led to mild deterioration in ASR accuracy when the accuracy of a complex, general purpose SI system was less than 70%.

5 Conclusion

The improvements in word and sentence recognition accuracies after online adaptation (using just one test utterance) shows that MLLR transform based speaker adaptation of speech models indeed decreases the recognition error by a factor of 0.19. This demonstrates that MLLR transform based adaptation transforms the acoustic models in such a way that the difference between test and train conditions is reduced, resulting in better performance. It is evident that it is possible to successfully adapt the system using just one sentence spoken by the speaker, provided sufficient computing resources are made available.

References

1. Rabiner, L.R.: A Tutorial on Hidden Markov Models and Selected Applications in Speech Recognition. Proceedings of the IEEE 77, 257–286 (1989)
2. Legetter, C.J.: Improved acoustic modeling for HMMs using linear transformations. Ph.D. Thesis. University of Cambridge (1995)
3. Leggetter, C.J., Woodland, P.C.: Maximum likelihood linear regression for speaker adaptation of HMMs. Computer Speech & Language 9, 171–185 (1995)
4. Doh, S.-J.: Enhancements to Transformation-Based Speaker Adaptation: Principal Component and Inter-Class Maximum Likelihood Linear Regression, PhD. Thesis, Carnegie Mellon University (2000)
5. CMU sphinx – Speech Recognition Toolkit, `http://www.cmusphinx.sourceforge.net`
6. Chan, A., et al.: The Hieroglyphs: Building Speech Applications Using CMU Sphinx and Related Resources (2003)
7. Samudravijaya, K.: Hindi Speech Recognition. J. Acoustic Society of India 29(1), 385–393 (2000)

ECG Feature Extraction Using Wavelet Based Derivative Approach

Anita P.[1] and K.T. Talele[2]

[1] U.M.I.T, Santacruz, Mumbai
anitanandu@rediffmail.com
[2] S.P.I.T, Andheri, Mumbai
talelesir@yahoo.com

Abstract. Many real-time QRS detection algorithms have been proposed in the literature. However these algorithms usually either exhibit too long a response time or lack robustness. An algorithm has been developed which offers a balance between these two traits, with a very low response time yet with performance comparable to the other algorithms. The wavelet based derivative approach achieved better detection. In the first step, the clean ECG signal is obtained. then, QRS complexes are detected and each complex is used to locate the peaks of the individual waves, including onsets and offsets of the P and T waves which are present in one cardiac cycle. The algorithm was evaluated on MIT-BIH Database, the manually annotated database, for validation purposes. The proposed QRS detector achieved sensitivity of 98.91% and a positive predictivity of 99.65% over the validation database.

Keywords: ECG, Beat Detection, P-QRS-T waves, Daubechies wavelets, Feature Extraction.

1 Introduction

The electrocardiogram (ECG) is a diagnostic tool that measures and records the electrical activity of the heart in exquisite detail. Interpretation of these details allows diagnosis of a wide range of heart conditions. A normal ECG can be decomposed in characteristic components, named the P, Q, R, S and T waves. The electrical activity of the heart is generally sensed by monitoring electrodes placed on the skin surface. The electrical signal is very small (normally 0.0001 to 0.003 volt). These signals are within the frequency range of 0.05 to 100 Hertz (Hz.) or cycles per second .Automatic analysis and classification of ECG wave can ease the burden of cardiologist and speed up the diagnosis. Hence the need for automation of this process to extract useful information in the intervals and amplitudes defined by its significant points (characteristic wave peaks and boundaries, especially for long recordings. The standard parameters of the ECG waveform are the P wave, the QRS complex and the T wave. But most of the information lies around the R peak. Additionally a small U wave (with an uncertain origin) is occasionally present. The different waves that comprise the ECG represent the sequence of depolarization and repolarization of the atria and ventricles [14].

K. Shah, V.R. Lakshmi Gorty, and A. Phirke (Eds.): ICTSM 2011, CCIS 145, pp. 239–247, 2011.

2 Filtering

The two major sources of noise are: biological and environmental. Biological noise includes muscle contraction/electromyography interference, baseline drift with, respiration where the sinusoidal component of frequency motion artifacts caused by changes in electrode skin impedance with electrode motion. Environmental includes power line interference which includes 50/60 Hz pick and its harmonics, electrode contact noise caused by loss of contact between the electrode and skin that disconnects the measurement system from subject. As stated by Chavdar Levkov, Georgy Mihov [1] subtraction procedure totally eliminates power line interferencefrom ECG signal. Subtraction procedure combined with wavelet having multiresolution characteristics and adaptive filtering gives a clean ECG signal.

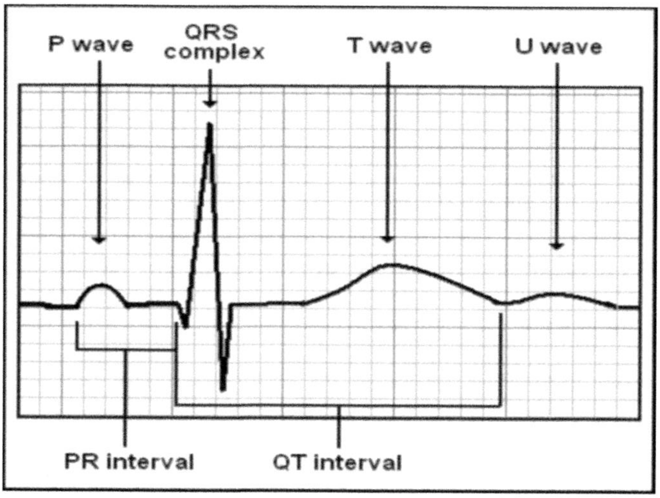

Fig. 1. Normal ECG signal

3 QRS Detection

A critical feature of many biological signals is frequency domain parameters. Time localization of these changes is an issue for biomedical researchers who need to understand subtle frequency content changes over time. To analyze the ECG signals, both frequency and time information are needed simultaneously. The short time Fourier transform gives a frequency-amplitude representation of the signal. The major drawback of the STFT is that it uses a fixed window width. For a compact representation using as few basis functions as possible, it is desirable to use basis functions that have a wider frequency spread as most biological signals do. Wavelet theory, which provides for wideband representation of signals [5, 6, 7, 8], is therefore a natural choice for biomedical engineers involved in signal processing. The Continuous Wavelet Transform (CWT) [3] is defined thus for a continuous signal, $x(t)$ or with change of variable as where $g(t)$is the mother or basic wavelet, * denotes

$$CWTx(\tau,a)=1/\sqrt{a}\int x(at)g^*(t-\tau)/a \quad (1)$$

a complex conjugate, *a* is the scale factor, and τ —a time shift. The Discrete Wavelet Transform (DWT), [5,6] which is based on sub-band coding, is found to yield a fast computation of Wavelet Transform. It is easy to implement and reduces the computation time and resources required. DWT is computed by successive low pass and high pass filtering of the discrete time-domain signal as shown in Figure 2. The signal is denoted by the sequence x[n], where n is an integer. The low pass filter is denoted by G0 while the high pass filter is denoted by H0. At each level, the high pass filter produces detail information; d[n], while the low pass filter associated with scaling function produces coarse approximations, a[n]. At each decomposition level, the half band filters produce signals spanning only half the frequency band. This doubles the frequency resolution as the uncertainty in frequency is reduced by half.

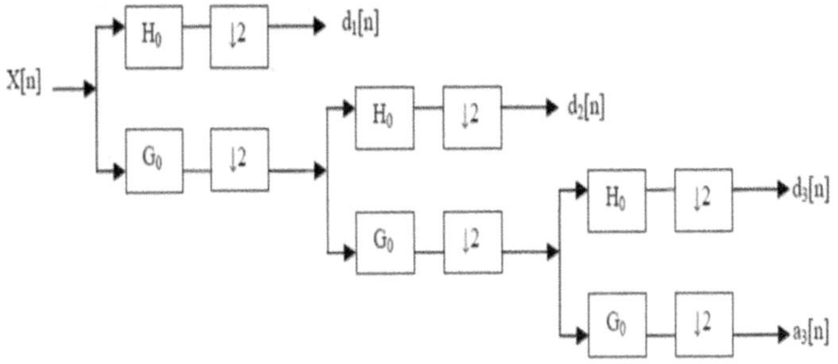

Fig. 2. Signal Decomposition using Discrete Wavelet Transform

The filtering and decimation process is continued until the desired level is reached. The maximum number of levels depends on the length of the signal. The DWT of the original signal is then obtained by concatenating all the coefficients, a[n] and d[n], starting from the last level of decomposition.

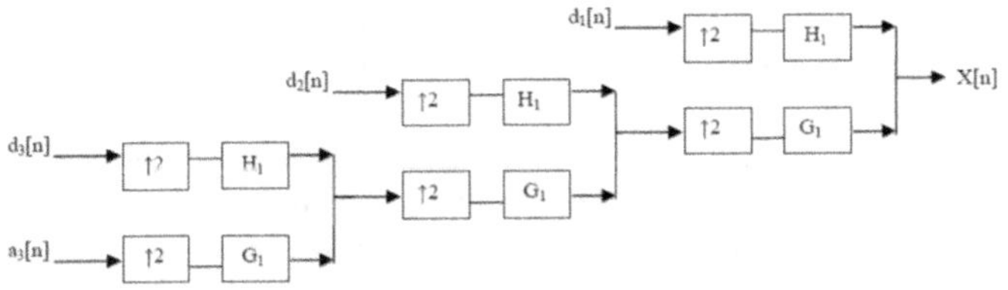

Fig. 3. Signal Reconstruction using Discrete Wavelet Transform

To determine the choice of wavelet, properties of the QRS were examined. There are three properties of ECG that are useful for detection of QRS complex.QRS has the highest slope; the shape of the signal is important; event is localized in time. Most of the energy of the ECG signal lies at scales 2^2, 2^3 & 2^4. For scales higher than 2^4, the energy of the QRS is very low. Only P and T waves have significant components at scale 2^5, but at this scale the influence of baseline wandering is very important. Consequently, we only use the first four scales $W_2\ kx[\mathrm{b}]$, $k = 1, 2, 3, 4$. A sharp change in the signal is associated to a line of maxima or minima across the scales. Q and S wave peaks have zero crossings associated in the WT, mainly at scales 2^1 and 2^2. P or T-like waves have their major component at scale $2^{4,}$ whereas artifacts produce isolated maximum or minimum lines, which can be easily discarded. If the signal is contaminated with high-frequency noise, like EMG noise ***(f)*,** the most affected scales are 2^1 and 2^2, but higher scales are hardly affected by this kind of noise. Baseline wander only affects slightly to scale 2^4 · The WT's across the scales show that the peak of the complex corresponds to the zero crossing between two modulus maxima of the WT at all the scales as shown in Figure 4.

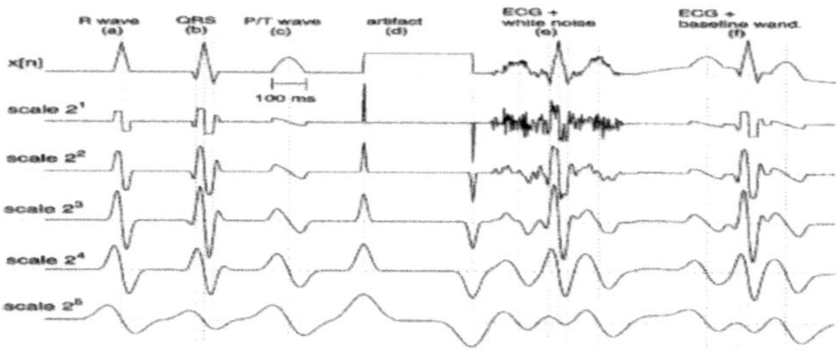

Fig. 4. Dyadic wavelet transform of a ECG –like simulated signal

For detecting the *R*-peak, the modulus maxima–minima pair is located for the lowest scale [7, 10], which is done by fixing a threshold for detection. The threshold is defined as the mean of extracted ECG signal levels. After reconstruction of the signals the signal is compared to the threshold value and signals exceeding this limit is defined as the R-peak. The frequency of the signal is calculated by dividing the length of the signal by the total time interval. The detected sample values of the R peaks and the calculated frequency are used to compute the time of occurrence of R peaks. The beat rate is derived as an inverse function of average R-R interval. Detection of R peaks in ECG signal is subsequently followed by application of the function of differentiation on the extracted QRS waves. In the first stage, the function is applied to QRS detected signal, generating a differentiated signal, namely y(n).The first and second order differentiated signal is used for extracting the onset and offset of QRS. The zero crossing for the wave detects the onset and offset points. QRS interval is thus found out. This gives an accurate estimation of widths of different QRS interval of the ECG.

4 P and T Wave Detection

The P wave and T wave can be found based on the location of the QRS complex.[13]. The slope threshold for P and T waves detection, pt-thresh, is determined by pt-thresh = pt-param * slope-thresh where pt-param is set between 0.1 to 0.3. The algorithm is to locate the onset, peak and offset of the P and T waves.The P wave offset is located by backward search in ECG data, starting from the QRS onset. The search distance is set based on the information that the PR interval is less than 0.2 second. If two consecutive data points having their slope (n) > pt-thresh, the first data point is identified as the P wave offset. The P wave onset is located by forward search from a far enough data point before the QRS onset to the just found P wave offset. If two consecutive data points having their slope (n) > pt-thresh, the first data point is identified as the P wave onset. Finally, between the onset and offset of the P wave, the peak can be easily identified by examining the amplitude of the ECG data. The T-wave's energy is mainly preserved between the scales 2^3 and 2^4. Therefore it was more appropriate to turn away from the dyadic scales and to choose the scale 10 for the WT. The next step consists of the search for modulus maxima. At scale 10 we analyze a signal and search for modulus maxima larger than a threshold. This threshold is determined by using the root mean square (RMS) of the signal between two R-peaks. J.P Martinez found that $\varepsilon = 0.25 \cdot RMS$ is suitable for detecting most of the T-peaks. When there are two or more modulus maxima with the same sign, the largest one is selected. After finding one or more modulus maxima, it is possible to determine the location and character of the T-wave. The first situation occurs when there is a modulus maxima pair with opposite signs. This indicates a small hill when the signs are +/- and a small inverted hill when the signs are -/+. When there is only one modulus maxima present, the + sign indicates a T-wave that consists only of a descend. When the sign is -, we see a T-wave formed by an ascend. The added advantage over other methods is that it can be easily extended to detect other abnormalities of the ECG signal.

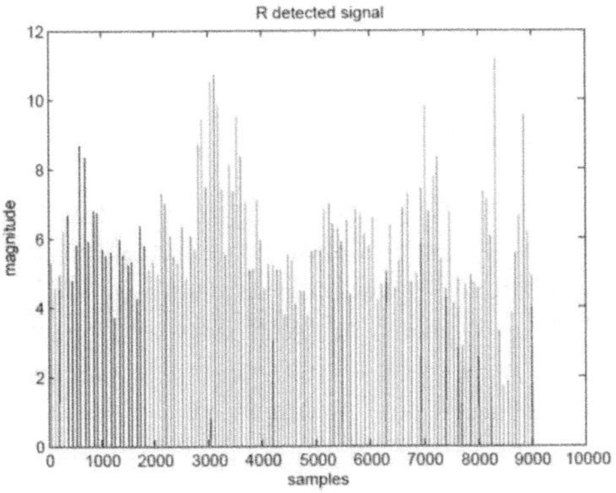

Fig. 5. R detected signal

5 Experimental Results

For validation purpose the MIT-Arrhythmia database was used. The test protocols described in EC57 standard [18], require that, for each record, the output of the device has been recorded in an annotation file (the "test annotation file"), in the same format as the reference annotation file for that record. The programs "bxb," was used to perform the comparisons between the test annotation files and the reference annotation file [18]. In order to test the beat detection algorithm which is related to R peak detection, an annotation file was generated. Three performance measures are commonly used to assess an analyzer performance: sensitivity, specificity, and positive predictive accuracy [38]. Sensitivity, *Se*, the ratio of the number of correctly detected events, to the total number of events is given by

$$Se=TP/TP+FN \quad (2)$$

Where TP (true positives) stands for correct identification of QRS complex present in the input test signal. *FN* (false negatives) represents a detector error of missed QRS complex identification that exists in the analyzed signal is the number of missed events. It gives the percentage of misdetections. The specificity, *Sp*, the ratio of the number of correctly rejected nonevents, *TN* (true negatives), to the total number of nonevents is given by

$$Sp=TN/TN+FP \quad (3)$$

where *FP* (false positives) is the number of falsely detected events or represents a detector error of QRS complex identification that doesn't exist in the analyzed signal. Positive predictive accuracy, *+P*, (or just positive predictivity) is the ratio of the number of correctly detected events, *TP*, to the total number of events detected by the analyzer and is given by

$$+P=TP/TP+FP \quad (4)$$

These performance measures actually are frequencies in a statistical sense and approximate conditional probabilities.

$$\text{Error rate: } ER= FN+FP\ /\ \text{Total no.of beats} *100\% \quad (5)$$

The positive predictivity for the wavelet based derivative approach yielded a value of 99.65.Only eight out of the 48 records tested did not achieve a predicitvity of 100% .Only record 232 attained a lowest predicitivity value of 95.18.The other records obtained values close to 98% and 99%. The QRS sensitivity gives the correctly detected QRS complexes. The sensitivity of record 203 is the lowest as the record is noisy and has QRS complexes of unusual form. Another low value is observed for record 207 .This record has an occurrence of atrial flutter and the QRS complex did not have a high slope. The R peaks for the other noisy records were obtained (record 108, 105) and record with high number of Premature Ventricular contractions (record 201 and 217) yielded good results. The average error rate was found to be 1.46. The wavelet based derivative method reduces the probability of error in the detection of QRS complex with less computational time as shown in Figure 5.3. Only record 203 and 207 showed a high error rate. Figure 5.4 shows the comparison between the false positives and false negatives .There were number of false negatives obtained as compared to the false positives .Only seven of the total records examined showed false positives. The number of false negatives observed were in the range of 0 to 2, except two records which showed false negative of 11(record 203) and false negative of 4 (record 207 & 232).

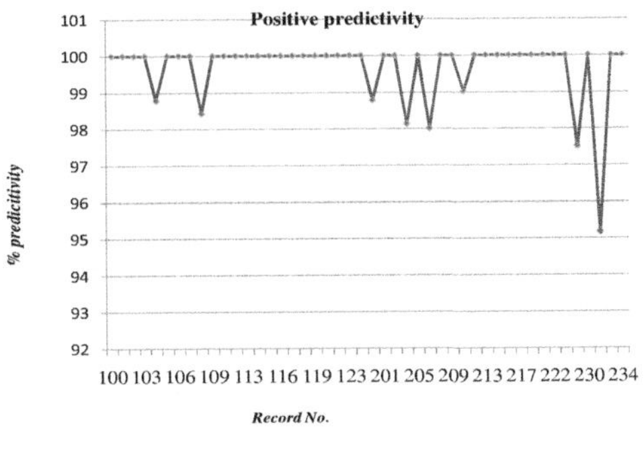

Fig. 6. Predictivity for wavelet based derivative approach

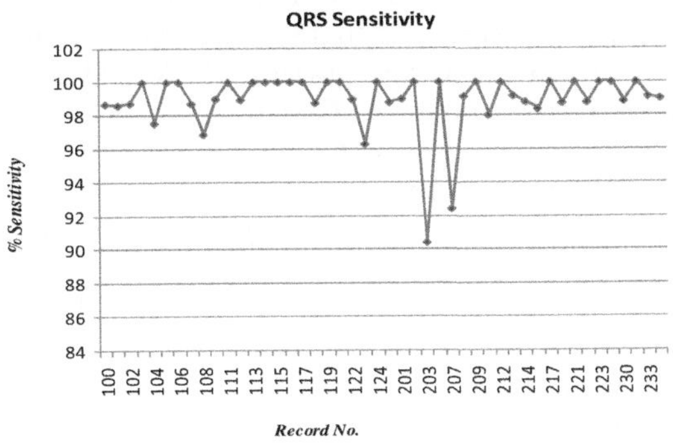

Fig. 7. Sensitivity for wavelet based derivative approach

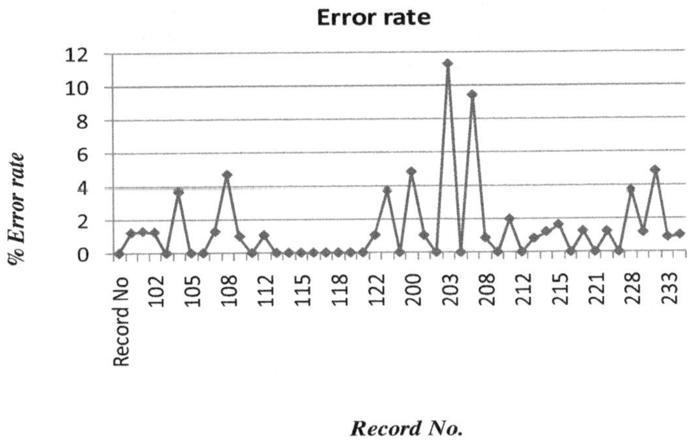

Fig. 8. Error rate for wavelet based derivative approach

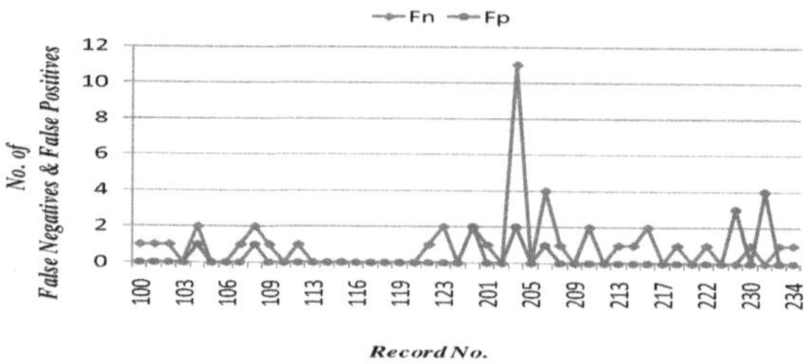

Fig. 9. False positives and false negatives for wavelet based derivative approach

6 Conclusion

Filtering of ECG signal corrupted with different sources of noise like power line interference, baseline wandering effect and high frequency noise incurred due to motion artifact and electromyogram signal was carried out. The subtraction procedure eliminates power-line interference from the ECG signal without affecting its spectrum. The procedure operates successfully even with amplitude and frequency deviations of the interference. The wavelet approach itself is very promising, because it is possible to evaluate different characteristics of the signal by using different scales of its wavelet transforms. . From the result, it can be seen that this method is having a comparatively higher sensitivity and nominal positive predictivity value. The algorithm treats each beat individually, hence the accuracy in measurement. The limitation of this method is that the computations required due to the calculation of WT. But the added advantage over other methods is that this can be easily extended to detect other abnormalities of the ECG signal, the wavelet based derivative approach has the least number of FN's indicating that it did not miss as many QRS complexes as other algorithms did. The main advantages of the method over existing techniques are its (1) robust noise performance and (2) its flexibility in analyzing non-stationary ECG data. Comparison shows clearly that QRS detection algorithms based on wavelet transform are very efficient .The QRS detector attained sensitivity of 98.91% and a positive predictivity of 99.65%. They are self-synchronized to the QRS steep slope and the heart rhythm, regardless of the resolution and sampling frequency used.

References

1. Levkov, C., Mihov, G., Daskalov, I., Dotsinky, I.: Removal Of Power-Line Interference from the ECG: A Review Of the Subtraction Procedure. Biomedical Engineering Online (2005)
2. Nunes, R.C., Filho, R.G., Rodrigues, C.R.: Adaptive ECG Filtering and QRS Detection Using Orthogonal Wavelet Transform, University of Santa Maria, CEP 97105900, Brazil (2005)
3. Sayadi, O., Shamsollahi, M.B.: Multiadaptive Bionic Wavelet Transform: Application to ECG Denoising and Basline Wandering Reduction. In: Biomedical Signal and Image Processing Laboratory, Sharif University of Technology, France (January 2007)

4. Kohler, B.U., Henning, C., Orglemeister, R.: The Principles of Software QRS Detection, Department of Electrical Engineering. In: Biomedical Electronics Group, Berlin University of Technology (February 2002)
5. Dinh, H.A.H., Kumar, D.K., Pah, N.D.: Wavelets for QRS Detection. In: Proceedings of the 23rd annual EMBS International Conference, Istanbul, Turkey (October 2001)
6. Kadambe, S., Murray, R., Bartels, G.F.: Wavelet Transform Based QRS Complex Detector. IEEE Transactions on Biomedical Engineering 46 (July 1999)
7. Kadambe, S., Murray, R.: Dyadic Wavelet Transform Based QRS Detector, Department of Electrical Engineering, University of Rhode Island, Kingston (1992)
8. Saritha, C., Sukanya, V., Murthy, Y.N.: ECG Signal Analysis Using Wavelet Transforms. Journal of Physics, 68–77 (2008)
9. Szilagyi, L., Szilagyi, S.M., Szlavec, A., Nagy, L.: On-Line QRS Complex Detection Using Wavelet Filtering. In: Proceedings of 23rd Annual EMBS International Conference, Istanbul, Turkey, October 25-28
10. Ranjith, P., Baby, P.C., Joseph, P.: ECG Analysis Using Wavelet Transform: application to myocardialischemia detection (January 2002), `http://www.sciencedirect.com`
11. Germán-Salló, Z.: Applications Of Wavelet Analysis in ECG Signal Processing, PhD Thesis, Technical University Of Cluj-Napoca (2005)
12. Jouck, P.P.H.: Application of the Transform Modulus Maxima method to T-Wave detection in cardiac signals, M.S. thesis, Maastricht University, December 20 (2004)
13. Tan, K.F., Chan, K.L., Choi, K.: Detection of QRS Complex, P Wave and T Wave in Electrocardiogram, Department of Electronic Engineering, City, University of Hong Kong
14. Clifford, G.D., Azuaje, F., McSharry, P.E.: Advanced Methods and Tools for ECG Data Analysis. Archtech House, London (2006)
15. Josko, A.: Discrete Wavelet Transform In Automatic ECG Signal Analysis. In: Instrumentation and Measurement Technology Conference, Warsaw, Poland (2007)
16. TheMIT-BIHArrhythmiaDatabase, `http://physionet.ph.biu.ac.il/physiobank/database/mitdb`
17. `http://www.physionet.org`
18. ANSI/AAMI EC57: Testing and reporting performance results of cardiac rhythm and ST segment measurement algorithms (AAMI Recommended Practice/American National Standard) (1998), `http://www.aami.org`; Order Code: EC57-293

Designing and Implementing Navigation and Positioning System for Location Based Emergency Services

Sonal N. Parmar

Electronics & Telecommunication Dept., MPSTME-NMIMS University
sonalnparmar2008@rediffmail.com

Abstract. This paper proposes and implements the Global Positioning System (GPS) in a very simple and efficient way for navigation, positioning and tracking. This system is designed to provide information about emergency services such as nearest Hospital, Police Station, Fire Station and Petrol Pump from the current location of the user. The presented GPS based hand held terminal is composed of low cost hardware components like Microcontroller and GPS Receiver. The main feature of this Navigation System is its ease of use along with the compact size and visual display of results. Subsequently a small computer interface can be added to the system to provide download of location details and superimposing the results on to a GIS mapping system to provide On-Line tracking.

Keywords: Context-aware services, Emergency based services, Location based services.

1 Introduction

The GPS is a network of satellites that continuously transmit coded information, which is useful to precisely identify location on earth by measuring distance from the satellites [1]. Our main aim is to design the GPS in an effective way that allows navigating, positioning and tracking and provide emergency services information. The system proposed in this paper utilizes the GPGGA string format captured by the GPS receiver for providing exact location.

Section 2 provides details of related work. System design is discussed in section 3 and section 4 shows the different modes of operation used for navigation. System execution methodology is described in section 5 and section 6 summarizes empirical results and analyses.

2 Related Works

Many researchers have recognized the advantages of GPS for various applications. However, many of them have focused on its application like Localization, Tracking

K. Shah, V.R. Lakshmi Gorty, and A. Phirke (Eds.): ICTSM 2011, CCIS 145, pp. 248–253, 2011.

and Navigation. The effectiveness of GPS electronic navigation has been proposed in [2]. A localization System based on Enhanced Software GPS receiver has been introduced in [3]. The Enhanced Software GPS receiver is making use of hardware for RF Front End and Analog to Digital Converter, and Software approach for acquisition, tracking and navigation solution calculation [4, 5, 6]. An Automobile localization system has also been designed using GSM services for transmitting the position and localization of automobile to the owner on his mobile phone as a short message (SMS) at his request [7,8, 9]. A simple expandable GSM and GPS systems simulator allows user to learn advantages of modern wireless technology and use it in many complicated solutions [10, 11].

3 System Design

Following Figure 1 shows the block diagram of GPS Based Hand Held Navigator System. The system is composed of hardware as well as software modules.

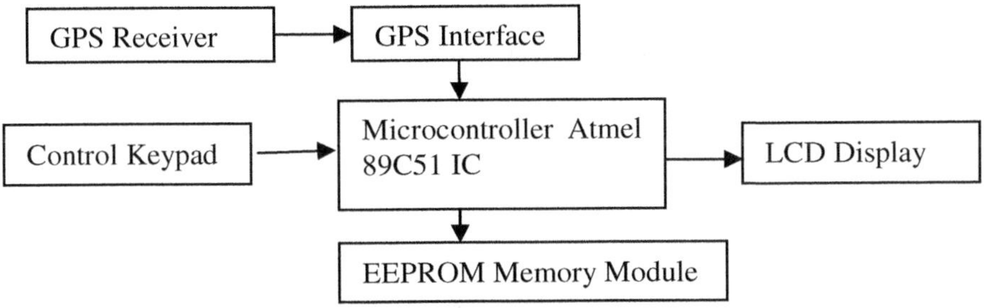

Fig. 1. GPS Receiver block diagram

- *Hardware Modules:* The hardware part is designed using following modules as described below

    - *Microcontroller*: All ports are used for the interfacing devices like GPS receiver, Data bus, keypad and latches
    - *EPROM Memory Module*: The memory module is used for the purpose of sorting location data along-with its positioning information
    - *LCD display and Keypad*: The LCD module is used for the purpose of showing the results as well as the display of menu screen to navigate through the system.
    - *GPS receiver and Interface*: The GPS receiver module generates information strings [12].
    - *Front Panel key interface:* Total 32 Key are connected to the microcontroller.

- *LCD Panel interface:* The LCD panel is connected to the Microcontroller through Latch 3 operating as LCD control port and Latch 4 operating as LCD Data port.

The development of embedded software required for implementing the required functionality is carried out in Embedded C and Assembly language of MC551 family.

4 Modes of Operation

The system has following modes of operation:

- *GPS Sense Mode- A Default Mode:* GPS receiver captures NMEA Strings carrying position information. User can switch between following types as tabulated in Table 1.
- *Calibration Mode:* The system records the location with its position co-ordinates. Figure 2 illustrates the program flow of GPS operations.

Table 1. Types used in GPS Sense Mode

Key No.	Type Name	Type Name used
0	Deleted Entry	DEL
1	Police Station	PST
2	Hospital	HSP
3	Bus Station	BUS
4	Railway Station	RLY
5	Fueling Station	FST

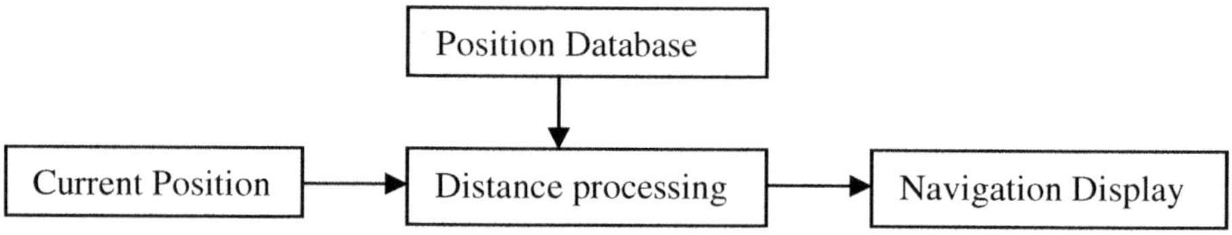

Fig. 2. GPS operational flow

5 System Execution Methodology

The Software main flow chart is shown in Figure 3. Figure 4, 5 and 6 illustrates the Pseudo Code for Test Entry, Calibration Mode and Run Mode.

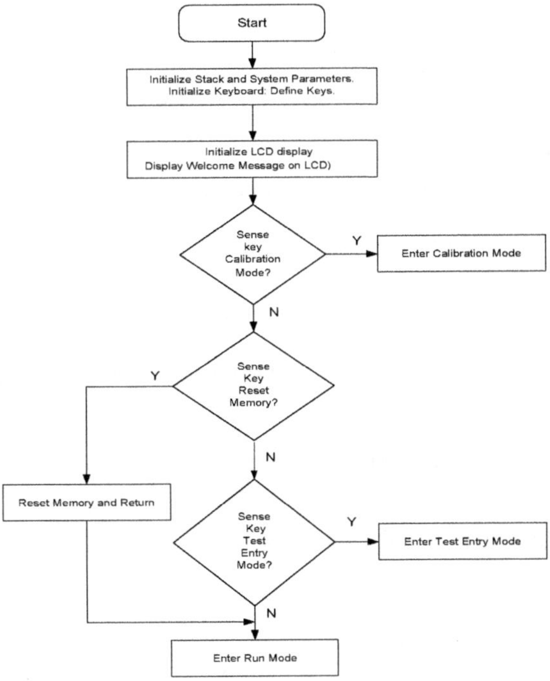

Fig. 3. System Execution Flow Sequence

The Pseudo Code for Test Entry Mode is as follows:

Initialize Variables for Test Entry Mode
Insert 16 Test Entries in the Database
End Test Entry Mode

Fig. 4. The Pseudo code for Text entry mode

The Pseudo Code for Calibration Mode is as follows:

Initialize Calibration parameters, Display Calibration mode input screen
Accept Point Details: Type
Accept Point Latitude/Direction
Accept Point Longitude/Direction
Accept Point Details: Name
Check for One More Entry?
Initialize Screen for next entry OR Exit Calibration Mode

Fig. 5. The Pseudo Code for Calibration Mode

Pseudo Code for Run Mode is as follows:

Initialize Parameters for Run Mode Display
Capture GPGGA String from GPS Receiver
Process the GPS String
Compare it with Memory Database Display information of the Point
Check for, Add this Point as Entry
Store the Point as New Entry and Continue Capturing GPGGA String from GPS Receiver OR Continue Run Mode
Check for Continue Run Mode
Continue Capturing GPGGA String from GPS Receiver OR End Run Mode :Shutdown

Fig. 6. The Pseudo Code for Run Mode

6 Empirical Results and Analyses

To evaluate the performance of the proposed handheld device, the real time results were taken by GPS based Hand Held Navigator System. The tested results are tabulated in Table 2. The locations tabulated in fourth column of table are real time locations but for brevity they are written in symbolic form.

This system can be used for geographical surveys for generating the Longitude, Latitude and Altitude information for the required locations. This system can be used for In-Motion Tracking system. With a front end application build on this, it can serve as Fleet Management system to manage large fleets of vehicles and keep a central control for tracking and guiding the vehicles across the country or anywhere in the world.

Table 2. Main performance index of GPS based Handheld Terminal

Sr. No.	Latitude North (N)	Longitude East (E)	Actual Point/ Location Name
1	21.091340	79.049504	A
2	21.110528	79.035799	B
3	21.152307	79.088947	C
4	21.157246	79.076565	D
5	21.121616	79.030762	E
6	21.113660	79.005641	F
7	21.139253	79.082424	G
8	21.122557	79.048824	H
9	21.124709	79.042980	I
10	21.130927	79.067218	J

References

1. Online Document The Global Positioning System, http://www.trimble.com/gps/index.html
2. Wright, M., Stallings, D., Dunn, D.: The Effectiveness of Global Positioning System Electronic Navigation. In: IEEE Southeastcon 2003. Bridging the Digital Divide. Renaissance Jamaica Granade Resort Ocho Rios, St. Ann, Jamaica WI (April 2003)
3. Lita, I., Visan, D.A., Popa, I.: Localization System Based on Enhanced Software GPS Receiver. In: ISSE 2006, 29th International Spring Seminar on Electronics Technology, Nano technologies for Electronics Packaging, May 2006, International Meeting Centre, St. Marienthal Germany (2006)
4. Rinder, P., Bertelsen, N.: Design of a single frequency gps software receiver. Aalborg university (2004)
5. Bao, J., Tsui, Y.: Fundamentals of Global Positioning System Receivers A Software Approach. John Wiley & Sons, Inc., Chichester (2000) ISBN 0-471-38154-3
6. Zheng, S.Y.: Signal acquisition and tracking for software GPS receiver. Blacksburg Virginia (February 2005)
7. Lita, I., Cioc, I.B., Visan, D.A.: A New Approach of Automobile Localization System Using GPS and GSM/GPRS Transmission. In: 29 International Conference Spring Seminar on Electronics Technology, ISSE 2006, May 2006, International Meeting Centre, ST. Marienthal, Germany (2006)
8. ETSI GSM 03.40: Digital cellular telecommunication system (phase 2+), Technical realization of the Short Message Service (SMS)
9. ETSI GSM 04.11: Digital cellular telecommunication system (phase 2+), Point to Point (PP) Short Message Service (SMS) support on mobile radio interface
10. Pochmara, I.J., Palasiewicz, J., Szablata, P.: Expandable GSM and GPS Systems Simulator. In: 17 International Conference on Mixed Design of Integrated Circuits and Systems, MIXDES 2010, June 2010, Wroclaw, Poland (2010)
11. Ardalan, A., Awange, J.L.: Compatibility of NMEA GGA with GPS Receivers Implementation 3(3), 1–3 (2000)

A New Method for Matching Loop Antenna Impedance with the Source for ISM Band

K.P. Ray[1], Uday P. Khot[2], and L.B. Deshpande[3]

[1] SAMEER, IIT campus, Powai, Mumbai-400 076, India
[2] Thadomal Shahani Engineering College, Mumbai-400 050, India
[3] S.B.M.Poly., Vile Parle (W), Mumbai-400056, India
kpray@rediffmail.com, udaypandit@rediffmail.com

Abstract. A new method for matching loop antenna impedance with the 50 ohm source for ISM band using parallel-wire transmission line has been proposed. Design of length and width of the parallel-wire transmission line for various loop configurations of 1λ perimeter at 915 MHz has been presented. Variation of impedance with frequency of these configurations has also been studied. Experiments have been performed to measure impedance, impedance bandwidth, and radiation patterns of these configurations, which tally well with simulated results. Proposed impedance matching method gives wide and symmetrical impedance bandwidth at VSWR = 2.

Keywords: Feed transmission line, ISM band, loop antenna.

1 Introduction

Many authors have presented analytical study of circular loop antenna. Measurements of its input impedance and radiation efficiency had been reported [1], [2]. Simulation data on input impedance and impedance bandwidth of polygonal loop antenna have been reported [3], [4]. However, data on experimental measurements have not been reported. This is probably because impedance matching using balun or matching network seems to fail as the small loop is highly inductive and difficult to match to the feed line [5]. Therefore, design and application of the polygonal loop antenna has received much less attention. However, with the operating frequencies of wireless communication moving into higher bands and the proposed impedance matching technique, the loop becomes a viable element for this application as the loop perimeter approaches 1λ.

In this paper, a new method for matching loop impedance with the 50 ohm source for ISM band using parallel-wire transmission line has been proposed. Design of length and width of the parallel-wire transmission line for various loop configurations of 1λ perimeter at 915 MHz using NEC has been presented. Variation of impedance with frequency of these configurations has also been studied. Experiments have been performed using the proposed impedance matching method to measure impedance, impedance bandwidth, and radiation patterns of these configurations, which agrees with simulated results. Unlike the balun or matching networks, the parallel-wire

K. Shah, V.R. Lakshmi Gorty, and A. Phirke (Eds.): ICTSM 2011, CCIS 145, pp. 254–259, 2011.

transmission line presented in this paper is simple and more attractive as it gives wide and symmetrical impedance bandwidth at VSWR = 2.

2 Parallel-Wire Transmission Line

A parallel-wire transmission line consists of two similar conducting circular rods separated by a dielectric (or air) as shown in Fig. 1. The characteristic impedance for such a parallel-wire line is given by [6],

$$Z_o = \frac{276}{\sqrt{\varepsilon_r}} \log \frac{2D}{d} \quad (1)$$

Since D is large compared to d, the characteristic impedance of this line are typically few hundred ohms. It is, therefore, proposed to use this line with an air dielectric to match the antenna impedance with the 50 ohm source.

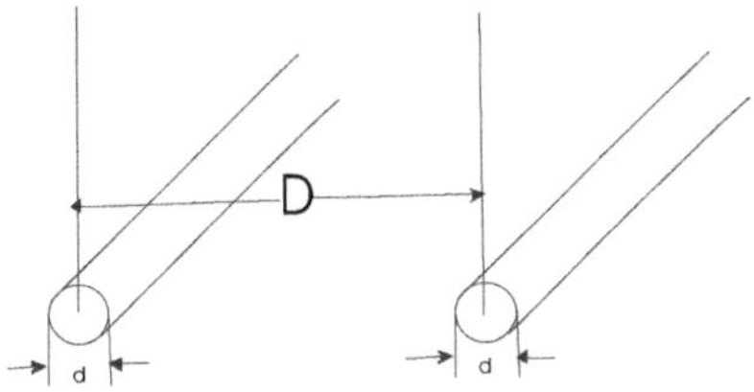

Fig. 1. Parallel-wire transmission line with air dielectric

3 Impedance Matching Using Parallel-Wire Transmission Line

A parallel-wire transmission line consisting of two similar copper wires separated by air dielectric as shown in Fig. 2 is used for matching antenna impedance (Z_A) with the 50 ohm source impedance. Using a parallel-wire transmission line of length `l' and width 'D' the load impedance $Z_L = Z_A$ is transformed into impedance Z_{in} such that $Z_{in} = 50$ ohms.

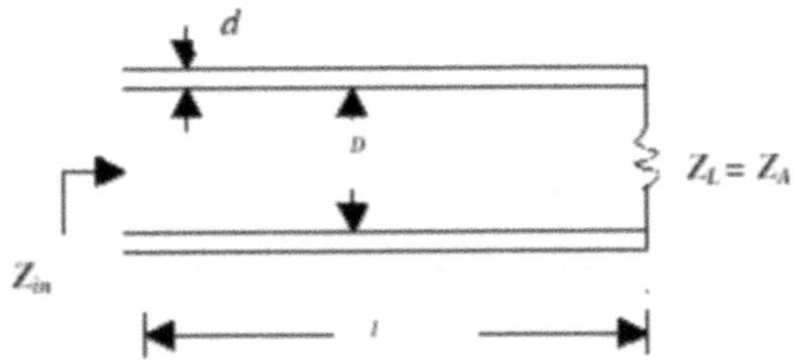

Fig. 2. Parallel-wire transmission line used for matching antenna impedance (Z_A) with source impedance of 50 ohms

4 Design of Parallel-Wire Transmission Line as an Antenna Feeder

The parallel-wire transmission line used as a part of antenna structure for feeding power to the loop antenna of 1λ perimeter is designed using the NEC software and is shown in Fig. 3.

Using NEC the length '*l*' and the spacing between the two parallel conductors '*D*' are computed for various loop shapes of 1λ perimeter at 915 MHz so that the antenna impedance Z_A at A-A' is transformed into Z_{in} = 50 ohms at B-B' with VSWR nearly equal to one. With VSWR nearly equal to one, Z_A is matched with 50 ohm source impedance and there is no reflection from the antenna. The design is verified using impedance transformation equations as well as experimentally, and shows good agreement.

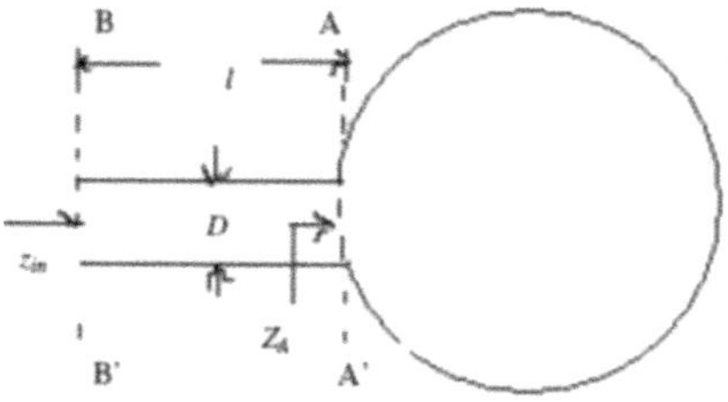

Fig. 3. Parallel-wire transmission line as a part of the antenna structure

The simulated values of length and width of transmission line simulated and measured values of VSWR and Z_{in} for various loop antennas of 1λ perimeter at 915 MHz are summarized in Table 1.

Table 1. Simulated values of length and width of transmission line, simulated and measured values of VSWR and Z_{in}

Sr. No.	Antenna Configuration	Length of trans. line '*l*'	Width of trans. line '*D*'	VSWR using NEC	Z_{in} using NEC	VSWR meas.	Z_{in} meas.
1	Circular loop	4 cm	1 cm	1.17	58.3 + j2.31	1.15	49.7 – j4.5
2	Rectangular loop with γ = 0.5	6cm	0.3cm	1.11	55 + j1.96	1.10	50.7+j1.18
3	Square loop	4.3 cm	0.4 cm	1.03	50.3 – j1.49	1.05	51.3 - j2.2

5 Experimental Verification

For carrying out experimental investigations, various loop configurations with 1λ perimeter (excluding feeder) at 915 MHz were fabricated using a copper wire of

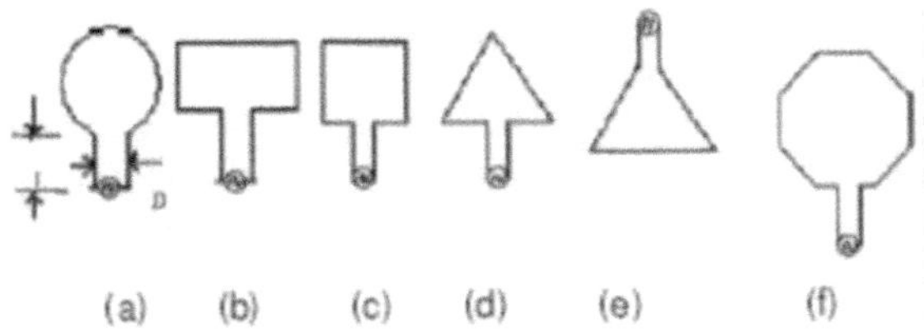

Fig. 4. Loop configurations fabricated for experimental investigations (a) Circular loop (b) Rectangular loop with γ = 0.5 (c) Square loop

2 mm diameter as shown in Fig. 4. The design parameters '*l*' and '*D*' of the feeder were computed using NEC.

An N-type connector was used for feeding power to the antenna from the source. Experimental investigations were carried out to determine the properties of loop configurations such as input impedance, impedance bandwidth, and radiation pattern and are shown in Fig. 4. Loop configurations shown in Fig. 4 were fabricated and excited using a Hewlet Packard source for the measurement purpose. Initially the source was calibrated by connecting open, short, and standard load.

5.1 Investigations on Input Impedance

The input impedance of each loop was measured by sweeping the frequency of source by 75 MHz on either side of the design frequency of 915 MHz i.e. span of 150 MHz. Measured input impedance plots for all the loops have been compared with simulated plots on the smith chart and are given in Fig. 5.

The simulated and measured values of input impedance at 915 MHz for all the three loop configurations are compared in Table 1. It is noticed that the simulated and measured values of Z_{in} tally with a high degree of accuracy and are very close to 50 ohms.

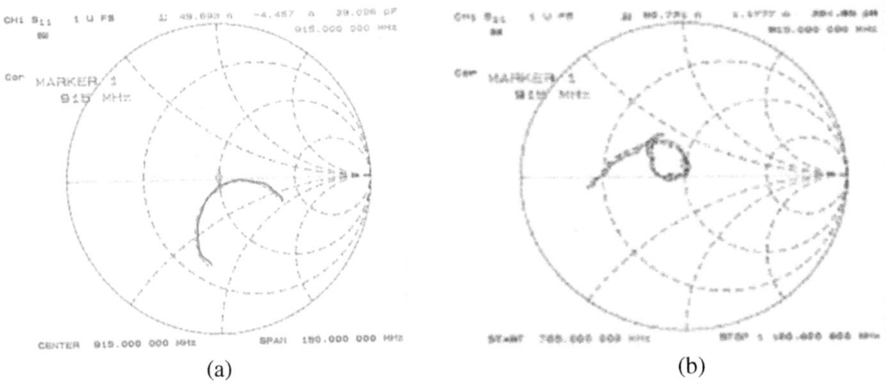

Fig. 5. Comparison of measured and simulated input impedance plots on smith chart (a) Circular loop (b) Rectangular loop with γ = 0.5(——Measured and --------- Simulated)

5.2 Investigations on Impedance Bandwidth

Impedance bandwidth at VSWR = 2 was measured for all the three configurations, which are compared with respective simulated bandwidth in Table 2. It is noticed that bandwidth measured agrees within ± 2% accuracy with simulated values of bandwidth.

Table 2. Comparison of VSWR = 2 bandwidth of various loop configurations with 1λ perimeter

Sr. No.	Antenna Configuration	Simulated Freq. range '*fs*' for VSWR< 2 (MHz)	Measured Freq. range '*fm*' for VSWR<2 (MHz)	%Error $\left\{\frac{fs-fm}{fm}\right\}*100$
1	Circular loop	870 to 957	880 to 967	-1.13 to -1.03
2	Rectangular loop with γ = 0.5	830 to 1120	845 to 1100	-1.77 to 1.81
3	Square loop	870 to 975	863 to 960	0.81 to 1.56

5.3 Investigations on Radiation Pattern

For doing investigations on radiation pattern, two loop antennas, one as transmitting and one as receiving, were placed at a height of 8 feet above the ground. The distance between the two antennas was kept to be 20 feet.

Normalized measured and simulated elevation and azimuth radiation patterns of all the three loop configurations with 1λ perimeter at 915 MHz are compared and due to the space constrain, the results for only circular loop have been shown in Fig. 6. It is

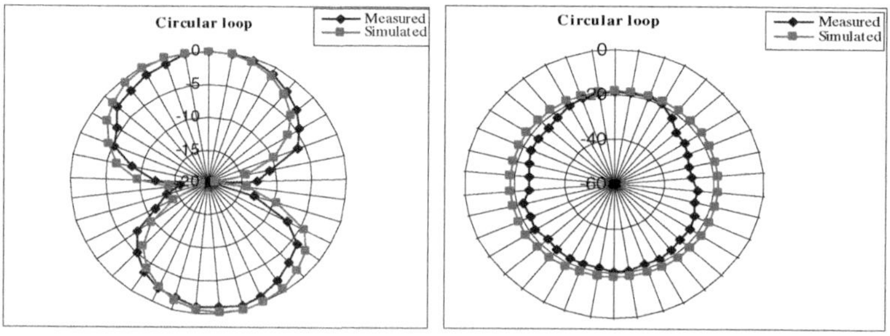

Fig. 6. Comparison of measured and simulated azimuth and elevation radiation pattern of loop antenna

observed that the elevation radiation pattern is omni directional, whereas the azimuth radiation pattern is close to figure of ‘8’ because of maximum radiation in broad-side direction i.e. at 90° to the plane of antenna as the perimeter is 1λ.

6 Conclusions

A new method for matching of antenna impedance to the 50 ohm source using parallel–wire transmission line as part of the antenna structure has been devised. Experimental results obtained using this method agrees well with simulated results. Impedance bandwidth at VSWR = 2 was measured for three configurations, which are compared with respective simulated bandwidth. It is noticed that bandwidth measured agrees within ±2% accuracy with simulated ones.

Normalized measured and simulated elevation and azimuth radiation patterns of all the three loop configurations with 1λ perimeter at 915 MHz are compared. It is observed that the elevation radiation pattern is omni directional, whereas the azimuth radiation pattern is close to figure of ‘8’ because of maximum radiation in broad-side direction i.e. at 90° to the plane of antenna as the perimeter is 1λ.

The measured radiation pattern for all the loop configurations agrees with the simulated ones. The minor differences in measured and simulated radiation pattern are because of reflections from the nearby objects.

References

1. Kennedy, P.A.: Loop Antenna Measurements. IRE Trans. Antenna Propagat, 610–618 (1956)
2. Boswell, A., Taylor, A.I., White, A.: Performance of A Small Loop Antenna in The 3-10 MHz Band. IEEE Antennas Propagat Magazine 47(2), 51–55 (2005)
3. Awadalla, K.H., Sharshar, A.A.: A Simple Method to Determine The Impedance of A Loop Antenna. IEEE Antennas Propagat Magazine 32(11), 1248–1251 (1984)
4. Tsukiji, T., Tou, S.: On Polygonal Loop Antennas. IEEE Trans. Antenna Propagat. 28(4), 571–574 (1980)
5. Jensen, M.A., Samii, Y.R.: Performance Analysis of Antennas for Hand-Held Transceivers using FDTD. IEEE Trans. Antenna Propagat. 12(8), 1106–1112 (1994)
6. Shevgaonkar, R.K.: Electromagnetic Waves, 1st edn. Tata McGraw-Hill, New York (2006)

Optimized Carry Look-Ahead BCD Adder Using Reversible Logic

Kanchan Tiwari, Amar Khopade, and Pankaj Jadhav

Department of Electronics & Telecommunication
M.E.S. College of Engineering, Pune-01, Maharashtra
kanchan.s.tiwari@gmail.com,
amarkhopade20@gmail.com,
jadhavpankaj275@gmail.com

Abstract. In this paper, we propose design of the BCD adder using reversible logic gates. Reversible Logic plays important role in CMOS, quantum computing, nanotechnology and optical computing. The proposed circuit can add two 4-bit binary variables and it transforms the result into the appropriate BCD number using efficient error correction modules. We show that the proposed reversible BCD adder has lower hardware complexity and it is much better and optimized in terms of number of reversible gates and garbage outputs with compared to the existing counterparts. Further we have validated our design by showing our synthesis and simulation results.

Keywords: Reversible Logic, quantum computing, BCD adder, garbage outputs, VHDL.

1 Introduction

With new technology nowadays conventional systems are becoming more complex due to rapid increase in number of transistors which results in increase in power dissipation. While computing logical operation information about input is lost. Landauer [1] has posed the question of whether logical irreversibility is an unavoidable feature of useful computers, arguing that it is, and has demonstrated the physical and philosophical importance of this question by showing that whenever a physical computer throws away information about its previous state it must generate a corresponding amount of entropy. Therefore, a computer must dissipate at least kTln2 of energy (about 3 X 10^-21 joule at room temperature) for each bit of information it erases or otherwise throws away, where k is Boltzmann's constant and T is absolute temperature[2]. A gate is reversible if Boolean function it computes is Bijective. Bijective means one-to-one and onto; or, for those of us who forget our mathematics terminology, there must be the same number of inputs as outputs, and for each output value there is a unique input value that leads to it. A reversible circuit should be designed using minimum number of reversible gates. One key requirement to achieve optimization is that the designed circuit must produce minimum number of garbage outputs. Also they must use minimum number of constant inputs [3, 4].

K. Shah, V.R. Lakshmi Gorty, and A. Phirke (Eds.): ICTSM 2011, CCIS 145, pp. 260–265, 2011.

Conventional irreversible systems lose information while computing. Erasure of is not necessary for computation. Reversible systems are used to conserve information. In this Paper we are dealing with design and implementation of carry look-ahead BCD adder using standard reversible logic gates.

- **Garbage Outputs:** outputs that are not used as primary outputs are termed as garbage outputs
- **Constants:** constants are the input lines that are either set to zero(0) or one (1) in the circuit's input side
- **Gate Count:** number of gates used to realize the system
- **Hardware Complexity:** refers to the number of basic gates (NOT, AND and EXOR gate) used to synthesize the given function

1.1 Synthesis of Reversible Circuits

Synthesizing a circuit with reversible gates is different from synthesizing an irreversible circuit. The main differences are the following

- The number of inputs and outputs are equal.
- We should try to produce minimum number of garbage outputs.
- Neither feedback nor fan-out is allowed in reversible logic; every output can be used only once.

Consequently, in the proposed circuits the constraints are carefully abided.

2 Basic Reversible Gates

2.1 Feynman Gate

It is very popular in field of quantum computing for performing XOR operation. Besides XOR it is used to make copies of signals which are often necessary to avoid fan out. We can obtain same signal at both outputs if we give zero to second input [5].

Fig. 1. Feynman gate

2.2 Toffoli Gate

Toffoli gate [6] has both applications of making copies as well as AND operation. In the main design of paper many Toffoli gates are used to perform XOR/AND operations.

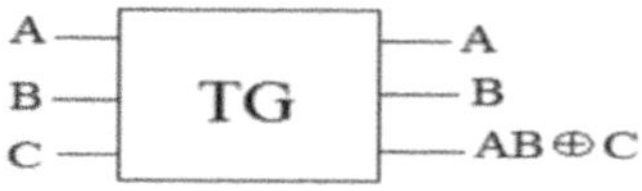

Fig. 2. Toffoli gate

2.3 Peres Gate

This gate can be used to form one bit full adder. Using Peres we are capable of performing both XOR/AND with less garbage outputs [7].

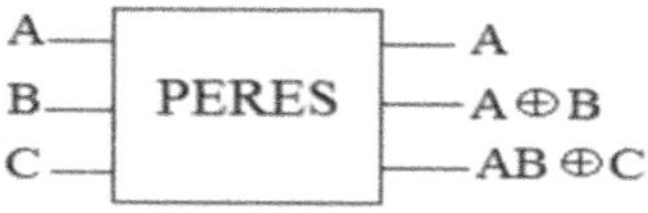

Fig. 3. Peres gate

2.4 HNG Gate

HNG gate [11] can also act as one bit adder. It is working as (4,4) reversible gate. Generally used when garbage values has to reduced and many functions are to be performed on single gate.

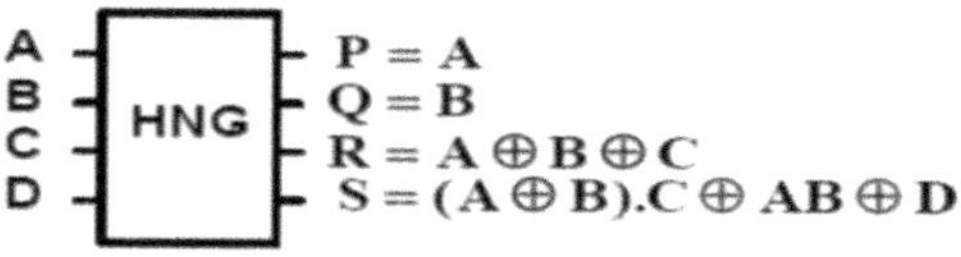

Fig. 4. HNG gate

3 Proposed Circuit

Peres and Feynman are used to get generate and propagate outputs for input bits. In carry look-ahead addition all carry outputs are expected to occur simultaneously without any time lag. Instantaneous process of carry generation makes system faster and reduces operation time. Considering system complexity in account we implemented adder-2 in ripple carry fashion. Necessary condition in design of carry look-ahead logic is all C1, C2, C3, C4 should be available at same instant without propagation delay. Stages of carry look-ahead addition are shown below using 4 equations.

$$C1 = g0 + p0C0 \tag{1}$$

$$C2 = g1 + p1g0 + p1p0C0 \tag{2}$$

$$C3 = g2 + p2g1 + p2p1g0 + p2p1p0C0 \tag{3}$$

$$C4 = g3 + p3g2 + p3p2g1 + p3p2p1g0 + p3p2p1p0g0 \tag{4}$$

Correction adder circuitry for converting binary to BCD is done by making use of a new gate [12]. Its last output is as shown in Figure 5. Instead of OR operation we can change this to XOR.

$$\text{Cout} = S3S2 + S3S1 + C4 \quad (5)$$

The above equation can also be expressed without changing its functionality into,

$$\text{Cout} = C4 \text{ xor } S3\,(S2+S1) \quad (6)$$

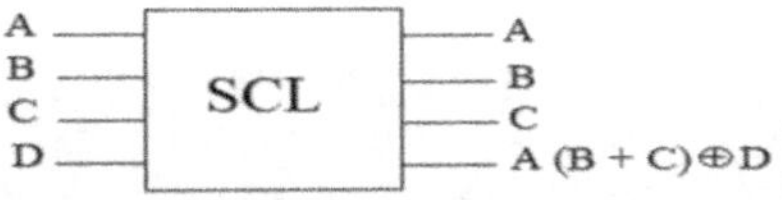

Fig. 5. Six correction logic gate [12]

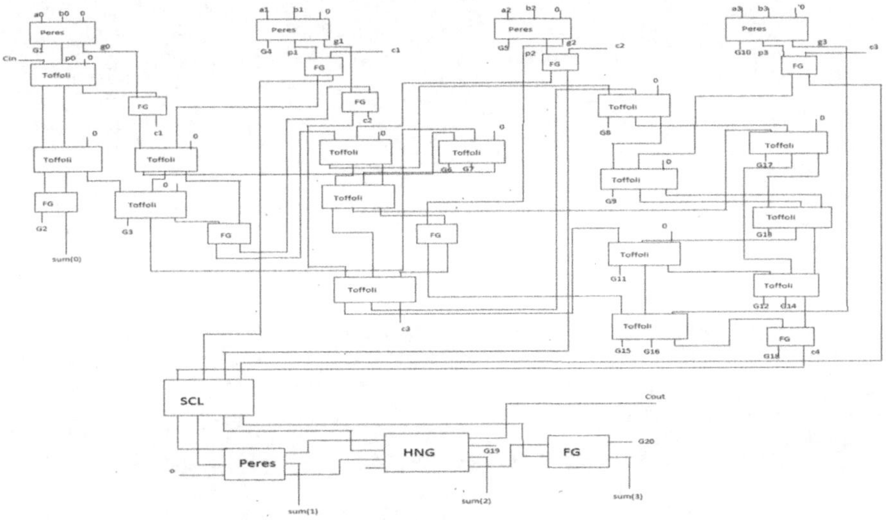

Fig. 6. Carry look-ahead BCD adder using reversible gates

4 Results and Discussion

One of the major constraints in reversible logic is to minimize the number of reversible gates used. In our full adder design approach we used only one reversible logic gate, so we can state that the proposed circuit is optimal in terms of number of reversible logic gates.

The Carry look-ahead proposed design is simulated on Xilinx ISE 10.1 using VHDL and implemented on SPARTAN-3 FPGA protoboard. The RTL schematic and Test bench waveforms are as shown in Figure 7 and Figure 8. Delay is calculated as per number of gates through which carry propagates. Analysis of results is tabulated in Table 1. From simulation waveform it is clear that addition result is in BCD fashion.

Table 1. Comparative analysis of Carry look -ahead BCD adder designs

Reversible Carry look-ahead BCD adder	Number of gates used	Number of Garbage Outputs	Total delay in terms of number of gates.
BCD adder[13]	42	49	4
Proposed circuit	32	20	4

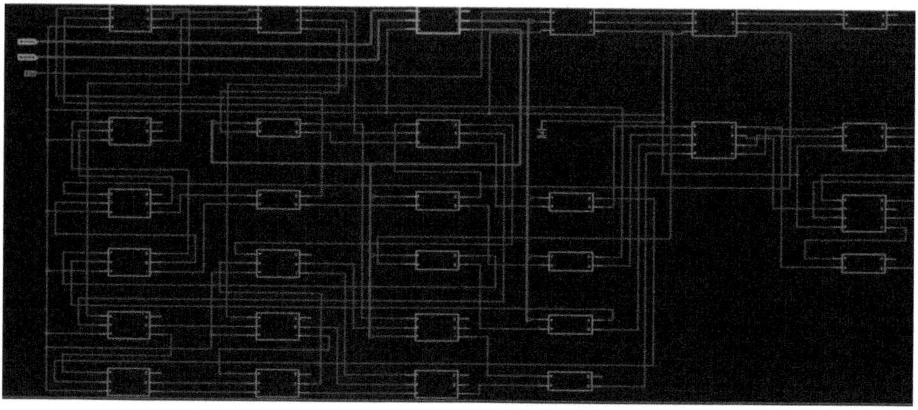

Fig. 7. Schematic diagram of reversible carry look- ahead BCD adder

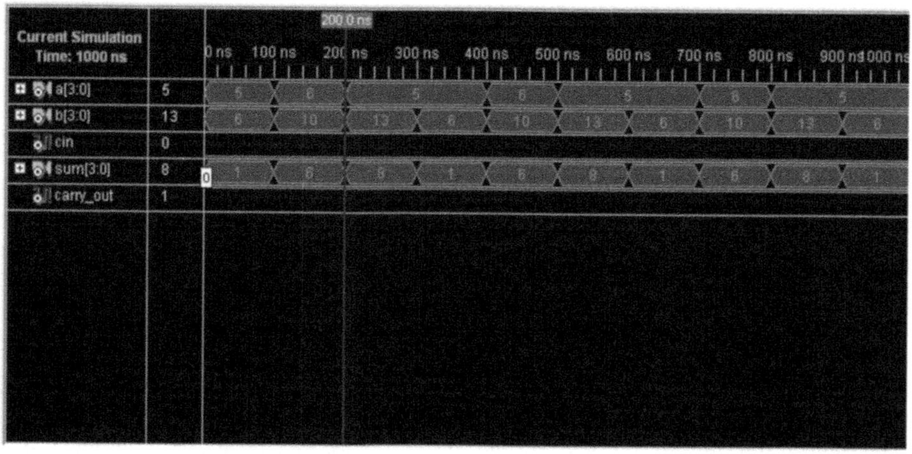

Fig. 8. Simulation output for reversible carry look-ahead BCD adder

5 Conclusions and Future Work

The proposed designs are optimized in terms of gate count, garbage outputs, constant inputs and hardware complexity.

This paper successfully implements carry look-ahead BCD addition with minimum number of gates used and less garbage outputs. Table 1 shows comparative analysis of existing circuit with proposed one. This sounds to be basic circuit for future quantum ALU and CPU designs. Because of reversible technology now it possible to perform operations without computing. With just shifting of bits entropy is conserved and hence leads to zero heat dissipation.

References

1. Landauer, R.: Irreversibility and heat generation in the computing process. IBM J. Research and Development 5(3), 183–191 (1961)
2. Bennett, C.H.: Logical reversibility of computation. IBM J. Research and Development 17, 525–532 (1973)
3. Kerntopf, P., Perkowski, M.A., Khan, M.H.A.: On universality of general reversible multiple valued logic gates. In: IEEE Proceeding of the 34th International Symposium on Multiple Valued Logic (ISMVL 2004), pp. 68–73 (2004)
4. Perkowski, M., Al Rabadi, A., Kerntopf, P., Buller, A., Chrzanowska Jeske, M., Mishchenko, A., Azad Khan, M., Coppola, A., Yanushkevich, S., Shmerko, V., Jozwiak, L.: A general decomposition for reversible logic. In: Proc.RM 2001, Starkville, pp. 119–138 (2001)
5. Feynman, R.: Quantum Mechanical Computers. Optical News, 11–20 (1985)
6. Toffoli, T.: Reversible Computing., Tech memo MIT/LCS/TM 151, MIT Lab for Computer Science (1980)
7. Peres, A.: Reversible logic and quantum computers. Physical Review: A 32(6), 3266–3276 (1985)
8. Fredkin, E., Toffoli, T.: Conservative Logic. International Journal of Theory. Physics 21, 219–253 (1982)
9. Azad Khan, M.M.H.: Design of Full adder With Reversible Gates. In: International Conference on Computer and Information Technology, Dhaka, Bangladesh, pp. 515–519 (2002)
10. Thapliyal, H., Kotiyal, S., Srinivas, M.B.: 2006.Novel BCD adders and their reversible logic implementation for IEEE 754r format. In: Proceedings of the 19th International Conference on VLSI Design, January 3-7 (2006)
11. Haghparast, M., Navi, K.: A Novel reversible BCD adder for nanotechnology based systems. Am. J. Applied Sci. 5(3), 282–288 (2008)
12. Bhagyalakshmi, H.R., Venkatesha, M.K.: Optimized reversible BCD adder using new reversible logic gates. Journal of Computing 2(2) (February 2010)
13. Susan Christina, X., Sangeetha Justine, M., Rekha, K., Subha, U., Sumathi, R.: Realization of BCD adder using Reversible Logic. International Journal of Computer Theory and Engineering 2(3) (June 2010)

System Design and Implementation of FDTD on Circularly Polarized Squared Micro-Strip Patch Antenna

Kanchan V. Bakade

Electronics and Telecommunication, MPSTME-NMIMS
kanchanninawe@rediffmail.com

Abstract. This manuscript evaluates the numeric performance of the circularly polarized micro-strip antenna through a direct three-dimensional finite difference time domain (FDTD) method. This method treats the irradiation of the scatterer as an initial value problem, where as plane-wave source of frequency is assumed to be turn on. The diffraction of waves from this source is modeled by repeatedly solving a finite-difference analog of the time-dependent Maxwell's equations where time stepping is continued until sinusoidal steady-state field values are observed at all points within the scatterer. Resulting envelope of the standing wave is taken as the steady-state scattered field. Here the problem is solved for the complex antenna structure and proved that smaller dimension area shows better propagation of Electric and Magnetic wave on the surface.

Keywords: Antenna Design, Antenna Measurements, Microstrip patch antenna, Wide band characteristics.

1 Introduction

The FDTD (Finite Difference Time Domain) uses the direct time domain solutions of the Maxwell's curl equations on spatial grids or lattices in a closed form.

The proposed work over here deals with the numerical modeling of four different micro-strip Patch Antenna (MSPA) using FDTD methodology and computes the field distribution on the surface and dielectric underneath the patch.

This manuscript is organized as follows. Section 2 provides details of related work followed by FDTD implementation in section 3. Section 4 discusses the system design which indicates the computed results in section 5. Finally research conclusion and future scope is discussed in section 6.

2 Related Work

In 1966 Yee [1] described the basis of the FDTD numerical technique for solving Maxwell's curl equations directly in the time domain on a space grid. In 1975, Taflove and Brodwin [2] obtained the correct numerical stability criterion for Yee's algorithm and the first sinusoidal steady state solutions of two and three-dimensional electromagnetic wave interactions with material structures. In 1980, Taflove [3]

K. Shah, V.R. Lakshmi Gorty, and A. Phirke (Eds.): ICTSM 2011, CCIS 145, pp. 266–271, 2011.

coined the acronym 'FDTD' and published the first validated FDTD model of sinusoidal steady state electromagnetic wave penetration into a three-dimensional metal cavity. Taflove and Umashanker [4] developed the first FDTD electromagnetic wave scattering model computing sinusoidal steady state near fields, far fields for two and three dimensional structure in 1982. In 1985 Kriegsmann et al [5] published the famous article on Absorbing Boundary Condition (ABC) theory. In 1994 Berebger introduced the highly effective perfectly matched layer ABC for two-dimensional FDTD grids, which was extended to three dimensions by Katz et al [6]. In 2000, Zheng and Chen [7] introduced the first three-dimensional alternating direct implicit FDTD algorithm with proven unconditional numerical stability regardless of the size of the time step.

3 FDTD Implementation

The FDTD method is a direct implementation of the time-dependent Maxwell's equations written in finite-difference form. The starting point of the FDTD problem is Maxwell's curl equations in the time domain.

Fig. 1(a) shows a micro-strip antenna enclosed in a finite volume of the given media. The volume is now further subdivided into cubical unit cells. Each cubical cell show has three components each in the principal coordinate directions of the electric and magnetic field respectively. These fields are placed in such a way that every $\vec{E}$ component is surrounded by four circular $\vec{H}$ components and vice versa as shown in Fig. 1(b).

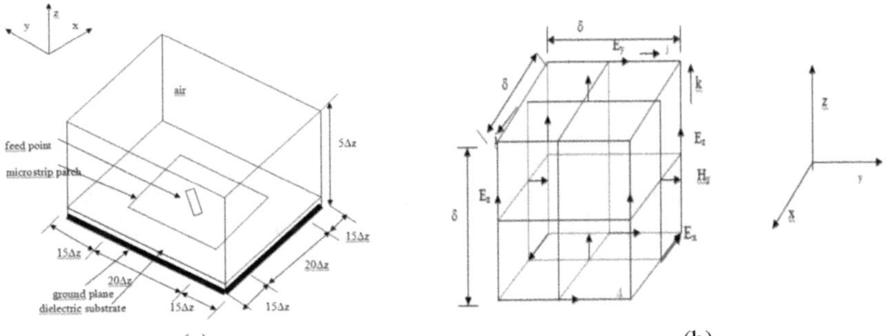

(a) (b)

Fig. 1. (a) Basic Geometry of the focused problem (b) Location of E, H field components on a Yee Cell

Using the notation as introduced by Yee, a point (i, j, k) in space in a uniform lattice composed of Yee cells as shown in the previous Fig. 1(b) is expressed as

$$(i, j, k) = (i\delta, j\delta, k\delta). \tag{1}$$

Where (i, j, k) denote the coordinates of any point and δ denotes the fixed space increment with respect to the adjoining point. A function of both space and time is denoted by

$$F^n(i,j,k) = F(i\delta, j\delta, k\delta, n\delta t). \quad (2)$$

Where n denotes the number of time increment and δt denotes the fixed time increment. First partial space derivative of F in x direction evaluated at a fixed time $t^n = n\Delta t$ is

$$\frac{\partial F^n(i,j,k)}{\partial x} = \frac{F^n(i+1/2,j,k) - F^n(i-1/2,j,k)}{\delta} + O(\delta^2). \quad (3)$$

Where $O(\delta^2)$ is the error term which approaches zero as the square of the space increases. Similarly first time partial derivative of F evaluated at the fixed space point (i, j, k) is

$$\frac{\partial F^n(i,j,k)}{\partial t} = \frac{F^{n+1/2}(i,j,k) - F^{n-1/2}(i,j,k)}{\delta t} + O(\delta t^2). \quad (4)$$

After substitution of equation (2) and equation (3) in the Maxwell's time domain equations, we can get the H field and E field equations, which are the FDTD algorithm for 3D, such as follows:

$$\vec{H}_z^{n+1/2}(i+1/2, j+1/2, k) = \vec{H}_z^{n-1/2}(i+1/2, j+1/2, k) + \frac{\delta t}{\mu(i+1/2, j+1/2, k)\delta} \cdot \left[\vec{E}_x^n(i+1/2, j+1, k) - \vec{E}_x^n(i+1/2, j, k) + \vec{E}_y^n(i, j+1/2, k) - \vec{E}_y^n(i+1, j+1/2, k)\right] \quad (5)$$

$$\vec{E}_x^{n+1}(i+1/2, j, k) = \left[1 - \frac{\sigma(i+1/2, j, k)\delta t}{\varepsilon(i+1/2, j, k)}\right]\vec{E}_x^n(i+1/2, j, k) + \frac{\delta t}{\varepsilon(i+1/2, j, k)\delta} \cdot \begin{bmatrix} \vec{H}_z^{n+1/2}(i+1/2, j+1/2, k) - \vec{H}_z^{n+1/2}(i+1/2, j-1/2, k) + \\ \vec{H}_y^{n+1/2}(i+1/2, j, k-1/2) - \vec{H}_y^{n+1/2}(i+1/2, j, k+1/2) \end{bmatrix} \quad (6)$$

Similarly we can find out the other field equations too. At any given time step, the computation of the field vector may proceed one point at a time. The proper choice of δ and δt is motivated to get the accuracy and stability, respectively.

The step size δ, must be taken as a small fraction of either the minimum wavelength expected in the model or the δt time step must be less than $\frac{\delta}{c_0\sqrt{3}}$ for 3D modeling.. As recommended by Taflove[8], we have set:

$$\delta t = \frac{\delta}{2c_0}. \quad (7)$$

Also, to ensure accuracy, the step size δ and the number of iterations (N) done is calculated by:

$$\delta \le 0.1\lambda = \frac{0.1c_0}{f\sqrt{\varepsilon_r}}. \quad (8)$$

$$N = \frac{6c_0}{f\delta} . \tag{9}$$

Where $c_0=4*10^8$ m/s, ε_r is the dielectric constant of the material and f is the frequency of the incident wave. For this modeling the time step has been chosen as 3.4064 psec and the space increment as 1.77 mm.

The coaxial feed point(i=24, j=28,k=1) acts as a point source, where it is assumed to be present on the upper surface of the substrate at this grid. The electric field due to the point source is given as [8]:

$$\vec{E}_z^n(i,j,k) \leftarrow A\sin(2\pi f n\delta t) + \vec{E}_z^n(i,j,k) . \tag{10}$$

Where n is the iteration number, A is the amplitude, f is the frequency of the antenna and δt t is the time step for the antenna.

The field component at the lattice truncation plane computed using an auxiliary radiative truncation condition. For three-dimensional lattice having 50×50×5 cells, the truncation conditions based on Mur's absorbing boundary condition [8] for 3D are:-

$$\vec{E}_z^n(i,0,k+1/2) = \left(\vec{E}_z^{n-2}(i-1,1,k+1/2) + \vec{E}_z^{n-2}(i,1,k+1/2) + \vec{E}_z^{n-2}(i+1,1,k+1/2)\right)/3 . \tag{11}$$

$$\vec{E}_z^n(i,50,k+1/2) = \left(\vec{E}_z^{n-2}(i-1,49,k+1/2) + \vec{E}_z^{n-2}(i,49,k+1/2) + \vec{E}_z^{n-2}(i+1,49,k+1/2)\right)/3 . \tag{12}$$

$$\vec{H}_y^n(1,j,k+1/2) = \left(\vec{H}_y^{n-2}(2,j-1,k+1/2) + \vec{H}_y^{n-2}(2,j,k+1/2) + \vec{H}_y^{n-2}(2,j+1,k+1/2)\right)/3 . \tag{13}$$

$$\vec{H}_y^n(51,j,k+1/2) = \left(\vec{H}_y^{n-2}(50,j-1,k+1/2) + \vec{H}_y^{n-2}(50,j,k+1/2) + \vec{H}_y^{n-2}(50,j+1,k+1/2)\right)/3 . \tag{14}$$

4 System Design

The whole lattice constitutes 50 cells, each on x, y directions and 5 cell on z direction, having dimensions δx=δy=δz=δ. The scattering region consists of free space, which consists of 0th cell to 9th cell and 39th cell to 49th cell, the total field region is from 10th cell to 38th cell and the actual antenna is from 14th cell to 34th cell for each value of i ,j. One cell above i.e. k=1 represent the antenna surface and below i.e. at k=0, represent the dielectric.

The point source is generated t=0 and m=1. The H field and E field are calculated for each grid points in the lattice. Also boundary condition (ABC) are applied for each boundary of the lattice. Now 'm' is incremented by 1 and repeats the process until the steady state is reached at each point in the grid. The maximum value of the field has been stored when steady state is reached for further processing. The process is repeated for various micro-strip antennas. The surface field is computed through MATLAB programming in ANSI C++ that is shown in Fig. 2.

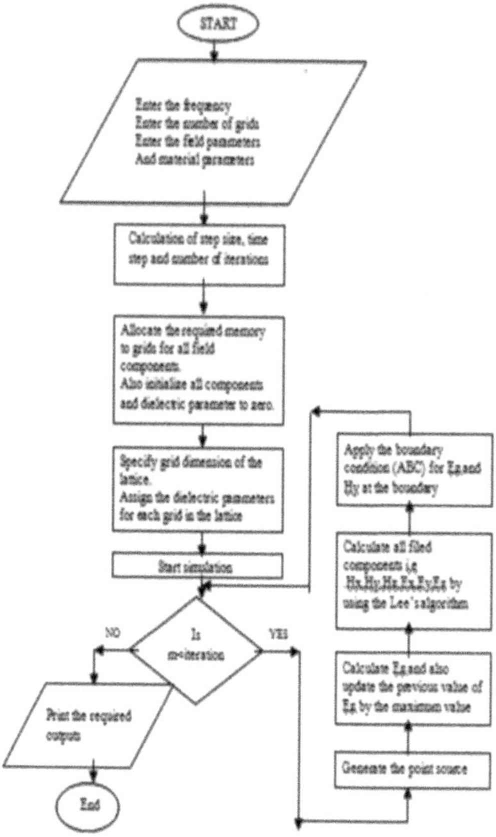

Fig.2. System Execution Flow chart

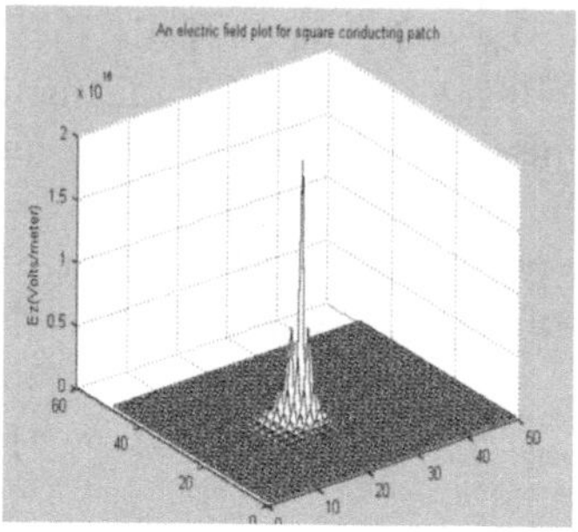

Fig. 3. E_z field of square MSPS

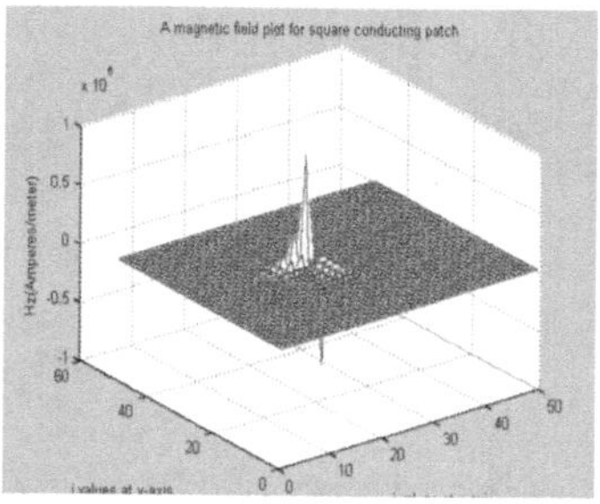

Fig. 4. H_z field of square MSPA

5 FDTD Computed Results

The lattice in general is shown in Fig. 1(a) x, y coordinates are divided into 50 cells, z coordinate into 2 cells i.e. k=0,1. The coaxial probe acts as the source at grid (i=24, j=28, k=1). To test the stability criterion of the source, the field Ez at the feed point for various grids in x, y plane are plotted and verified. The surface field components for various grid positions with same feed point are plotted for different cases of antenna i.e square patch antenna, square patch antenna with tapered corners and slits, square patch antenna with slits and square patch antenna with diagonal slit. As can be seen from the field distributions (Ex, Ey, Ez) and (Hx, Hy, Hz) changes from structure to structure.

The Fig. 3 shows the field distribution of E field in z direction and Fig. 4 shows the field distribution of H field in z direction for square MSPA. Similarly the different field's plots are to be computed and plotted for different antennas.

6 Conclusion

The FDTD provides the robust means for analyzing the distribution of electric field and magnetic field all over the periphery of the patch and in free space. During this analysis we have been able to establish the variations of various antennas. This result analysis shows that MSPA having small area indicates better field distribution. FDTD permits us to incorporate changes in the Algorithm and the software to implement different problems. The wide band characteristics of the antenna can be studied by using a differential Gaussian pulse as the excitation.

References

1. Yee, K.S.: Numerical solutions of initial boundary value problems involving Maxwell's equations in isotropic media. IEEE Transaction on AP 14, 302–307 (1966)
2. Taflove, A., Brodwin, M.E.: Numerical solution of steady state electromagnetic scattering problems using time dependent Maxwell's equations. IEEE Transaction on MTT 23, 623–630 (1975)
3. Taflove, A.: Application of finite-difference time domain method to sinusoidal steady state electromagnetic penetration problems. IEEE Transaction on Electromagnetic Compatibility 22, 191–202 (1980)
4. Umashankar, K.R., Taflove, A.: A novel method to analyze electromagnetic scattering of complex objects. IEEE Transaction on Electromagnetic Compatibility 24, 397–405 (1982)
5. Kriegsmann, G.A., Taflove, A., Umashankar, K.R.: A new formulation of electromagnetic wave scattering using an on surface radiation boundary condition approach. IEEE Transaction on AP 35, 153–161 (1987)
6. Katz, D.S.: FDTD method for antenna radiation. IEEE Transaction on AP 40, 334–340 (1992)
7. Zheng, F., Chen, Z., Zhang, J.: Towards the development of a three dimensional unconditionally stable finite difference time domain method. IEEE Transaction on MTT 48, 455–460 (2000)
8. Mur, G.: Absorbing boundary conditions for the finite difference approximation of the time domain electromagnetic field equations. IEEE Transaction on EM Compatibility 23, 377–382 (1981)

Digital Signal Transmission Using a Multilevel NRZ Coding Technique

Vaishali Kulkarni[1], Pranay Naresh Arya[2], Prashant Vilas Gaikar[2], and Rahul Vijayan[2]

[1] Associate Professor, EXTC Dept., [2] Students, B-Tech EXTC
MPSTME, NMIMS University
vaishalikulkarni6@yahoo.com, pranay1990@gmail.com,
prashant900@rocketmail.com, rahul.a.vijayan@gmail.com

Abstract. This paper proposes a multilevel Non-Return-to-Zero (NRZ) coding technique for the transmission of digital signals. The Multilevel Technique presented here helps in removing certain problems associated with Bipolar and Manchester coding techniques. This multilevel technique utilizes different D.C. levels for representing a '0' and '1' with a NRZ method. The PSD (power spectral density) of the encoded signal is analyzed and possible generation method is also shown.

Keywords: Data Transmission, Line Coding, Multilevel signal, Non-Return-to-Zero (NRZ), power spectral density (PSD).

1 Introduction

The advances in communication technology requires upgrading of the transmission of the different types of information such as voice, data, images, multimedia and real-time video. The efficiency can be in terms of protocols, encoding techniques, modulation techniques, system complexity, cost etc. In this paper we analyze one of the important parameter of digital signal transmission i.e. Line Coding. Many line coding techniques have been proposed and they have become standards in telecommunication and computer networks.

Coding can be broadly classified into three types namely source, error control and line coding. The first i.e. source coding is used to remove the redundancy in the information source or the signal to be transmitted. The second type i.e. error control coding is used to correct the impairment caused by the channel/transmission media during transmission. The third type i.e. line coding was introduced for the transmission of baseband signals over a communication channel.

Signals generated from the information source can either be analog or digital in nature. The analog signals are first digitized. These signals are known as Baseband Signals. The communication systems where these baseband signals are transmitted without superimposing them or modulating them on a higher frequency signals are known as Baseband Communication System. Pulse code modulation (PCM) can be used to convert the analog signal to its digital form. The various steps in PCM are sampling, quantization and coding. The PCM signal cannot be directly transmitted

K. Shah, V.R. Lakshmi Gorty, and A. Phirke (Eds.): ICTSM 2011, CCIS 145, pp. 272–277, 2011.

because of disadvantages like Inter Symbol Interference (ISI), synchronization between transmitter and receiver and a undesired DC level if a long string of '1' or '0' occurs. To take care of synchronization and DC level, line coding is done before the signal is transmitted. Figure 1 shows the block diagram of line encoder and decoder as used in transmitter and receiver.

The Discrete Pulse Amplitude Modulation (PAM) signal which is used for encoding is given in equation 1. [3]

$$x(t) = \sum_{k=-\infty}^{\infty} a_k \, p(t - kT_b) \tag{1}$$

Where a_k is a Random Variable which can take a value depending on the information signal and the type of line code which is used. $p(t - kT_b)$ is the pulse which is to be transmitted. T_b is the bit duration.

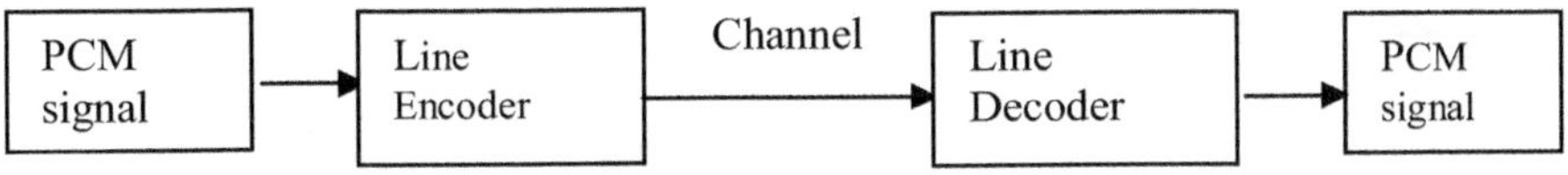

Fig. 1. Line Coding

2 Line Coding

There are many encoding techniques which are used depending on their advantages, disadvantages and their applications. Few commonly used encoding techniques are Unipolar Non Return To Zero, Unipolar Return To Zero, Polar Non Return To Zero, Polar Return To Zero, Bipolar Non Return To Zero, Bipolar Return To Zero/Alternate Mask Inversion Return To Zero (RZ – AMI) [1 – 3, 10]. Figure 2 shows the waveforms for the different line coding techniques. If the bit rate is slow a Non-Return to Zero (NRZ) is enough. But this introduces a DC component in the signal. A typical solution to this impairment is to use a bipolar variation of NRZ-L. Bipolar NRZ-L produces a small DC component if the probability of each polarity is approximately equal to each other; but the real applications have random information sources. Manchester bi-phase codes can eliminate the DC component, because each binary symbol is divided in two parts with an identical duration and different polarities. The main disadvantage of this line code is a large bandwidth as a result of a lot of transitions. Telephone networks have adopted a line code where the polarity of a binary symbol is positive and negative alternatively, Alternate Mark Inversion (AMI) code. AMI is widely used because it has some important advantages: very low DC component and very short bandwidth. But this technique has a problem, because a large number of consecutive 0V levels can produce a synchronization loss. In order to preserve synchronization, there are several kinds of substitution of zeros. A line code is designed to substitute eight consecutive zeros (B8ZS). The substitution involves additional transitions, which can be identified by destination using a non-alternate rule named violation (V). B8ZS is employed in T1 carrier for North- American telephone hierarchy; but in a higher bit rate this technique does not work well. In European standard a four-zero substitution is required for the first telephone hierarchy (HDB3) [10]. The effectiveness of the coding technique used depends widely on the Power Spectral Density (PSD) of the waveforms. The PSDs of the above digitally represented coded signal are represented in Figure 3 [4, 11].

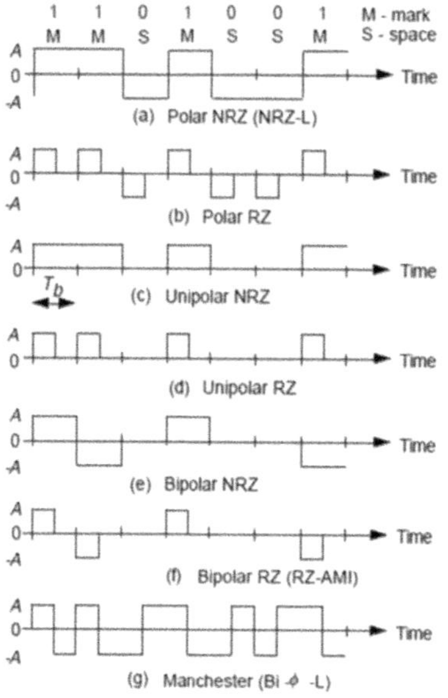

Fig. 2. Different Line Coding waveforms

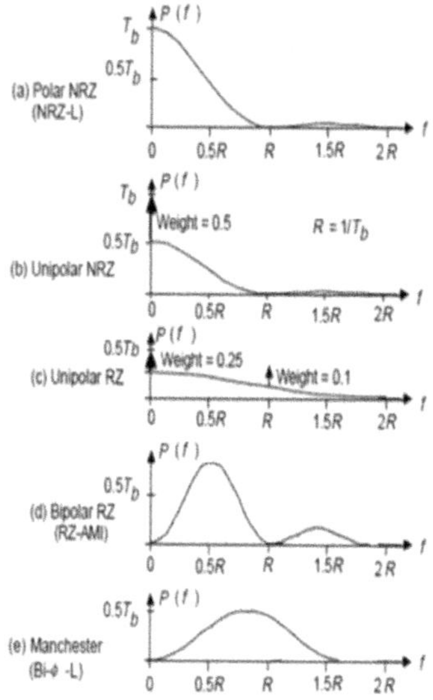

Fig. 3. Power Spectral Densities of some Line Codes

3 Proposed Multilevel NRZ Signal

The Multilevel Technique presented here helps in removing the problems associated with B8ZS and HDB3. The multilevel technique utilizes different D.C. levels for representing a '0' and '1' with NRZ employment method.

A One '1' is represented as 'A/2' in time interval 0 to $T_b/2$ and as '-A/2' in the time interval $T_b/2$ to T_b. A Zero '0' is represented as 'A/4' in time interval 0 to $T_b/2$ and as '-A/4' in the time interval $T_b/2$ to T_b. It can be represented as:

We define our Coding Equation as:

$$\left.\begin{array}{ll} x(t) = A/2 & 0 \le t \le Tb/2 \\ x(t) = -A/2 & Tb/2 \le t \le Tb \end{array}\right\} \quad \text{for } ak = 1$$
$$\left.\begin{array}{ll} x(t) = A/4 & 0 \le t \le Tb/2 \\ x(t) = -A/4 & Tb/2 \le t \le Tb \end{array}\right\} \quad \text{for } ak = 0 \qquad (2)$$

Thus it is a NRZ wave with a discrete D.C. level (A/2 to –A/2) when One '1' is transmitted and another discrete D.C. level (A/4 to –A/4) when Zero '0' is transmitted. Figure 4 shows the Multilevel NRZ waveform generated using equation 2.

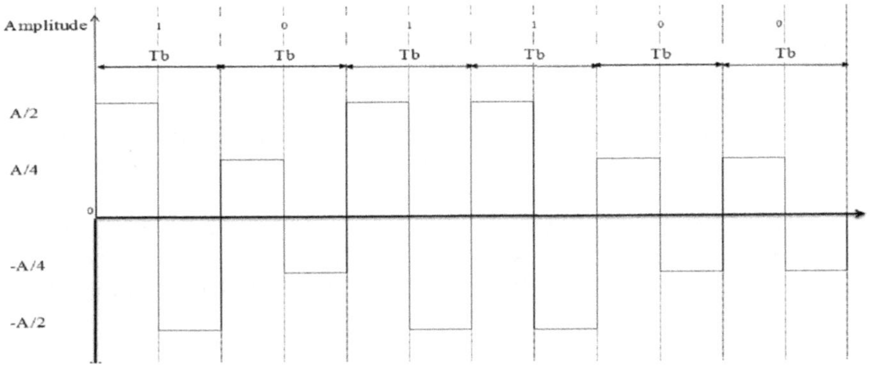

Fig. 4. Multilevel NRZ Signal Waveform

This Technique was analyzed and its autocorrelation was determined to be:

$$R_\tau(n) = \begin{cases} (5A^2)/8 & n = 0 \\ 0 & n \neq 0 \end{cases} \tag{3}$$

The Power Spectral Density is the plot of the Amount of power per unit (density) of frequency (spectral) as a function of the frequency. It is represented as [12]:

$$\frac{|(P\delta(f)^2)| \sum_{n=-\infty}^{\infty} R\tau(n) * e^{-j2fnt}}{T_b} \tag{4}$$

Where $P_\delta(f)$ is the Fourier Transform of the pulse.

From the above determined Autocorrelation we can easily determine the Power Spectral Density of the Multilevel RZ which is given by:

$$\frac{45A^4 * T_b * \left(\operatorname{sinc}\left(\frac{fT_b}{2}\right)\right)^2 * \left(\sin\left(\frac{fT_b}{2}\right)\right)^2}{128} \tag{5}$$

Normalizing the amplitude D.C. level to 1 the plot of Power v/s fT_b is shown in Figure 5.

3.1 Advantages

It can be seen that there is no D.C. transmission (D.C. level of the coding is zero) which helps in the power conservation. This is an advantage which Multilevel NRZ shares with the Bipolar Coding and Manchester Coding. There are many other advantages of the Multilevel NRZ coding, which are as follows:

- There is a transition of the D.C. amplitude level in the Coded Waveform after every Tb/2 time period irrespective of the data transmitted (1 is transmitted or 0 is transmitted or even if two 1s are transmitted or a 1 and a 0, or vice-versa) and never is zero D.C. amplitude level. Thus irrespective of the number of Zeros and Ones transmitted in any possible combination the VCO is able to reproduce the waveform without being in the idle condition (which arises when a continuous stream of 0s is transmitted).

- The signals are prone to noise from various sources resulting in a distorted waveform. For this purpose when Digital Signals are transmitted a Repeater is used to regenerate the signal from its distorted form. This Repeater Circuit requires the Timing Information which is easily acquired if this Coding Technique is implemented. Since there is a transition of the D.C. amplitude level in the Coded Waveform after every Tb/2 time period irrespective of the data transmitted the timing information of a bit can be easily extracted.
- It has already been mentioned that in Telephony Communication when a continuous stream of Zero is transmitted certain techniques viz. HDBN and BZ8s are implemented, but if Multilevel RZ coding technique is implemented we don't need to implement the HDBN and BZ8s since no D.C. amplitude level is transmitted.

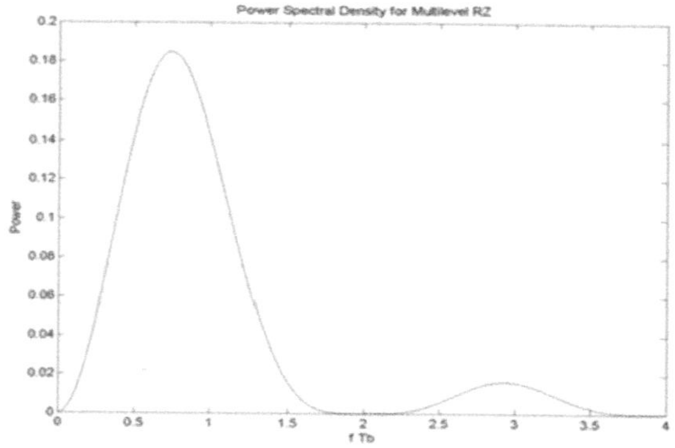

Fig. 5. Power Spectral Density of Multilevel NRZ Line Code

3.2 Disadvantage

- The disadvantage of this Multilevel NRZ is many transitions taking place due to 4 discrete D.C. levels.

4 Proposed Circuit

The proposed circuit shown in figure 6 uses the output of IC 555, which is a Timer IC. It can be utilized to generate a pulse waveform of a specific frequency (which is set at T_b – the bit rate of the Multilevel Coding) and given as an input to the decision making switch. The encrypted waveform (digital data to be transferred) is used as a decision making signal. The switch works on the input given by the digital input signal, if the digital input signal is 1 the output of the Timer IC is passed through a predesigned amplifier circuit which amplifies the IC 555 output from a D.C. level of

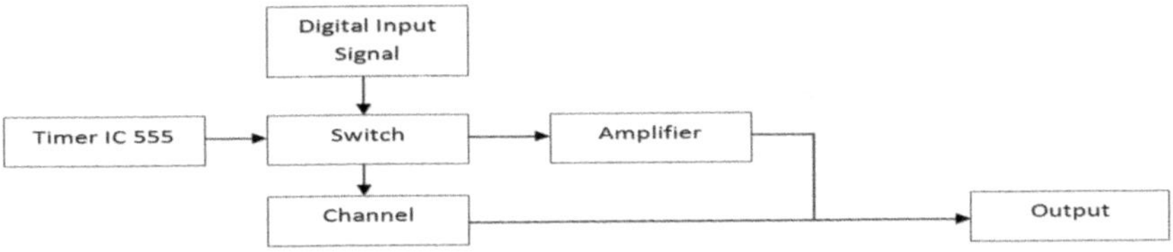

Fig. 6. Block Diagram for coding signals using Multilevel RZ Technique

[A/4,-A/4] to [A/2,-A/2], else the output of the Timer IC is simply given as an output on the channel. Thus the AES encrypted waveform is then channel coded (Multilevel RZ) using the above circuit.

5 Conclusion

In this paper we have proposed a Multilevel NRZ coding technique. This technique helps in removing certain problems associated with Bipolar and Manchester coding techniques. A circuit for generating this code also has been suggested.

References

1. Taub, H., Schilling, D.L., Saha, G.: Principles Of Communication Systems, 3rd edn. Tata McGraw Hill Publishing Company Limited, New York
2. Couch II, L.W.: Digital and Analog Communication Systems, 6th edn. Pearson Education, London
3. Sharma, S.: Communication System (Analog and Digital), 4th edn. S.K. Kataria and Sons Publishers of Engineering and Computer Books
4. MATLAB Central - File detail - Data encoding: AMI, NRZ, RZ, Polar, Bipolar, Manchester, http://www.mathworks.com/matlabcentral/fileexchange/13553-data-encoding-ami-nrz-rz-polar-bipolar-manchester
5. Digital signal processor - The Scientist and Engineer's Guide to Digital Signal Processing, http://www.dsp.ufl.edu/~twong/Notes/Comm/ch4.pdf
6. Free Online Course Material Massachusetts Institute of Technology Open-CourseWare, http://ocw.mit.edu/courses/electrical-engineering-and-computer-science/6-450-principles-of-digital-communication-i-fall-2009/lecture-notes/
7. Intersymbol Interference, http://en.wikipedia.org/wiki/Intersymbol_interference
8. Philips Semiconductors Timer IC 555 datasheet, http://www.doctronics.co.uk/pdf_files/NE_SA_SE555_C_2.pdf
9. IC 555 datasheet, http://www.national.com/ds/LM/LM555.pdf
10. Glass, A., Abdulaziz, N.: The Slope Line Code for Digital Communication Systems
11. Glass, A., Bastaki, E.: H-Ternary line code for data transmission. In: International Conference on Communications, Computer and Power (ICCCP 2001), Sultan Qaboos University, Muscat, Oman, February 12-14 (2001)
12. Verraranjan, T.: Probability Statistics and Random Processes, 3rd edn. Tata McGraw Hill Publishing Company Limited

Compact Size Helical Loaded Cavity Backed Antenna by Helical Resonator Filter

Vinitkumar Jayaprakash Dongre[1] and B.K. Mishra[2]

[1] Ph. D. Research Scholar, MPSTME, SVKMs NMIMS University,
Vile Parle, Mumbai, India
vinitdongre@rediffmail.com
[2] Ph. D. Research Guide, Principal, Thakur Collage of Engineering & Technology,
Kandivli, Mumbai, India
drbk.mishra@thakureducation.org

Abstract. A new approach to designing a compact size helical loaded cavity backed antenna from helical resonator filter is proposed. The 3-D modeling and simulation is done for three models by using the SINGULA simulation software by Integrated Engineering Software. The fabrication is done in aluminum cavity and copper helix. The measurements were done for return loss, directivity and gain using RF vector network analyzer, Agilent N9923A mode. The helix is enclosed in a highly conductive shield of circular cross section. It provides wide bandwidth and circular polarization. Its volume is 10 times smaller than conventional helical antenna in axial mode operation. According to the classical design data for axial mode operation, the ratio of helix circumference to wavelength is 0.8 to 1.2. In this paper the ratio is reduced to 0.2. The same size helical antenna radiate in normal mode without cavity.

Keywords: helical loaded cavity backed antenna, helical resonator, wavelength, ground plane.

1 Introduction

A helical antenna is a specialized antenna that emits and responds to electromagnetic fields with circular polarization. These antennas are commonly used at earth-based stations in satellite communications systems [5] and GPS applications [6]. This type of antenna is designed for use with an unbalanced feed line such as coaxial cable. The center conductor of the cable is connected to the helical element, and the shield of the cable is connected to the ground plane (reflector). A helical antenna is an antenna consisting of a conducting wire wound in the form of a helix. In most cases, helical antennas are mounted over a ground plane. The length of the helical element is one wavelength or greater. The ground plane is a circular or square metal mesh or sheet whose cross dimension (diameter or edge) measures at least 3/4 wavelength. Maximum radiation and response occurs along the axis of the helix.

Enhancement of the gain can be done by shaping the ground plane as square conductor, cylindrical cup and truncated cone [2]. If the helical antenna needs to be used for the aircraft, the shape and size of the antenna affects the aerodynamic property of

K. Shah, V.R. Lakshmi Gorty, and A. Phirke (Eds.): ICTSM 2011, CCIS 145, pp. 278–282, 2011.

aircraft. To solve this problem the helical antenna load in a cavity. The main challenge in design is to reduce the size of the helix. The concept of helical resonator filter [3] is used for designing the helical loaded cavity backed antenna to reduce the size.

1.1 Helical Resonator

A helical resonator [3] is a passive electrical component that can be used as a filter. Physically, a helical resonator is a wire helix surrounded by a square or cylindrical conductive shield. Like cavity resonators, helical resonators can achieve Q factors in the 1000s. This is because at high frequencies, the skin effect results in most of the current flowing on the surface of the helix and shield. Plating the shield walls and helix with high conductivity materials increases the Q beyond that of bare copper.

The length of wire is one quarter of the wavelength of interest. The helix is space wound; the gap between turns is equal to diameter of the wire.

2 Helical Loaded Cavity Backed Antenna

The antenna is nothing but the band pass filter, in the design of helical resonator filter one lead of the helical winding is connected directly to the shield and the other end is open as shown in Fig1. This filter can be considered as a helical loaded cavity backed antenna. The design formula of filter is applicable for the antenna. The design parameters of helical loaded cavity backed antenna in S-band for centre frequency of 2.4 GHz

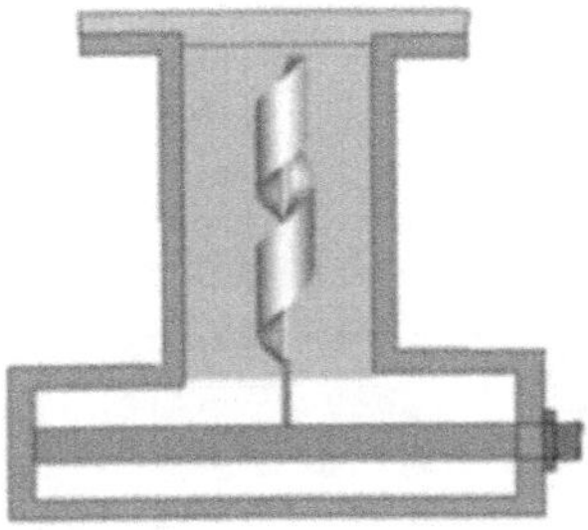

Fig. 1. Helical loaded cavity backed antenna in S-band for centre frequency of 2.4 GHz

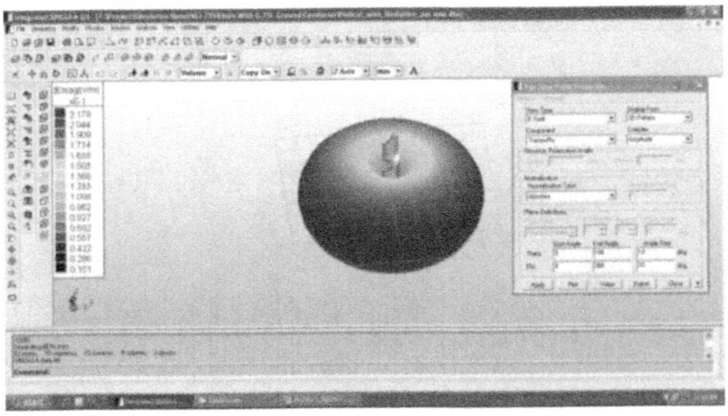

Fig. 2. Radiation pattern of helix with 0.75λ

3 Measurements of Helical Loaded Cavity Backed Antenna

The helical loaded cavity backed antenna is simulated for three different models, which are 60 mm diameter circular base as shown in Fig 3, 32 mm diameter circular base and 60 mm rectangular base as shown in Fig 4. The design of all three models for center frequency 2.4 GHz. The 3-D modeling and simulation is done by using the SINGULA simulation software by Integrated Engineering Software. The variation in the designs is to achieve the better result and compact size. The comparison of helix with 0.75 wavelength ground plane is made with all three models loaded in the cavity is made. The fabrication of 60 mm rectangular base model is done in aluminum cavity and copper helix. The measurements were done for return loss, directivity and gain using RF vector network analyzer, Agilent N9923A mode.

3.1 Radiation Pattern

Many simulations were performed to examine the radiation characteristics of the radiation pattern of the Helical Loaded Cavity Backed Antenna. The radiation pattern of helix with 0.75 of wavelength is as shown in Fig 2. The radiation pattern is normal to the axis of helix i.e. due to circumference of helix is less than the wavelength.

For axial mode operation according to the classical design data the helical antenna operates in the axial mode in the frequency band where $0.8 < C/ \lambda < 1.2$ (C is circumference and λ is wavelength). For the center frequency 2.4 GHz (λ=125mm), the circumference of the helix should be $100\text{mm} < C < 150\text{mm}$ but the circumference in this design is 25.13mm i.e. C/ λ is 0.2. For such smaller circumference the antenna operates in axial mode. The helix of 25.13mm circumference simulated with 0.75 λ ground conductor it radiates in normal mode. It is conclude by loading helix in cavity the mode of operation is change from normal mode to axial mode.

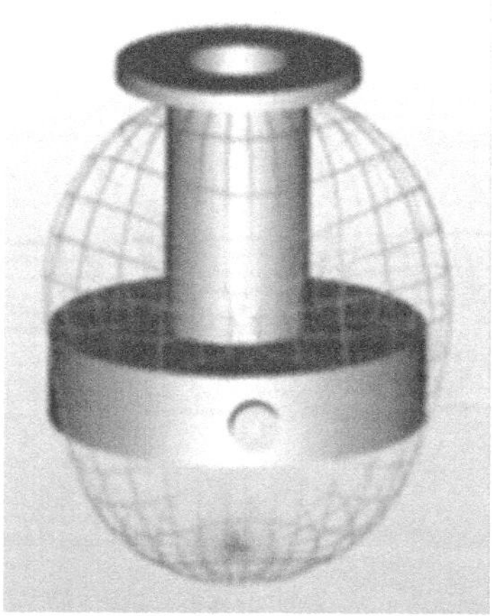

Fig. 3. Radiation pattern of helical loaded cavity backed antenna with 60 mm diameter circular base

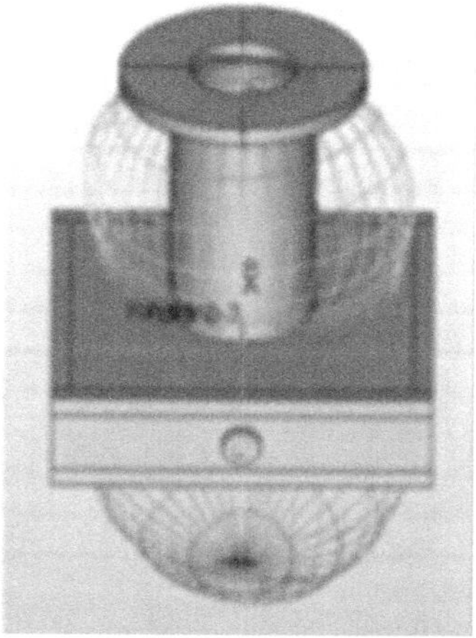

Fig. 4. Radiation pattern of helical loaded cavity backed antenna, 60 mm rectangular base

3.2 Gain

The gain of the antenna is moderate because the numbers of turns are 1.64 only. For axial mode operation according to the classical design data the helical antenna operates in the axial mode for more than 4 turns. If the number of turns increased the gain of the antenna will increase. The Helical Loaded Cavity Backed Antenna, 60 mm rectangular base 5.5 dB is the measured gain. As shown in Fig 5.

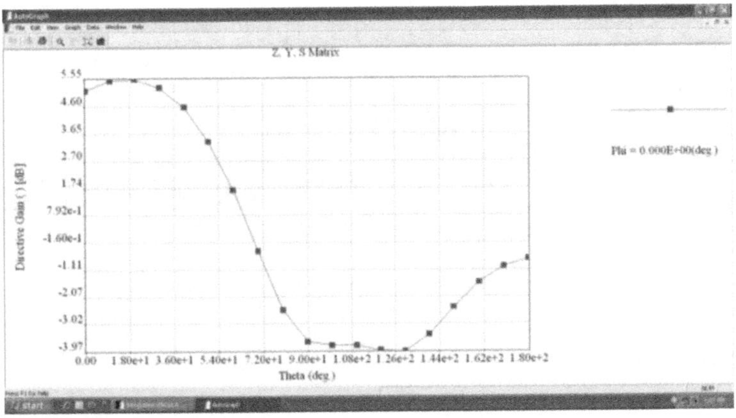

Fig. 5. The Directive Gain of Helical Loaded Cavity Backed Antenna with 60 mm diameter circular base, Rectangular plot

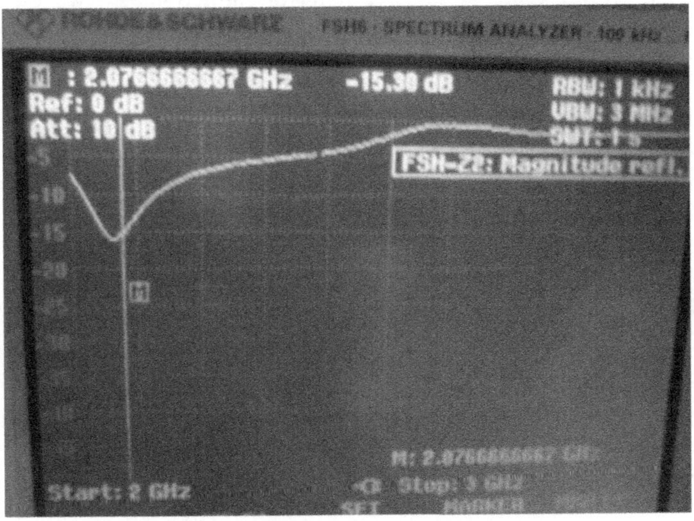

Fig. 6. Measurement of return loss S11=-15.30 dB

4 Results and Discussion

This antenna has only 1.64 turns which is unlike the conventional concepts of the helical antenna. The measured return loss S11=-15.30 dB for center frequency and 120 MHz is measured bandwidth of antenna, as shown in Fig 6. The Directive Gain of 5.55dB is achieved. The helix of 25.13mm circumference simulated with 0.75 λ

ground conductor it radiates in normal mode. It is conclude by loading helix in cavity the mode of operation is change from normal mode to axial mode.

Fig. 7. helical loaded cavity backed antenna, 60 mm rectangular base

5 Conclusions

The novel antenna that provides moderate gain and circular polarization over a wide bandwidth is fabricated and tested as shown in Fig 7. This new antenna occupies a much smaller volume. The size of the proposed antenna is reduced by a factor of 10, compared to a conventional helix. The helix circumference to wavelength ratio is reduced to 0.2, according to the classical design data for axial mode operation it is 0.8 to 1.2. The same size helical antenna radiate in normal mode without cavity. The compact size makes this antenna very attractive for satellite communications and aerospace applications.

By increasing the number of turns the gain of the antenna can be increase. The cavity is circular, the rectangular cavity can be considered and the optimum shape and size can be found out.

References

1. Kraus, J.D.: Antennas. McGraw Hill, New York (1988)
2. Djordjevic, A.R., Zajic, A.G.: Enhancing the gain of helical antennas by shaping the ground conductor. IEEE Antennas and Wireless Propagation Letters 5, 138–140 (2006)
3. Zverev, A., Blinchikoff, H.: Realization of a Filter with Helical Components. Transactions on IRE 8(3), 99–110 (1961)
4. Colline, R.E.: Foundation for Microwave Engineering. McGraw Hill, New York (1992)
5. Colantonio, D., Rosito, C.: A spaceborne telemetry loaded bifilar helical antenna for LEO satellites. In: Microwave and Optoelectronics Conference, pp. 741–745. IEEE Press, New York (2009)
6. Rabemanantsoa, J., Sharaiha, A.: Small-folded, printed quadrifilar helix antenna for GPS applications. In: 14th International Symposium on Antenna Technology and Applied Electromagnetics & the American Electromagnetics Conference, pp. 1–4. IEEE Press, New York (2010)

A Review for Supplier Selection Criteria and Methods

Ashish J. Deshmukh[1] and Archana A. Chaudhari[2]

[1] SVKMs NMIMS Mukesh Patel School Of Technology Management & Engineering, Mumbai
ashish.deshmukh@nmims.edu
[2] Atharva College Of Engineering, Mumbai
arch2309@rediimail.com

Abstract. Supplier selection is a multicriteria problem which includes both qualitative and quantitative factors. In order to select the best suppliers it is necessary to make a Trade off between these tangible and intangible factors.It has been well focused by Dickson since 1966. Dickson ranked the importance and placed 23 criteria of purchasing agents and managers. Weber *et al.* reviewed 74 articles from 1966 to 1991 concerning supplier selection criteria and methods. In this study, we have collected 49 articles from 1992 to 2007 and made a review according to Weber *et al.*'s method. Supplier selection criteria both in Dickson's 23 criteria and new developed ones are reviewed and compared with Weber *et al.*'s (1991) study. Supplier selection methods are summarized. Finally, conclusions future research is presented.

Keywords: selection of **supplier**, methods and criteria, SCM.

1 Introduction

In today's competitive operating environment it is impossible to successfully produce low cost, high quality products without satisfactory vendors. Thus one of the important purchasing decisions is the selection and maintenance of a competent group of suppliers. Selecting a good set of suppliers to work with is crucial to a company's success. In supplier selection decisions, two issues are of particular significance. One is what criteria should be used, and the other, what methods can be used to compare suppliers.

The objectives of the paper are twofold: One is to summarize the literature on supplier selection issues from 49 articles published during 1992 to 2007, particularly on supplier selection criteria and approaches. Based on that, the other is to compare with Weber *et al.*'s work (1991) to identify the differences under different circumstances. In the next section, a review of criteria is presented; also a comparison with Weber *et al.*'s (1991) study. In the third section, development of supplier selection criteria is discussed. Supplier selection methods are described in the fourth section. Conclusions and future research are given in the final section. The 49 articles examined in this paper are searched using key words "vendor selection" and "supplier selection", "SCM" from 1992 to 2007.

Over the years, the significance of supplier selection has been long recognized and emphasized. Lewis (1943) suggested that of all the responsibilities that related to

K. Shah, V.R. Lakshmi Gorty, and A. Phirke (Eds.): ICTSM 2011, CCIS 145, pp. 283–291, 2011.

purchasing, no one was more important than the selection of a proper source. England and Leenders (1975) made the same point by stating "supplier selection is purchaser's most important responsibility". Later, Weber *et al.* (1991) wrote, More recently, with emergence of the concept of Supply Chain Management (SCM), more and more scholars and practitioners have realized that supplier selection and management was a vehicle that can be used to increase the competitiveness of the entire supply chain (Lee *et al.*, 2001). Weber *et al.* (1991) pointed that supplier selection decisions were complicated by the fact that various criteria must be considered. Meanwhile, different approaches could be employed to make the selection. The analysis of the two issues of supplier selection have attracted the attention of many academicians and purchasing practitioners since the 1960's. Dickson (1966) identified 23 supplier selection criteria, which deeply influenced later researches in this area. In 1991, Weber *et al.* reviewed, and classified 74 related articles which had appeared since 1966. Specific attention was given to the impact on supplier selection of Just-In-Time (JIT). Weber *et al.*'s work provided an explicit overview on issues of supplier selection up to 1991. Since then, a large number of articles concerning supplier selection have been published. However, no reviewing work as Weber *et al.*'s (1991) has been undertaken.

2 Review of Criteria in Supplier Selection

Before the review, Dickson's study regarding the importance of the 23 criteria for supplier selection is firstly presented in Table 1.

Some observations made from Table 2 lists the number of articles in which each criterion was addressed, along with the rank and rating of the criteria in Dickson's (1966) study. The criteria mentioned in these two studies seem to have the analogical distributions. Table 2 indicates that net price, quality and delivery were discussed in 90%, 86% and 76% of the articles reviewed in this study respectively, distinctively more than any other criteria. They also occupied the top three positions in Weber *et al.*'s study with relatively smaller frequencies. These three criteria were rated as having 'extreme importance' or 'considerable importance' by Dickson. It implies that these three criteria have invariably been the most important ones in supplier selection decisions, and that there maybe more agreement on recognition of their importance. Production facilities and capacity takes the 4th position in both Weber *et al.*'s study and ours. Below that, there are differences in positions of the remaining 19 criteria, which may indicate changes in attentions of researchers and practitioners over the years. For example, financial position was discussed in only 9% of the articles in Weber *et al.*'s study, but 31% in this study.

The reason may be that in today's business environment, firms and their suppliers may form a more close relationship to enhance the whole supply chain's profit. Thus, supplier's financial position and stability may become more important. Another example is geographical location which was discussed in 11% of the articles in this study but 21% in Weber *et al.*'s study. The main reason is that with the economic globalization, suppliers are selected from the whole world, especially in the developing countries for their low cost and with the development of global logistics,

Table 1. Dickson's supplier selection criteria

RANK	FACTOR	MEAN RATING	EVALUATION
1	Quality	3.508	Extreme Importance
2	Delivery	3.417	
3	Performance History	2.998	
4	Warranties & Claims Policies	2.849	Considerable Importance
5	Production Facilities and Capacity	2.775	
6	Price	2.758	
7	Technical Capability	2.545	
8	Financial Position	2.514	
9	Procedural Compliance	2.488	
10	Communication System	2.426	
11	Reputation and Position in Industry	2.412	
12	Desire for Business	2.256	
13	Management and Organization	2.216	
14	Operating Controls	2.211	
15	Repair Service	2.187	
16	Attitude	2.120	Average Importance
17	Impression	2.054	
18	Packaging Ability	2.009	
19	Labor Relations Record	2.003	
20	Geographical Location	1.872	
21	Amount of Past Business	1.597	
22	Training Aids	1.537	
23	Reciprocal Arrangements	0.610	Slight Importance

Dickson (1966), 0=no importance, 4=extreme importance.

geographical location is no longer a major criterion for supplier selection. Yet another criterion which is of large difference between Weber *et al.*'s study and ours is communication system, 3% versus 11%. Information share is nowadays critical to the success of both parties in a supply chain. Thus building an effective communication system is important. Most other criteria are mentioned in less than 4% of the articles in both studies. Of the 49 articles, 29 or 60% were published in the last seven years. It seems that supplier selection issues got more attention in the competitive environment. It can also be seen that net price, quality and delivery have received the greatest amount of attention in the last seven years, appearing in 90%, 86% and 76% of the articles published in that time frame, respectively.

Table 2. Criteria discussed in annotated bibliography

Dickson (1966)		Criteria	This study			Weber *et al.*	
Rank	[a]Rating		[b]Rank	No	%	No.	%
6	1	Net Price	1	44	90	61	80
1	1A	Quality	2	42	86	40	53
2	1	Delivery	3	37	76	44	58
5	1	Production Facility & Capacity	4	22	45	23	30
7	1	Technical Capability	5	16	39	15	20
8	1	Financial Position	6	15	31	7	9
20	2	Geographical Location	7	5	11	16	21
13	2	Management & Organization	7	5	11	10	13
3	1	Performance History	7	5	11	7	9
14	2	Operating Controls	7	5	11	3	4
10	2	Communication System	7	5	11	2	3
11	2	Reputation & Position in Industry	12	3	7	8	11
15	2	Repair service	13	2	4	7	9
18	2	Packaging Ability	13	2	4	3	4
22	2	Training Aids	13	2	4	2	3
9	2	Procedural Compliance	13	2	4	2	3
19	2	Labor Relations Record	13	2	4	2	3
4	1	Warranties & Claims Policies	13	2	4	0	0
16	2	Attitude	19	1	2	6	8
23	3	Reciprocal Arrangement	19	1	2	2	3
17	2	Impression	21	0	0	2	3
12	2	Desire for Business	21	0	0	1	1
21	2	Amount of Past Business	21	0	0	1	1

[a] Ratings: 1A = Extreme importance. 1 = Considerable importance.
2 = Average importance. 3 = Slight importance.
[b] Rank: Based primarily on frequency in our study.

3 Development of Supplier Selection Criteria

There are top three criteria for supplier selection quality, delivery and net price. For quality, Dickson (1966) defined it as the ability to meet quality specifications consistently. In the new development, some specifies are added, for example, quality staff , experimentation and inspection (Choy and Lee, 2002, 2003) ISO9001 system (Lee *et al.*, 2003). Delivery as the ability to meet specified delivery schedules. Its meaning is extended into criteria such as delivery capacity (Karpak *et al.*, 1999); freight terms (Min, 1994); lead time (Youseef *et al.*, 1996); cycle time and JIT delivery capability (Bevilacqua and Petroni, 2002); shipment quality (Choy and Lee, 2002, 2003).

The meaning of net price is extended more. In Dickson's study, net price was defined as price offered by each vendor including freight and discounts charges. These include, fixed cost (Current and Weber, 1994); design cost and supplier cost (Gupta and Krishnan, 1999); quality cost, technology cost and after-sales service cost (Bhutta and Huq, 2002); inventory holding cost and fixed ordering cost (Tempelmeier, 2002);. In recent years, total cost of ownership (TCO) has become increasingly important. In addition to proliferation of the three core criteria, some new criteria are generated with the advance of management philosophy. First one is the product design and development by Person and Ellram (1995), and Chan (2003); product development and product improvement by Choy and Lee (2002, 2003); commitment to continuous improvement in product and process by Kannan and Tan (2003), another one is flexibility, including production flexibility and responsiveness to customers (Mummalaneni *et al.*, 1996); response to changes and process flexibility (Ghodsypour and Brien, 1998); flexibility in changing the order (Verma and Pullman, 1998); flexibility (Masella and Rangone, 2000); flexibility of response to customer's requirements (Bevilacqua and Petroni, 2002); reverse capacity or the ability to respond to unexpected demand (Kannan and Tan, 2003); quota flexibility (Kumar *et al.*, 2003). The third one is relationship between the buying firms and the suppliers. Mummalaneni *et al.* (1996) regarded highly quality of relationship with suppliers.

Another trend is global sourcing. Buying firms search their suppliers all over the world for cost reduction. In this circumstance, criteria such as political stability, foreign exchange rate and tariff and customer duties should be considered (Min, 1994; Motwani *et al.*, 1999). Also, environment issues start to be considered in supplier selection process in recent years (Zhu and Geng, 2001; Humphreys *et al.*, 2003).

4 Approaches to Supplier Selection

Weber *et al.* (1991) grouped the quantitative approaches to supplier selection into three categories: mathematical programming models, linear weighting models, and statistical / probabilistic approaches. As their study, in linear weighting models, usually a weight is placed on each criterion (typically subjectively determined) to reach a total score for each supplier by summing up his performance on the criteria multiplied by these weights. Mathematical programming models include linear programming, mixed integer programming, and goal programming. Statistical approaches include methods such as cluster analysis and stochastic economic order quantity (EOQ) model.

4.1 Linear Weighting Models

Linear weighting models used Analytic Hierarchy Process (AHP). The AHP is a 'modern multi-criteria decision making method' first structures the problem in the form of a hierarchy, to capture the criteria, subcriteria, and alternatives. All the criteria are compared fairly to determine their relative weights. Then, the alternatives are compared fairly with regard to each criterion. The final outcome of the procedure is a score for each alternative. AHP avoids the main drawback of the traditional linear scoring model, which assigns weights and scores arbitrarily. At the same time, it can make trade-off between the quantitative and qualitative criteria. It is a relatively practical method in supplier selection. Interpretive structural modeling(ISM) is another technique that has been applied to supplier selection. Its main goal is to identify and summarize relationships among items and form a structural model of the problem (Mandal and Deshmukh, 1994). Min (1994) used multi-attribute utility approach for international supplier selection, which is another method to determine the relative weights for the attributes. Kwong *et al.* (2002) combined scoring method and fuzzy expert system for supplier selection. Ozden Bayazit and Birsen Karpak(2005) used AHP for real-world case and presented along with sensitivity analysis.

4.2 Mathematical Programming Models

The purpose of a mathematical optimization method is to select several suppliers in order to maximize an objective function subject to supplier/buyer constraints. The objective function can be a 'single' criterion (classical optimization models) or 'multiple' criteria (goal programming or multi-objective programming) (Lee et al., 2003).

In this study, most of the 15 articles used mixed integer program, including single-objective and multi-objective. Weber and Current (1993) used multi-objective mixed integer program to minimize the total purchase price, late deliveries and rejected units. Chaudhry *et al.* (1993) used linear and mixed binary integer programming models to solve cost minimizing problems of vendor selection with price breaks. Rosenthal *et al.* (1995) developed a mixed integer linear program found the purchasing strategy for the buyer to minimize the total purchase cost. Ghodsypour and O'Brien (2001) presented a mixed integer non-linear programming model to solve the multiple sourcing problem, which took into account the total cost of logistics, including net price, storage, transportation and ordering costs. Buyer limitations on budget, quality, service, etc. Dahel (2003) presented a multi-objective mixed integer programming approach to simultaneously determine the number of vendors to employ and the order quantities.

Data envelop analysis (DEA) is a mathematical programming method for assessing the comparative efficiencies of decision-making units (DMUs). Weber *et al.* (1998), Braglia and Petroni (2000) and Liu *et al.* (2000) used this approaches for supplier selection problem. Kumar *et al.* (2003) used a fuzzy mixed integer goal programming for vendor selection problem that included three primary goals: minimizing the net cost, minimizing the net rejections, and minimizing the net late deliveries subjected to realistic constraints regarding buyer's demand, vendors' capacity, vendors' quota flexibility, purchase value of items, budget allocation to individual vendor, etc, because some of the parameters are fuzzy in nature.

4.3 Statistical Approach

Statistical approaches, were used in some articles. Mummalaneni *et al.* (1996) used conjoint analysis to find Chinese purchasing managers' preferences and trade-offs in supplier selection and performance evaluation. Verma and Pullman (1998) used discrete choice analysis to examine the choice of suppliers. Tracey and Tan (2001) employed confirmatory factor analysis and path analysis to examine empirically the relationships among supplier selection criteria, supplier involvement on design teams and in continuous improvement programs, customer satisfaction, and overall firm performance.

4.4 Other Methods

There are also some other methods employed in supplier selection problem solving. One is cost based method, including activity based cost (ABC) approach (Roodhooft and Konings, 1996), total cost of ownership (TCO) (Degraeve *et al.*, 2000), and transaction cost theory (Qu and Brocklehurst, 2003). Fuzzy logic approach is also employed (Bevilacqua and Petroni, 2002). Ghodsypour and Brien (1998) integrated AHP and linear programming for this problem, and Karpak *et al.* (1999) employed visual interactive goal programming.

5 Conclusion and Future Research

Supplier selection have attracted the interest of researchers since the 1960s. Dickson (1966) ranked the importance placed on 23 criteria. On the basis of this, Weber *et al.* (1991) reviewed 74 articles from 1966 to 1991 concerning supplier selection criteria and methods. We have collected 49 articles from 1992 to 2007 and made a review using similar methodology.

Of the 49 papers, except 4 in which no concrete criteria were presented, all other articles discussed more than one criterion. This shows that supplier selection decisions are of inherently multi-objective nature, as Dickson (1966) and Weber *et al.* (1991) stated. Dickson's (1966) 23 criteria seem to have the analogical distributions in this and Weber *et al.*'s (1991) study. Net price, quality and delivery were the most important criteria cited and were discussed in majority of the articles in this study.

Production facilities and capacity, technical capability and financial position, which were categorized as having 'considerable importance' by Dickson, were discussed in large proportion of the articles we reviewed. Financial position and communication system as supplier selection criteria got more attention from the literature reviewed in this study than that in Weber *et al.*'s (1991) study, while the reverse happened to geographical location. Other criteria, such as warranties and claims policies, procedural compliance, labor relation record, packing ability, training aids, attitude, reciprocal arrangement, impression, desire of business and amount of past business were much less mentioned in the articles reviewed in both studies. Since 1994, new criteria have been presented in the supplier selection articles. Some of them are extensions of Dickson's original criteria; some are generated with the development of management philosophy. While delivery and quality remain much the same, cost seems to be substituting net price. Product design and development, flexibility, and relationship with the suppliers are three newly developed criteria along with the development of supply chain management (SCM).

AHP is most used in linear weighting models. Mixed integer programming is dominant in mathematical programming models. For all these mathematical programming articles, computational cases were presented and solved with computer software, which indicates that with the technology development, the mathematical programming model can be easily simulated and solved by computer.

For future research, two aspects, supplier selection criteria and methods, will continue to be the focus. For supplier selection criteria, combining supply chain performance measurement and supplier selection seems to be an important area. Some new criteria to reflect the whole supply chain performance should be developed in the process of supplier selection. The methods mentioned in this study have shortcomings in dealing with the selection problem. New method to simulate the process of human decision making, such as neural network, seems to be promising, and the computer programming for supplier selection should also be developed.

References

1. Amid, A., Ghodsypour, S.H., O' Brien, C.: Fuzzy multiobjective linear model for the Supplier Selection in Supply Chain. International Journal of Production Economics 105, 394–407 (2006)
2. Bevilacqua, M., Petroni, A.: From traditional purchasing to supplier management: a fuzzy logic-based approach to supplier selection. International Journal of Logistics 5(3), 235–255 (2002)
3. Bevilacqua, M., Ciarapica, F.E., Giacchetta, M.: A Fuzzy- QF approach to Supplier Selection. Journal of Purchasing and Supply Management 12, 14–27 (2006)
4. Bhutta, K.S., Huq, F.: Supplier selection problem: a comparison of total cost of ownership and analytical hierarchy process approach. Supply Chain Management: An International Journal 7(3), 126–135 (2002)
5. Chan, F.T.S., Kumar, N.: Global supplier development considering risk factor using Fuzzy extended AHP- based approach. Omega 35, 417–431 (2007)
6. Chopra, S., Meindle, P.: Supply Chain Management: Strategy, Planning, and Operation, pp. 1–24. Prentice Hall, Inc., Upper Saddle River (2001)
7. Current, J., Weber, C.: Application of facility location modeling constructs to vendor selection problems. European Journal of Operation Research 76, 387–392 (1994)
8. Dahel, N.E.: Vendor selection and order quantity allocation in volume discount environments. Supply Chain Management: An International Journal 8(4), 335–342 (2003)
9. Dickson, G.W.: An analysis of vendor selection systems and decisions. Journal of Purchasing 2(1), 5–17 (1966)
10. Ghodsypour, S.H., O'Brien, C.: A decision support system for supplier selection using an integrated analytical hierarchy process and linear programming. International Journal of Production Economics 56-57, 199–212 (1998)
11. Ghodsypour, S.H., O'Brien, C.: The total cost of logistics in supplier selection, under 16 conditions of multiple sourcing, multiple criteria and capacity constrains. International Journal of Production Economics 73, 15–27 (2001)
12. Gupta, S., Krishnan, V.: Integrated components and supplier selection for a product family. Production and Operation Management 8(2), 163–182 (1999)
13. Kumar, M., Vrat, P., Shankar, R.: A fuzzy goal programming approach for vendor selection problem in a supply chain. Computers & Industry Engineering (2003); Available online December 5, 2003
14. Lee, H., Wellan, D.M.: Vendor survey plan: a selection strategy for JIT/TQM suppliers. Industrial Management and Data Systems 93(6), 8–13 (1993)
15. Mandal, A., Deshmukh, S.G.: Vendor selection using interpretive structural modeling (ISM). International Journal of Operations & Production Management 14(6), 52–59 (1994)
16. Sevkil, M., Lennykoh, S.C., Demirbag, S.M., Tatoglu, E.: Hybrid analytical hierarchy process model for supplier selection. Industrial Management and Data system 108, 122–142 (2007)

17. Motwani, J., Youssef, M., Kathawala, Y., Futch, E.: Supplier selection in developing countries: a model development. Integrated Manufacturing System 10(3), 154–161 (1999)
18. Patton III, W.E.: Use of human judgment models in industrial buyers' vendor selection decisions. Industrial Marketing Management 25, 135–149 (1996)
19. Rosenthal, E.C., Zydiak, J.L., Chaudhry, S.S.: Vendor selection with bundling. Decision Science 26(1), 35–48 (1995)
20. Satty, T.H.: How to make a decision: the analytic hierarchy process. Interface 24(6), 19–43 (1994)
21. Ting, S.-C., Cho, D.I.: An Integrated Approach For Supplier selection and Purchasing decisions. Supply Chain Management: An International Journal 13/2, 116–127 (2007)
22. Verma, R., Pullman, M.E.: An analysis of the supplier selection process. Omega 26(6), 739–750 (1998)
23. Weber, C.A., Current, J.R., Benton, W.C.: Vendor selection criteria and methods. European Journal of Operational Research 50, 2–18 (1991)
24. Weber, C.A., Current, J.R.: A multiobjective approach to vendor selection. European Journal of Operational Research 68, 173–184 (1993)

Petri Net Model for Knowledge-Based Value Chain

U.N. Niranjan, Salma Itagi, and Biju R. Mohan

Department of Information Technology,
National institute of Technology Karnataka,
Surathkal, India
{un.niranjan,salma03.itagi}@gmail.com, biju@nitk.ac.in

Abstract. In this paper a novel theoretical model is presented in which the dynamics in a knowledge-based value chain is modeled using Petri nets. From the generic scheme of a knowledge-based value chain, the various components are individually modeled. The theory of Petri nets aptly captures the evolution of knowledge in a system and the process is usually highly interactive in nature. The properties and analysis methods of various classes of Petri nets can be conveniently used to check various constraints while designing the system.

Keywords: Knowledge-Based System; Value Chain; Petri Nets.

1 Introduction

Technologies such as the intranets [1], extranets, and groupware have facilitated extensive knowledge transfer in the e-world. Knowledge portals play a key role in knowledge transfer. Various theoretical frameworks [3, 5] have been proposed for the modelling and analysis of knowledge-based systems. Theoretical approaches are popular in the context of semantic web [7] and digital library [2] also.

Various classes and numerous variants of Petri nets [6, 8] have been used to model supply chains, and other processes. In this paper, some of the general ideas and techniques are adopted to model the generic knowledge-based value chain, which is henceforth referred to as the "system".

In Section 2, some essential definitions are recalled and the schematic diagram of a generic knowledge-based value chain is presented. In Section 3, the modelling of individual components of the system is done using Petri nets. Section 4 presents some further directions for research. Section 5 concludes this paper.

2 Knowledge-Based Value Chain

A comprehensive discussion of the concepts of value chains, knowledge transfer, and related concepts can be found in [4]. The following definition provides a suitable starting point.

Definition 1: A value chain is a way of organizing the primary and secondary activities of a business so that each activity provides productivity to the total operation of the business.

The formal definition [6] of a Place-Transition (PT) net is recalled in brief.

K. Shah, V.R. Lakshmi Gorty, and A. Phirke (Eds.): ICTSM 2011, CCIS 145, pp. 292–296, 2011.

Definition 2: A PT net is a 5-tuple $N=\{P, T, I^-, I^+, M_0\}$ where P is a nonempty set of places, T is a nonempty set of transitions, $I^-, I^+: P\times T\rightarrow N_0$ are the backward and forward incidence functions, and M_0 is the initial marking. Tokens capture any quantity that flows in the system. A transition is enabled only when at least as many tokens as given by the arc weight are located on that place. Firing will destroy exactly this many tokens from that place and create in the receiving place as many number of tokens as given by the weight on the arc from the transition to the next place.

In any system, especially with respect to business, raw materials are taken as inputs. Value is added at each stage of processing. Then, the output is sold to the customers. The knowledge-based value chain is a useful way of looking at an organization's knowledge activities and at how various knowledge exchanges add value to adjacent activities and to the company in general. The schematic picture of a generic knowledge-based value chain [4] is given in figure 1.

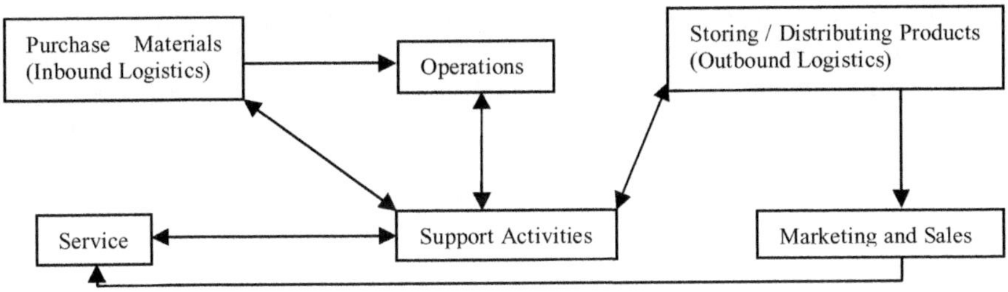

Fig. 1. A generic knowledge-based value chain

The knowledge in the system is abstracted by tokens and the flow is described by a firing sequence. In figure 1, the inbound logistics is marked initially. It is to be noted that the support activities are very "interactive" components of the system and consequently enhance the connectivity in the system. These may themselves be composed of subsystems like infrastructure, resources, and other technology [9].

3 Petri Nets for Individual Components

3.1 Inbound Logistics

The inbound logistics serves as the starting point which can be marked initially. For simplicity of modelling, the influx of raw materials is denoted by a single token which is present in this place initially. If more details are available from the system design, the generation of tokens and also the weights on the arcs can be accordingly varied. For instance, it may well be considered as a source element with respect to the system in which case the PT net of the inbound logistics cannot be both live and bounded [6].

Here, the n sources of input are denoted by n places of the PT net. If a certain quantity of some raw material is required, the constraint can be incorporated in the weights on the arcs. Also, the places may take feedback from other components, especially the support activities.

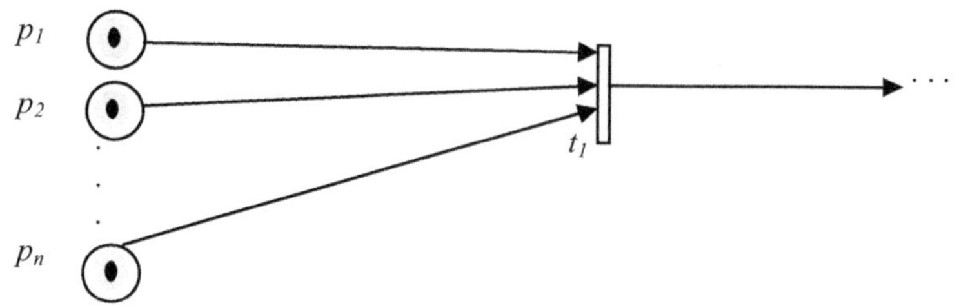

Fig. 2. PT net for inbound logistics

3.2 Operations

The operations denote that phase in the knowledge-based value chain that accounts for the processing parts. This is initially unmarked. A token appears in this place when an enabled transition preceding it fires, i.e., the inbound logistics phase enables its output transition. When the transitions succeeding these places in this phase are fired, it signifies the end of processing, and passing of the token (knowledge activities) to the support activities. Note that the output may go to different transitions since a bunch of support activities may require the results from processing a certain input. In the present context, this makes the PT net of this phase belong to the class of simple or SPL nets [6]. The following definition is recalled in simple words.

Definition 3: Simple nets are those PT nets which do not have multiple places that may enable a single transition.

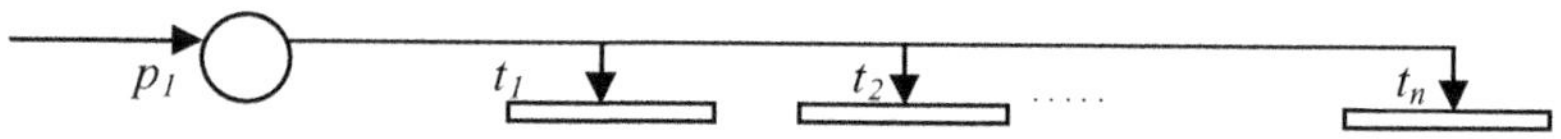

Fig. 3. PT net for operations

3.3 Outbound Logistics

The outbound logistics serves as the stage which interacts with the external environment surrounding the organization. Initially, there are no tokens corresponding to the distribution of end products since this phase waits for input from the preceding phases that carry out the processing. It may take the services of some support activities and usually directs the results to the marketing and sales. Hence, the classes of state machines and marked graphs [6] would suffice to characterize this phase.

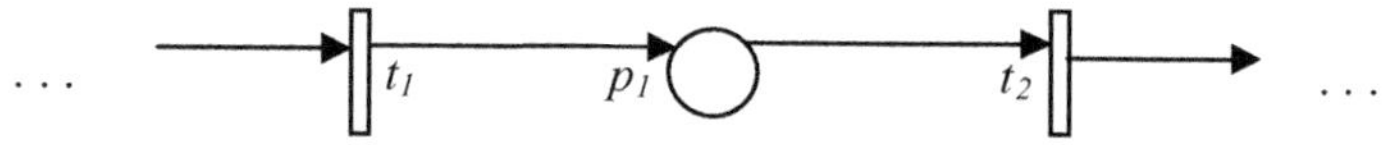

Fig. 4. PT net for outbound logistics

3.4 Marketing and Sales

The marketing and sales phase varies in intensity with respect to the knowledge activities of the particular organization under consideration. For small scale systems, it

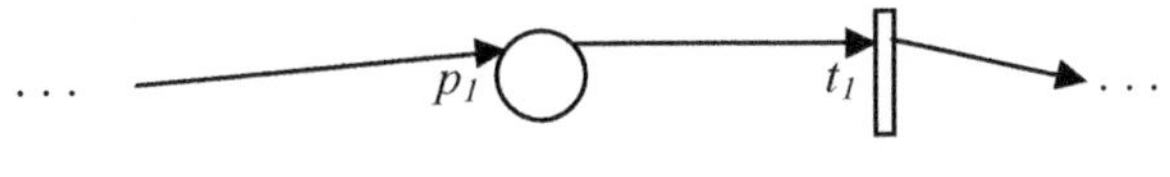

Fig. 5. PT net for marketing and sales

may also be optional since the functionalities of this phase can be integrated into other phases, thus reducing the overall complexity of the system.

3.5 Service

The services serve as the stage which actively interacts with the support activities. If the marketing and sales phase is separately mentioned in the requirements, then services need input from the marketing phases also, as shown in figure 1. Again the weights on the arcs may vary depending on the nature of the support activities. It has a structure similar to that of the operations phase.

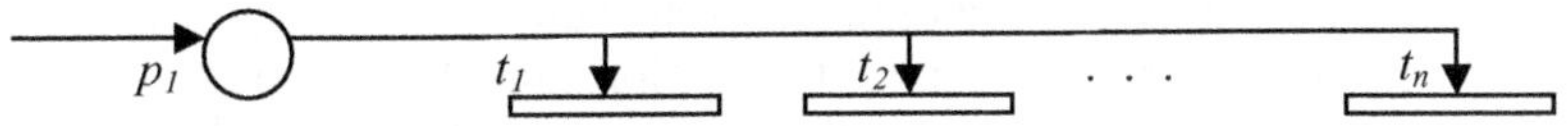

Fig. 6. PT net for service

3.6 Support Activities

The support activities may be considerably complex and may themselves be composed of subsystems demanding individual modeling. However, in the present context, for simplicity it is considered as an atomic component, which cannot be decomposed further. It is to be noted that support activities usually cater to the requisites of most other components and hence, interact with other units forming the central hub of the system. Since the interaction of support activities with other components is bidirectional, it can be viewed to be composed of two places, one having arcs directed toward it and the other having arcs away from it. These may be synchronization points for the various knowledge activities in the system. They can be combined if stochastic behavior is imposed and the probabilities on the arcs would filter the enabling of transitions. Such details may be obtained from previously available statistics or from the requirements elicitation phase.

Fig. 7. PT net for support activities

4 Future Implications

From the previous section, the design of the individual components can be used to obtain the complete system design. The weights on the arcs determine the enabling of various transitions. Depending on the specific requirements of the organization, the system can be designed using other extensions of Petri nets also. For example, continuous Petri nets obtained by "fluidification" [8] of their discrete counterparts provide various advantages. The computational complexity of the analysis and control of complex systems may be significantly reduced. Dimension of the state space does not become unwieldy and the approximation strategy does not introduce too many errors.

Analogous to the modelling presented here, the subsystems in support activities can also be modelled depending on the specifications that arise in the context. Hybrid models [8] of Petri nets may again be suitable for certain components of the system. Since Petri nets are popular tools in the field of concurrency control, in the present context, the co-ordination of knowledge activities within an organization can be potentially modelled in terms of constraints related to concurrency control.

5 Conclusion

In this paper, a generic theoretical framework for capturing the essence of knowledge-based value chains has been proposed by using Petri nets. An approach such as the one adopted in this paper can be used to develop knowledge management tools and automation techniques for knowledge engineering based on this. The advantage with this approach is that the dynamics is clearly described by a Petri net formalism and the well-developed techniques of analysis such as those based on reachability set and coverability tree in Petri nets can be easily applied.

References

1. Anton, K.: Effective Intranet Publishing: Getting Critical Knowledge to Any Employee, Anywhere. Intranet Design Magazine, 1–5 (August 12, 2000)
2. Parirokh, M., Daneshgar, F., Fattahi, R.: A Theoretical Framework for Development of a Customer Knowledge Management System for Academic Libraries. In: World Library and Information Congress: 75th IFLA General Conference and Council, Milan, Italy (2009)
3. Lai, H., Chu, T.: Knowledge management: A review of Theoretical Frameworks and Industrial Cases. In: Proc. of the 33 Hawaii International Conference on System Sciences (2000)
4. Awad, E.M., Ghaziri, H.M.: Knowledge Management. Pearson Education Inc., London (2004)
5. Kuo, H., Chen-Burger, Y., Robertson, D.: Knowledge Management using Business Process Modelling and Workflow Techniques. Advanced Knowledge Technologies (2004)
6. Bause, F., Kritzinger, P.S.: Stochastic Petri Nets: An Introduction to the Theory, 2nd edn. Vieweg & Sohn Verlagesselschaft mbH, Braunschweig/Wiesbagen (2002)
7. Li, Y., Thompson, S., Tan, Z., Giles, N., Gharib, H.: Beyond ontology construction; ontology services as online knowledge sharing communities. In: Fensel, D., Sycara, K., Mylopoulos, J. (eds.) ISWC 2003. LNCS, vol. 2870, pp. 469–481. Springer, Heidelberg (2003)
8. Kordic, V. (ed.): Petri Net: Theory and Applications. I-Tech Education and Publishing (2008)
9. Turati, C., Dino Ruta, C.: Technology in Knowledge-Based Value Chain. In: Proc. of the Portland International Conference on Management of Engineering and Technology, vol. 1, p. 82 (2001)

A Data-Driven View of the Evolving IT Infrastructure Technologies and Options

Tapati Bandopadhyay[1], Pradeep Kumar[2], and Anil K Saini[3]

[1] NMIMS University, Bangalore Campus, India
[2] Visiting Professor, Canada
[3] USMS, GGS Indraprastha University, Delhi, India
tapti.bandopadhyay@nmims.edu,
pkgarg12@yahoo.com,
aksaini1960@gmail.com

Abstract. For effective service delivery on any infrastructure and SOA-based environment, data is an essential building block both as inputs as well as in monitoring the output. Data is an integral part of any formation delivery and/or service delivery in an IT-enabled business environment. For the new-age IT infrastructure models with cloud and virtualization technology options and web-based service-delivery models, recording of the characteristic data parameters that define or govern the run-time service delivery environment is crucial both for client businesses and provider organizations to be cost-viable and quality-effective. In this paper, we create a data-driven framework of the new-age IT Infrastructure. Operationally, this framework can be used for IT-user organizations as well as service providers, to track their delivery concerns. Strategically, this framework may serve as a baseline template to plan for a minimal set of parameters recoded as operational data-points from the real-time run-time IT infrastructure operations environment.

Keywords: IT infrastructure, cloud, data parameters.

1 Introduction

SOA (Service oriented Architecture) is a reality now. Modern IT infrastructure and operations are getting modelled for SOA, even though understanding of SOA and maturity of SOA application practices are still in a nascent stage. Cloud computing and SPI-as-a-Service (SPI: Software/Platform/Infrastructure) are fast becoming realized dreams, based on SOA. Modern enterprises are aggressively pursuing cloud IT strategies for making their IT infrastructure and operations agile and lean. SOA's basic concepts, for example the Enterprise Service Bus (ESB) can be thought as the equivalent of a ubiquitous service provider channel or pipeline. ESB is the newest category of enterprise infrastructure software. Broadly speaking, ESBs attempt to provide a service-oriented, high performance, standards-based business integration infrastructure. The Enterprise Service Bus is an integral concept of SOA- Service Oriented Architecture. Service Oriented Architecture concepts open up loads of promises and possibilities [6] in terms of integration of services and flexible, dynamic composition of services.

K. Shah, V.R. Lakshmi Gorty, and A. Phirke (Eds.): ICTSM 2011, CCIS 145, pp. 297–304, 2011.

SOA has been inclusive of various other approaches like:

- EAI (Enterprise Application integration),
- ASPs (Application Service Providers)
- ITIL and ITSM (IT services management) standards

In a broad-based, flexible, all-encompassing fashion, it includes all resource aspects of the IT infrastructure of any organization (e.g. as proposed by ITIL- Information Technology infrastructure Library) [9] and is primarily driven by the ever-increasing business needs of integration and flexibility of service compositions.

Delivery of any services, be it IT-enabled or any other traditional forms of business services, depend on data. [3,5] Therefore, data integration is crucial for SOA to be successful.

Consequently, various data integration approaches have been suggested to achieve maximum functional and/or services integration, including UDWH (Universal Data warehouse), DKM (Distributed Knowledge Management) architecture and so on.
In this context of service-based software and IT delivery, service level management and supplier relationship management become very crucial for a user-business to survive both economically and operationally. Therefore, measurability becomes an imperative for better controls. For ensuring quality of measurability, right definitions of metrics that are business-relevant, are must. In this paper, this particular facet of service-based delivery is developed, with example parameters and functions, which encapsulate the business impacts and operational criticalities of the service-based delivery models.

2 The New-Age IT Infrastructure Model with Possible Technology Options

2.1 Evolving IT Infrastructure

IT infrastructure and operations are gradually moving towards SOA (Service Oriented Architecture), providing on-demand and usage-based service delivery modes. Everything in the world of application software even ERP systems are s attempted to be conceived and delivered as a service. ERP-as-a-Service has become a reality by Ramco offering ERP on the clouds for SMB's (Small and Medium Businesses) that could not afford an expensive ERP suite. SAP has followed this trend and has made a flexible variant of My SAP to be offered in a hosted cloud environment, even though not as mature, popular of cost-effective as Ramco On Demand ERP 2.0 (RODE 2.0) that can be deployed within 7 days, and is a pay-per-use model. The IT operations delivery space has already seen Software as a Service, Storage as a service, Platform as a service, Infrastructure as a service and Analytics as a service, etc. Therefore, XaaS i.e. X as anything as a service, is catching up as a lean, agile, flexible, on-demand, cost-efficient platform with minimum legacy obligations of IT user industries.

In the technology front, this value proposition stands on three legs, namely:

1. Virtualization
2. Cloud computing
3. SOA based SaaS

These options are making lean IT a realistic possibility for IT service consuming industries, incl. SMBs. According to a latest industry report, >67% user industries are optimistic about server virtualization, 56% on apps consolidation and rationalization, 51% bullish on Blade servers. VMWare's vCloud Service Director and success of Project Redwood, on vSphere as a virtualization management layer have made the dream of a truly virtual platform a reality. For a new VM(virtual machine), processes or even data-sharing or communication threads can move seamlessly between private – public cloud infrastructure: a value proposition that some people call as hybrid [Hybrid cloud has been broadly defined little differently though i.e. a static combination of private, public, community cloud. But here, it assumes that hybrid clouds are dynamic with vMotion i.e. virtual movement of service assets]. Similarly, Virtual Desktop Infrastructure- VDI View 4.5 with Virtual profile feature and form acquired from RTO software, are enhancing VD functionalities to integrated media convergence and bleeding-edge platforms. PC-over-IP, security servers are ensuring security over virtual desktop environment. Network virtualization and media convergence with VMworld TV and network virtual chassis platform, network OS(as an engine), and hypervisor(like a Gearbox) are all available options now, not any work of fiction or dreams of IT user industries. These are offering learn options while remaining focused on operations, backup-recovery, and critical BCP-DRP (Business Continuity Plans, Disaster Recovery Plans) policies.

2.2 Emerging Business Models

Based on the new technology options available for IT infrastructure, new business models are also emerging fast. For example:

1) Cloud Brokerage, integration and third-party cloud management services. For example, Cap Gemini specializes in this for governance-security- integrators domain.
2) Cloud service brokers, cloud infra providers (value-added distributors, value-added resellers), cloud apps providers (like Google apps, Facebook, Amazon Relational Database Services) etc. IBM by October-end 2010 has tried Infospehere information on demand in their Cloud computing lab at Hursley UK on a Test cloud.
3) cloud tech providers (VMWare, NetApps, Microsoft Hyper V, Cisco, then CA, BMC, EMC2 for SLM).
4) specialist vertical clouds (e.g. for Auto OEMs, for healthcare, for retail), TCS offering 'Bank-in-a-Box, ITaaS (Nano of IT services)- PaaS- SaaS for its BFSI product offered on cloud for Grameen Bank, SMB's.
5) specialist horizontal clouds e.g. finance/ HR (Infosys using HR product for hosting salary-apps)/ CRM(Salesforce.com) / SCM (i2 will be available).
6) hybrid clouds.
7) Specialist legacy vendors to support legacy management!

Technologies SP's (Service Providers) are making these options available. With hyper-competition leading to consolidation, these are fast becoming the most viable alternative realities for businesses. For example Oracle after acquiring Hyperion has stopped selling Oracle Apps Services offerings, as Hyperion has better application suites that oracle, in these product lines. Essbase is also a proven multi-dimensional Database platform, ideal for data warehousing. So Oracle is keeping both Essbase and its own Oracle 11g DWH platforms, but is consolidating apps based on Hyperion. Now, for a client with legacy Oracle apps will have to still be supported for them, for that APIs (Application Programs for Interfaces) to integrate communication channels between Hyperion apps and Oracle apps are being built. Now these API-based services can be offered by Oracle On-demand as Services. This creates a win-win situation for both Oracle as well as its clients.

3 The New-Age IT Infrastructure and Parameters to Be Monitored

The services and their constituent processes along with the technology options available to deliver them has been discussed in section 2. In section 3, the mapping of these technology and services options is presented. Then the data elements that are required to support smooth running of this kind of IT infrastructure, are given.

These parameters are only indicative and not exhaustive. The pre-defined processes for capturing the data parameters can be used as a starting point or a baseline, to be incremented by:

1) Each unique organization's IT infrastructure and sourcing models
2) Their business priorities.
3) Their IT operational criticalities and business impacts
4) Their data governance policies
5) Their service providers or partners relationships in terms of operational-level data sharing agreements
6) On their enterprise data architecture and the enterprise requirements that are getting outsourced to the cloud.

Therefore, these can be thought as a minimal set of generic data parameters that are commonly relevant for any IT-enabled enterprise.

By having these generic set of data parameters, organizations have a start-point and a template that can lead towards the right directions in terms of a data-driven IT infrastructure and operations management.

From figure 1, it can be seen that the generic architecture of IT infrastructure with the latest technology options has five main layers:

1) Layer 1: the underlying data layer – this is very similar to any enterprise system also, for example the 3-tier architecture of SAP R/3 with the data-application (business logic)-presentation (transactional interfaces, forms and reports)
2) Layer 2: An intelligent network layer named here as the 'Thinking Network'. This is a multi-gateway, multi-protocol network that has embedded intelligence in form of 'services-on-the-fly' modules for various services e.g. Data Cleaning, Deduplication, Compression and Encryption as Services, and various optimizers and accelerators as APIs.
3) Layer 3: The application layer hosting the business apps, transactional apps and information apps etc.

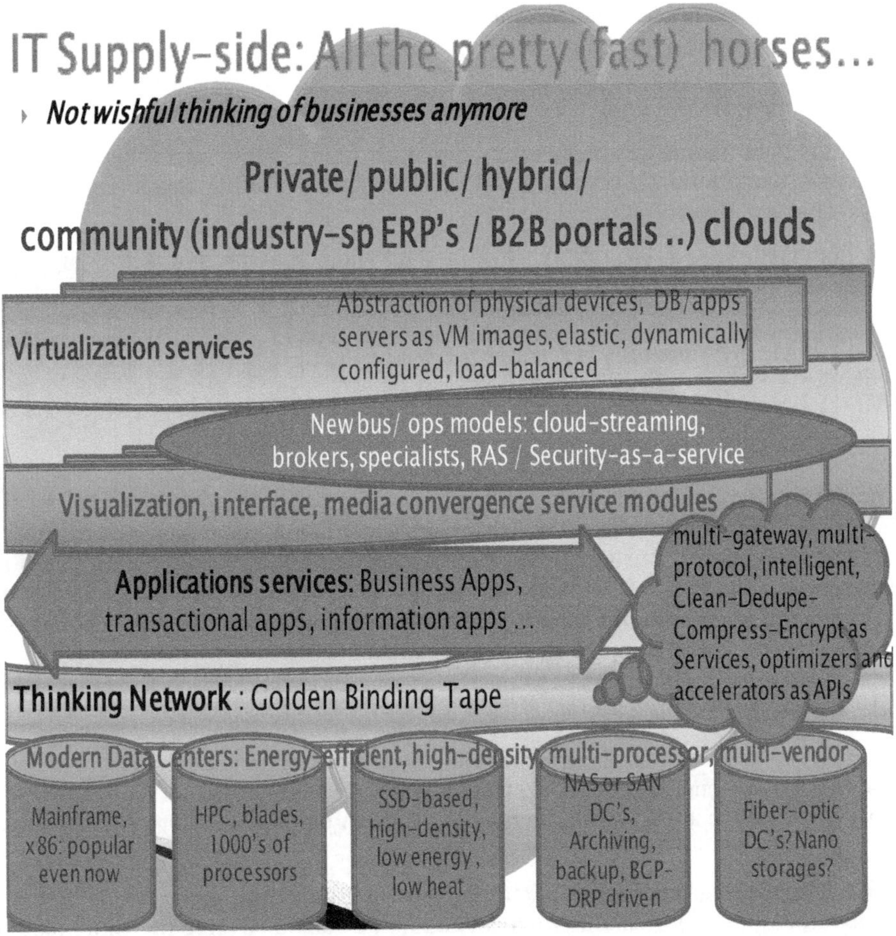

Fig. 1. New-age IT Infrastructure and technology options

4) Layer 4: Presentation and visualization layer – where content and media convergence like integration of channels including video/audio/telephone/digital can happen as services as the processing happening on the fly.
5) Layer 5: the virtualization technology actually is not just the top layer, but it is like a base, like the basic cement that holds the entire infrastructure together. At every level there will be virtualization e.g. virtual DCs(Data Centers), virtual network management by abstraction of networking devices and managing them seamlessly, virtual desktops as interfaces etc.

Given the scenario in figure 1 as an alternative reality that is available today, it is imperative that each of these service offerings need to be measured in terms of:

1. Quality
2. Reliability
3. Availability
4. Serviceability

5. Business impact(of downtime for example)
6. Cost efficiency
7. Capacity utilization
8. Other SLA (Service Level Agreements) compliance issues

From the data parameters captured as shown in figure 2 [not an exhaustive list, but indicative and a minimal set as a baseline], the service offering characteristic metrics can be measured as follows:

$$\text{Overall Quality-of-Service(QoS)} = \text{function\_of\_parameters } \{2, 4, 5, 6, 7, 8, 9, 10\} \quad (1)$$

$$\text{Reliability} = \text{function\_of\_parameters } \{2, 4, 6, 7, 9, 10\} \quad (2)$$

$$\text{Availability} = \text{function\_of\_parameters } \{1, 3, 4, 5, 6, 7, 9\} \quad (3)$$

$$\text{Capacity utilization} = \text{function\_of\_parameters } \{1, 2, 3, 9\} \quad (4)$$

- As indicative examples.

The functions represented in (1), (2), (3) and (4) can be either empirical i.e. like MTTF and MTBF, or specified by businesses according to their criticalities, priorities and business impact. Business impact in fact becomes part of each metric definition. The scope of the template as suggested in this paper is to identify the necessary data parameters that describe the new-age IT infrastructure service offering adequately so that the functions can compute using these available data.

3.1 Examples in Applications

Example 1: In case of an un-optimized or ill-intended code is blocking a VM Image instance for too long, [parameter 4] that indicates 'Gobbling' up the VM resources i.e. other processes that are waiting in queues or are in pipeline are not getting access to the VM instances.

=> If happening across multiple instances at the same time, this can be a severe capacity problem in a virtual environment and may potentially bring the entire platform down.

=>Hence, parameter 4 has to be continuously tracked and monitored and alerts must trigger some action the moment an event is sensed or is predicted to happen.

=> This implies that functions (1), (2) and (3) are to be invoked and used in such SLM environments.

Example 2: When a reporting application required very high quality data, it can request for data check service for source authentication, service for cross-referencing of values, and services for statistical validation of data that may require studying each variable- its distribution characteristics, detection and transformation as required to 'cure' any data from any statistical property violations i.e. 'abnormalities'.

=>All these services are likely to fail if parameters 6-10 do not work within their operational limits.

=> This implies that functions (1) and (2) must be invoked as they include most of the parameters from 6 to 10. Function (3) can be selectively invoked as this contains 3 of the critical five parameters.

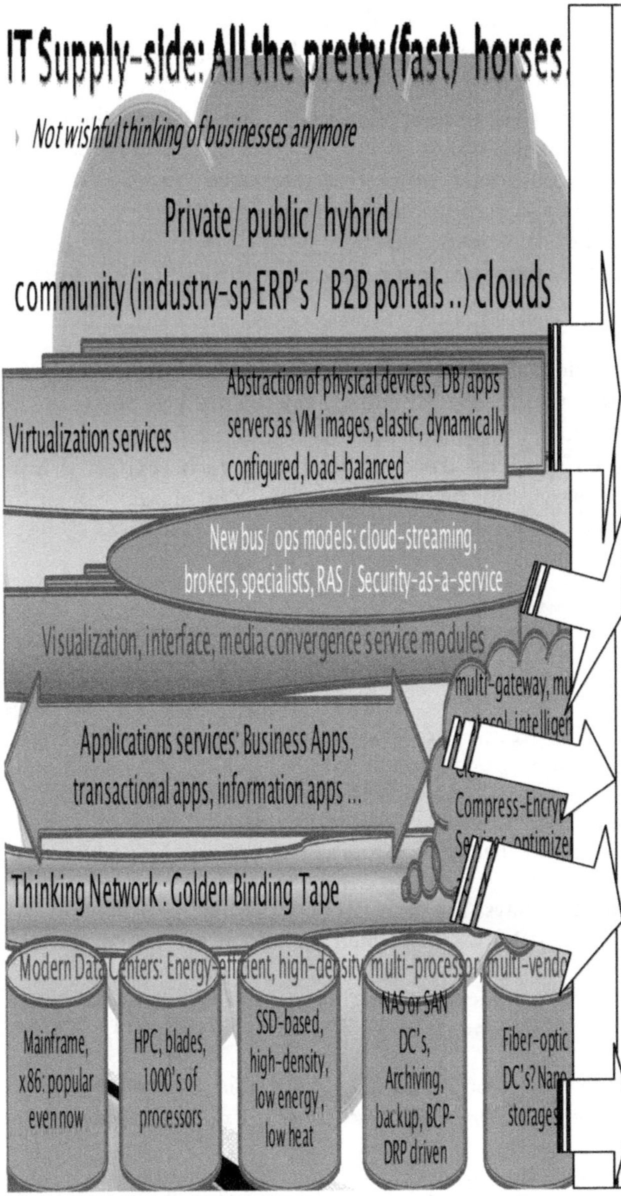

Pre-defined processes for collecting and monitoring following example data parameters:

1) Average VM images/ instances running during peak-load, medium-load, and lean-load
2) Distribution of durations- with mean, range and std. deviation- of VM's running several processes and time taken per process
3) No. of VM clones and templates and their utilization metrics
4) No. of instances where a process locks in a VM resource for more than the set acceptable threshold limit
5) No. of escalations per brokerage service providers per month, and their corresponding business impacts due to delay/ data loss/ opportunity losses
6) No. of apps failure instances and their business impacts- study of distributions
7) No. of network failures with root-cause identified and attributed, analyzed
8) No. of logic failures of network apps and their business impact
9) No. of data requests turned down or failed
10) No. of data errors and business impact

Fig. 2. Pre-defined data parameters and processes for capturing them

4 Conclusion

This paper presents and explains a data-driven approach for managing the new-age IT infra-structure operations and available technology options. The advantages of a simple parameters repository like the one proposed and explained with examples here, are manifold, for example:

1) This gives a holistic, integrated view of IT infrastructure management from the viewpoint of businesses more than of technology. For example, virtualization technologies like VMWare have their own logs and metrics that compute efficiencies of that technology platform in particular, but it cannot extend the impact calculations to business loss due to failure of running other processes, or network failures etc. Together, in an integrated view. This integrated view is extremely crucial for business-oriented IT operations management.
2) The template is simple and easily stretchable and can be easily contextualized or customized based on specific organizational requirements from IT infrastructure and operations.

The concept therefore can be practically extended to application in various organizations for a more holistic evaluation of IT infrastructure's operational efficacy issues.

It can also be extended technically for example while designing the APIs (Application Programming Interfaces) for the data parameters to be extracted from the run-time environments, or integrating them in a common XML-based format for better utilization, analysis and failure prevention.

References

1. Arkin, A., Askary, S., Bloch, B., Curbera, F., Goland, Y., Kartha, N., Liu, C.K., Thatte, S., Yendluri, P., Yiu, A.: OASIS Web Service Business Process Execution Language V2.0 Working Draft (May 2005), http://www.oasis-open.org/committees/download.php/12791/wsbpel-specification-draft-May-20-2005.html (accessed on May 2009)
2. Chappel, D.A.: Enterprise Service Bus. O'Reilly, Sebastopol (2006)
3. Cox, D., Kreger, H.: Management of the Service-Oriented-Architecture Life Cycle. IBM Systems Journal 44(4), 709–726 (2005)
4. Forrester Research, Inc.: Survey Data Says: The Time For SOA Is Now (April 2006)
5. Laguna, M., Marklund, J.: Business Process Modelling, Simulation and Design. Prentice Hall, Englewood Cliffs (2005)
6. De Paul, W., Lei, M., Pring, E., Villard, L., Arnold, M., Morar, J.F.: Web Services Navigator: Visualizing the Execution of Web Services. IBM Systems Journal 44(4), 821–846 (2005)
7. Portier, B., Budinsky, F.: Introduction to Service Data Objects, September 28 (2004), http://www-106.ibm.com/developerworks/java/library/j-sdo/ (accessed on May 2009)
8. Sadtler, C., Cotignola, D., Crabtree, B., Michel, P.: Patterns: Broker Interactions for Intra- and Inter- Enterprise. IBM Redbooks, SG24-6075, OpenDocument (2004), http://www.redbooks.ibm.com/redbooks.nsf/0/532fca172da15c6c85256d6d0046192e? (accessed on May 2009)
9. Schmidt, M.-T., Hutchison, B., Lambros, P., Phippen, R.: The Enterprise Service Bus: Making Service-Oriented Architecture Real. IBM Systems Journal 44(4), 781–798 (2005)

Impact of Information Technology on Supply Chain Capabilities – A Study of Indian Organizations

Srinivasan Sundar and Ganesan Kannabiran

Bharathidasan Institute of Management, Trichy
National Institute of Technology, Trichy

Abstract. Superior supply chain capabilities have become a potentially strategic route to gaining competitive advantage. This study examines the influence of Information technology usage on supply chain capabilities of organizations. The current study hypothesized that the usage of IT resources, facilitate supply chain capabilities. A Structural equation model was developed to test the relationships between IT usage, IT Advancement and IT alignment and Supply chain capabilities Findings are drawn from the analysis of the primary data collected from 307 managers from companies across eight manufacturing industry segments. The research findings support the hypothesis that the usage of IT has led to greater IT Advancement (advanced technology utilization) and IT Alignment (investment in IT to align with channel partners). The results indicate that such IT resources have a significant influence on supply chain capabilities.

Keywords: IT Advancement, IT Alignment, Supply chain capabilities, Indian organizations.

1 Introduction

A supply chain is a system of people, activities, information, and resources involved in producing a product and then moving it to reach the end-customer. Many organizations attempt to integrate and closely coordinate the various elements of their supply chains in order to enhance efficiency. They have invested heavily in Information Technology in the supply chain with the principal belief that they will gain competitive advantage in today's highly dynamic and changing business market. This study seeks to study the usage of IT in downstream supply chain processes manufacturing firms and its influence on the supply chain capabilities of firms.

2 Theoretical Background and Research Hypothesis

The usage of Information Technology (IT) in managing the supply chain processes has drawn increasing attention from the corporate world. Forrester research indicates that manufacturers are increasingly dependent on IT to improve supply chain agility, reduce cycle time, achieve higher efficiency and deliver products to customers in a timely manner [1]. Indian manufacturing companies have made substantial investments

K. Shah, V.R. Lakshmi Gorty, and A. Phirke (Eds.): ICTSM 2011, CCIS 145, pp. 305–311, 2011.

in IT in the supply chain typically in applications like Enterprise resource planning (ERP), Supply chain management (SCM), Customer relationship management (CRM) and invested in real time linkage with their branch offices to bring about process improvements that will have a definite impact on supply chain capabilities. Since Indian manufacturing organizations have made such massive investments in IT, it has become essential to evaluate whether such investments have led to enhanced supply chain capabilities.

2.1 IT Advancement

IT advancement is defined as the extent to which a firm adopts the most sophisticated available technology [2]. IT advancement is likely to be an important firm resource as the literature argues that firms with advanced technology outperform their competitors [3]. The IT enabled SCM helps the firm to gain a competitive position in the market by increasing efficiency in information and product flow across channel members, from the inception to distribution of the product.

The advancement of IT can result in stronger supply chain capabilities in four ways. First advanced IT can help to enhance the speed, quality and quantity of information transferred [4]. It can help to achieve better coordination and reduce transaction cost between partners [5]. It can help to improve interfirm collaboration between partners. Each collaborating partner focuses on its unique competency and, working together, the partners can achieve operational excellence that synergistically creates value [6]. Finally IT advancement can enhance greatly the ability of channel partners to respond to market changes in a timely manner [7].

2.2 IT Alignment

IT alignment in this study is defined as the extent to which a firm's IT investment is made to ensure compatibility with that of its channel partners. IT alignment reflects the degree of embeddedness of IT across the supply chain, and it requires channel partners to coordinate and align their business processes with each other in order to achieve efficiency [8]. Investment in IT by the organization and the channel partners is needed to maintain this alignment as new technological investments are made. The advancement and alignment of IT are equally important for the functional adequacy of the supply chain system [9].

2.3 Supply Chain Capabilities

The research literature on supply chain management has consistently identified four dimensions that have been positively affecting supply chain capabilities viz. information sharing, coordination, collaboration and responsiveness. These four dimensions have been discussed individually in past literature. Tracey et al. [10] have identified three types of SCM capabilities – outside-in capabilities (e.g. inbound transportation, warehousing), inside-out capabilities (e.g. packaging, outbound transportation) and spanning capabilities (e.g. customer order processing, information dissemination) and have thus typically used logistic capabilities.

Since the capabilities identified by Tracey et al. [10] are too narrow in their focus – mainly a logistics perspective, it is proposed to go with the four capability dimensions identified by Wu et al [2] – Information sharing, Coordination, Collaboration and Supply chain responsiveness - for the present research. Wu et al. [2] proposed to refer to these four dimensions together as Supply chain capabilities - a higher order construct encompassing all these four capabilities.

The above discussion leads us to the following hypotheses which are tested in this study:

H1: Greater the extent of IT usage, higher the level of IT advancement in firms.
H2: Greater the extent of IT usage, higher the alignment of the firm with channel partners.
H3: IT advancement has a positive impact on IT alignment.
H4: IT advancement in firms will influence supply chain capabilities positively.
H5: IT alignment in firms will influence supply chain capabilities positively.

3 Methodology

The study has adopted single cross sectional survey research design.In keeping with the scope of the research framework; a survey questionnaire was designed to capture the responses of respondents on the perceived impact of usage of IT on supply chain capabilities. The survey questionnaire was mailed to 975 manufacturing organizations across various industry segments in India having an asset base of Rs.500 crores and above (Information obtained from CMIE Prowess Data Base) which is used as the sampling frame for this study. A total of 307 responses were obtained from the downstream supply chain managers. These 307 respondent managers belong to eight broad industry segments. The respondents were asked to fill out the questionnaire where quantitative responses were measured using a five-point scale. The data has been analyzed through Structural Equation Modeling technique with the help of Amos 18.0 package.

4 Analysis and Results

Structural equation modeling (SEM) is a popular statistical data analysis technique to develop and test theory as well as construct validation. SEM takes a confirmatory (i.e., hypothesis-testing), rather than an exploratory, approach to the data analysis. Confirmatory factor analysis (CFA) allows researchers to specify relations between the observed variables and the underlying constructs *a priori*, and to analyze the causal relationships between the latent constructs that are measured by observed variables.

Instrument assessment is an important part of testing the theoretical research model.. CFA is performed by assessing the constructs for unidimensionality, reliability and validity. When the unidimensionality of a construct has been established, reliability and construct validity can be subsequently investigated. Unidimensionality mean that a set of measured variables (indicators) has only one underlying construct. After assessing unidimensionality, the measures are investigated for reliability and construct validity. Reliability refers to the consistency of the measurement. That is,

Table 1. Unidimensionality, Reliability and Validity results for constructs

Factor/ construct	No. of items	Cronbach Alpha	Composite Reliability	AVE	CFI*	NFI**
Extent of IT Usage	4	0.854	0.903	0.700	1.00	0.997
IT Advancement	4	0.822	0.884	0.656	0.997	0.993
IT Alignment	4	0.867	0.910	0.718	0.984	0.981
Supply chain capabilities	4	0.863	0.907	0.711	0.992	0.989
1. *CFI=Comparative Fit Index >0.90 indicates Unidimensionality						
1. Cronbach alpha >.60 confirms construct reliability.						
1. ** NFI=Normed Fit Index(Bentler-Bonnet Index) >0.90 & Composite reliability >.70 indicates Convergent Validity						

the degree to which the set of indicators in a latent construct are internally consistent in their measurements. The constructs are examined for convergent validity which implies that the indicators for a specific construct should converge or share a high proportion of variance in common.

The results of the tests for Unidimensionality, Reliability and Validity are presented in Table 1. The AVE values in Table 1 were found to be greater than the squared correlation estimates among the variables in the latent construct and thus also provided good evidence of discriminant validity.

A structural equation model consists of two major sub-models – a measurement model and a structural model. The measurement model describes the relationships between a latent variable or theoretical construct and its observed variables or indicators. The structural model defines the relationships or paths among latent constructs. Adopting this two step approach in SEM, the measurement model was first tested for fit and the results are indicated in the Table 2. The fit indices are all above .90 indicating good fit of the measurement models.

The structural model to assess the effects of IT usage, IT Advancement and IT Alignment on supply chain capabilities were assessed. The Standardized path

Table 2. Measurement model fit indices results

S. No.	Construct	Chi square	GFI	NFI	CFI	RMSEA
1.	Extent of IT Usage(ITU)	1.686	0.997	0.997	1.000	0.000
2.	IT Advancement (ITAD)	3.217	0.995	0.993	0.997	0.045
3.	IT Alignment(ITAL)	12.002	0.981	0.981	0.984	0.042
4.	Supply chain capabilities (SCC)	6.796	0.989	0.989	0.992	0.089

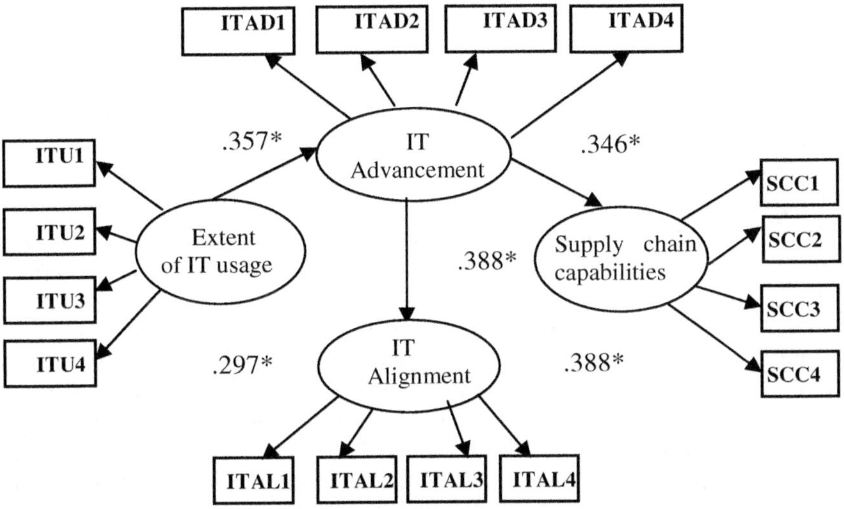

Chisquare:149.68 df:99 CMIN/df=1.513. GFI: .944 NFI: .942 CFI: .979 RMSEA: .041 ***p<.001.

Fig. 1. Impact of IT usage on Supply chain capabilities – test results

coefficients and the overall fit statistics which show excellent model fit for the structural model are shown in Figure 1.

Regarding the hypothesized relationships, the SEM analysis finds all the relationships significant and the entire hypotheses are supported. Table 3 presents the results of the SEM analysis for the five hypotheses proposed in this study.

Table 3. SEM Analysis – Results of tests of Hypothesis

Structural Path	**Hypothesis**	**Beta Coefficient**	**t-value**	**Decision**
Extent of IT Usage → IT Advancement	H1	0.370	5.321***	Supported
Extent of IT Usage → IT Alignment	H2	0.449	6.423***	Supported
IT Advancement → IT Alignment	H3	0.388	5.391***	Supported
IT Advancement → Supply chain Capabilities	H4	0.357	5.670***	Supported
IT Alignment → Supply chain capabilities	H5	0.424	6.599***	Supported
*****p < 0.001 (Significant at 99.9% Confidence level)**				
Beta => Structural Path Coefficient / Standardized Partial Regression Coefficient				

5 Discussion and Implications

The study provides a theoretical research framework that identifies positive and significant relationships between constructs that were developed for IT resources and supply chain capabilities. The study has confirmed the impact of usage of IT on Supply chain capabilities. The greater the extent of IT Usage leads to the higher level of IT Advancement in firms as they strive to seek a differential advantage over competing firms in their industry. The present study also accepts that IT usage leads to greater usage of advanced technology through *Hypothesis 1* which is supported in this study.

Greater IT usage leads simultaneously to IT Alignment to achieve higher system compatibility and integration between channel partners. When newer and more effect technologies are introduced and used by organizations, there is pressure to adopt it across the trading partners and thus greater use of Information Technology and IT Advancement leads to greater levels of IT Alignment across downstream channel partners. Thus the current study accepts the previous research findings in this area through *Hypothesis 2 and 3* which are supported in this study.

Proactive use of the latest IT solutions which are more advanced than the competition helps firms to gain better supply chain capabilities and thus confirms the prior research. A crucial factor that influences Supply Chain capabilities along with IT Advancement is IT resource alignment across channel partners. Firms can raise their supply chain capabilities not only by deploying advanced technologies but when its channel partners invests in having a compatible technology in the context of supply chain IT investments. Thus the research confirms that IT advancement and alignment positively influence supply chain capabilities through Hypothesis 4 and 5 which find support in the SEM analysis.

The study and the findings provide practical implications for managers in the downstream supply chain. The findings of this research assure supply chain managers that IT investments in downstream supply chain have a positive impact on supply chain capabilities. It gives a valuable guidance for managers in their approach to IT investments. Investment in IT advancement should not be concentrating merely on technology upgrades. They should find out if the advancements in technology would lead to improved functional capabilities in the supply chain. Along with such investments, the study demonstrates that IT alignment with channel partners is necessary for creating good supply chain capabilities. Organizations can gain good levels of supply chain capabilities when its channel partners have compatible technology with that of the organization. Alignment of IT resources along with IT advancement leads to superior supply chain capabilities.

6 Conclusion

The antecedents to creating good supply chain capabilities arise not only from extensive IT usage but through advanced IT usage and well aligned IT deployment. Traditionally, defending market share through product superiority and pricing was the strategy. Today however competitive advantage for a firm depends on anticipating and responding quickly to changing market needs. The creation of superior competencies is achieved by creating these four dimensions of capabilities in the supply chain. Future research can focus on studying the impact of supply chain capabilities on firm performance.

References

1. Rajdou, N.: U.S. manufacturers' supply chain mandate: increase IT funding and unleash innovation. World Trade (December 2003)
2. Wu, F., Yeniyurt, S., Kim, D., Cavusgil, S.T.: The impact of information technology on supply chain capabilities and firm performance: A resource-based view. Ind. Mark. Mgmt. 35, 493–504 (2006)
3. Rogers, D.S., Daugherty, P.J., Stank, T.P.: Enhancing service responsiveness: The strategic potential of EDI. Log. Inf. Mgmt. 6, 27–32 (1993)
4. Booth, M.E., Philip, G.: Technology driven and competency driven approaches to competitiveness: are they reconcilable? J. Info. Tech. 11, 143–159 (1996)
5. Clemons, E.K., Row, M.C.: Information Technology and Industrial Cooperation: The Changing Economics of Coordination and Ownership. J. Mgmt. Info. Sys. 9, 9–28 (1992)
6. Bowersox, D.J., Closs, D.J., Drayer, R.W.: The Digital Transformation: Technology and beyond. Sup. Chn. Mgmt. Rev. 9, 22–29 (2005)
7. Stank, T.P., Daugherty, P.J., Autry, C.W.: Collaborative planning: supporting automatic replenishment programs. Sup. Chn. Mgmt. 4, 75–85 (1999)
8. Powell, T.C.: Organizational alignment as competitive advantage. Strat. Mgmt. J. 13, 119–134 (1992)
9. Hausman, A., Stock, J.R.: Adoption and implementation of technological innovations within long-term relationships. J. Bus. Res. 56, 681–686 (2003)
10. Tracey, M., Lim, J.S., Vonderembse, M.A.: The impact of supply-chain management capabilities on business performance. Sup. Chn. Mgmt. 10, 179–191 (2005)

Describing a Decision Support System for Nuisance Management of Urban Building Sites

Pierre Hankach[1], Mohamed Chachoua[2],
Jean-marc Martin[1], and Yann Goyat[1]

[1] IFSTTAR
Route de Bouaye, BP 4129, 44341 Bouguenais France
{pierre.hankach,yann.goyat,jean-marc.martin}@lcpc.fr
[2] Ecole des Ingénieurs de la Ville de Paris
15 rue Fénelon 75010 Paris France
chachoua@eivp-paris.fr

Abstract. In this paper, a decision support system for managing urban building sites nuisances is described. First, the decision process for nuisance management is studied in order to understand the use context of the decision support system. Two levels are identified where decision support is appropriate: at the territorial level for the administrator of the public space and at the building site level for the project owner. The decision support system at the former level is described. The interaction interface and the different functionalities of the system are detailed, showing the decision aid brought to the user and the type of processing associated with it.

1 Introduction

Construction, reconstruction or maintenance works on urban road networks are necessary but are usually associated with a deterioration of the quality of residents' life because of the produced nuisance. Therefore, in order to maintain the generated nuisance at an acceptable level, these works need to be managed efficiently.

But managing the nuisances of urban building sites is very complex. In this context, the project FURET, sponsored by the ANR (French national research agency), was set up to deal with this problem. The goal is to supply decision makers with decision support systems in order to meet the social demand to minimize the nuisances, therefore improving the acceptability of building sites. Nuisances of all kinds are concerned: the traffic and the induced congestion, noise, vibration, air pollution, dust, odours, obstructing the movement of residents, economic losses, etc.

The objective of decision support systems developed in FURET is to evaluate the nuisance of urban building sites and to promote the choice of better intervention strategies. In this paper, we describe these decision systems, their features and functionalities, as well as their context of use.

The first challenge of building a decision support system is to understand the context of its use. In order to do so, we need to identify the levels of action on the nuisance. Two levels have been indentified (section 2) depending on the skills of the decision makers. The first is the territorial level where the decision maker is the administrator of

K. Shah, V.R. Lakshmi Gorty, and A. Phirke (Eds.): ICTSM 2011, CCIS 145, pp. 312–320, 2011.

the public space. At this level, acting on the nuisance is done by organizing and coordinating all the building sites on the whole territory. The second is the building site level where the decision maker is the project's owner. At this level, acting on the nuisance is done while choosing the methods and means of intervention.

Therefore, two dedicated systems are built, one for each level of action. In this paper, we mainly describe the decision support system addressed to the administrator of the public space (section 3). This system is described in terms of its use cases, different features, the sophistication of these features, etc. The decision support offered through the system is a hybrid approach of two categories. In the first, the decision maker is supported by supplying nuisance information in an adequate form (adapted interface, alerts ...) so she can choose the appropriate actions. This type of decision support requires the integration of nuisance estimators (traffic, noise ...) to the system. The second involves active participation of the system in the choice of actions. It requires, in addition to the models for estimating the nuisances, to develop methods of multi-criteria decision aid to compare and classify actions.

2 Decision-Making in the Nuisance Management Area

Before building a decision support system, it is essential to understand the process of human decision-making for managing nuisances. By doing so, the use context of the system is better understood and the needs of the decision makers are better addressed.

In the following, we investigate the types and context of decisions made in the nuisance management area. In section 2.1, the decision makers and their interactions are indentified. Based on this, two decision levels are indentified in section 2.2: the territorial level and the building site level. The former is further developed in section 2.3.

2.1 The Decision Makers and Their Interaction

Since the first steps of a construction project and until its delivery, it goes through several stages where different actors get involved and interact ([10] and [11]):

1. The project is defined by the project owner after identifying the needs
2. The project owner contacts the administrator of the public space to get the necessary permits
3. After getting the necessary authorizations, the technical aspects of the construction project are defined
4. A tender is organized to select an execution company
5. Beginning of construction works.

2.2 Two Decision Levels

Decision support for nuisance management has been deemed necessary at two distinct levels (figure 1):

- The first level where the administrator of the public space validates the work permit requests is called territorial level. The stake at this level is to optimize nuisance production for a set of building sites on a territory. Nuisance management is addressed by coordinating all building sites through the validation process.

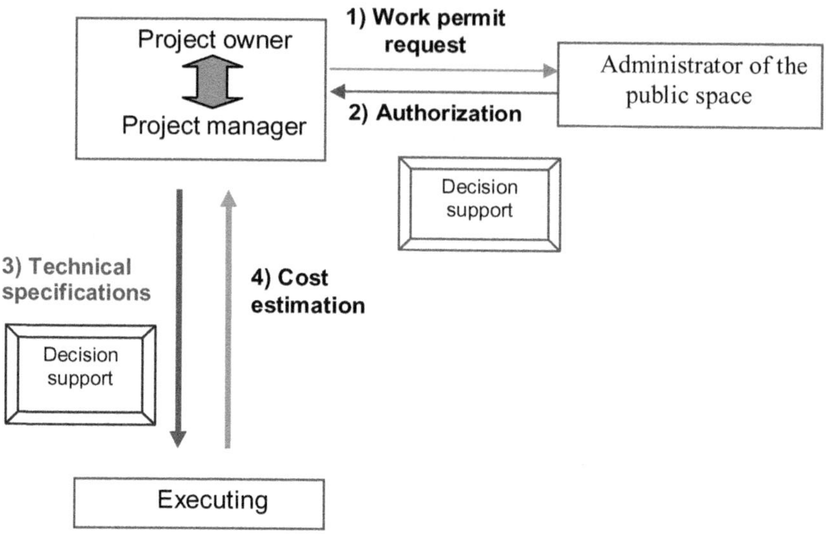

Fig. 1. Simplified interaction schema between decision makers in construction works. Decision support for nuisance management is marked where a need for it has been identified.

- The second level where the technical specifications of a construction project are defined is called the building site level. It primarily concerns to the project's owner. The stake at this level is to optimize the nuisance production of one building site in its context. At this level, nuisance management is addressed by altering technical specification in order to meet the desired goals in terms of nuisance reduction.

As one can see, at each of these two levels, we have a different decision maker with a different visibility, different data and different levers. Therefore at each level, a specialized decision system is put in place.

2.3 The Decision Level of the Administrator of the Public Space

In this paper, we focus on decision support supplied to the administrator of public space to manage the nuisances at the territorial level. The administrator of public space is faced with a continuous flow of work permit requests from project owners, and is entitled to accept or refuse these requests, proposing changes in order to achieve a better coordination with other accepted projects.

Such decisions by the administrator have a significant impact on nuisance outputs. Organizing projects on the territory in two different configurations produces different amounts of nuisance. Thus, the administrator's objective is to find the configuration that produces the least harm.

The administrator's decision context is characterized by an overall vision of the territory which allows her to coordinate various projects to improve the stealth of the whole. On the other hand, the data available at this level are limited and relates to the location of operations, their general description and the deadlines.

In conclusion, at this level the administrator manages nuisance by granting permits and formulating constraints on project owners. These decisions have a great influence on the state of nuisance on the territory. Notably, the impact on traffic whose state depends on the interference of all the building sites in place is important.

3 Decision Support for Urban Work Nuisance Management at the Administrator's Level

As it has been shown in our previous studies [2], the decision process allows choosing an action among a set of possible actions. The chosen action allows producing the desired outcome. To this goal, two main categories of approaches are possible:

1. The first category consists of extracting and presenting, for each action, the most relevant information to the human decision maker. So, by using these relevant information, the human can choose the better action, thus make the best decision.
2. In the second approach information about actions are processed automatically and all possible actions are classified based on some decision criteria. In this case, the best decision is the most highly ranked action of the ordered set.

For the decisional system described in this paper, a hybrid approach (that we have developed in [2] and that allows among others to process qualitatively the uncertain information) of the above two categories is used.

3.1 Decision Support by Extracting and Presenting Relevant Information

To assist the administrator in his manual programming of building sites, the system computes adequate information and presents them through a specially tailored GIS (geographic information system).

In fact, the administrator is solicited with a continuous flow of work permit requests. For each request one of the following actions can be taken by the administrator: the request is accepted; the request is refused; the request is accepted but with constraints (there are two types of constraints: constraints on indicators such as thresholds on the nuisances and constraints on accompanying measures for operations).

In the following, the different necessary components for information presentation and extraction to assist the administrator are discussed.

3.1.1 The Interface

The interface (figure 2) allows adequate visualization of relevant information by administrators so that they are in the best dispositions for taking decisions. It permits to its users to:

- visualize building sites on the territory as well as an estimation of associated nuisances (traffic, noise…)
- respond to work permits requests according to their repercussions on nuisances
- receive alerts corresponding to anomalies and thresholds set previously

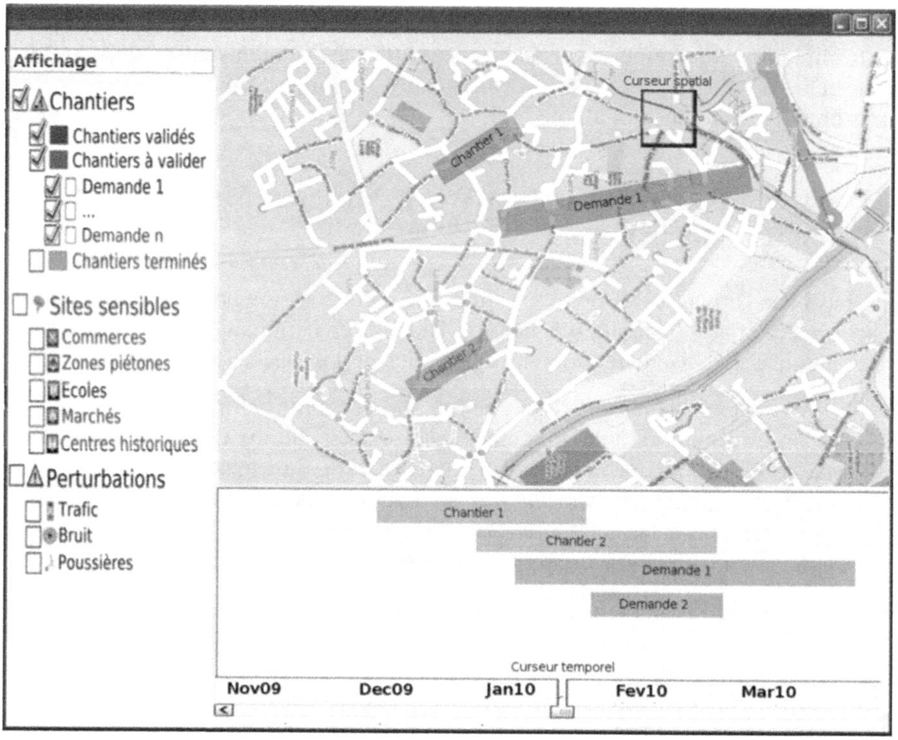

Fig. 2. The decision support system interface. In this example, building sites and work permit requests are displayed: green is used for validated permits; red for requests not validated yet and yellow for completed projects.

In order to respond to the administrator's needs, spatial and temporal information about building sites and their associated nuisances need to be displayed and effectively navigated. To do so, the interface is defined with the following components:

1. Control window: allows the selection of objects (building sites, nuisances…) for display in the spatial and temporal windows.
2. Spatial window: displays on a map the selected objects in the control window.
3. Temporal window: displays in a Gantt chart selected objects in the control window
4. Map cursor: indicates the position on the map of local objects (such as noise estimations) displayed in the temporal window. By moving this indicator, the display of some objects in the temporal window is changed.
5. Time Slider: indicates the temporality of the map displayed in the spatial window. By setting this indicator to a different date, the display in the spatial window is changed. The new map represents the state of the territory at the new date.

3.1.2 Nuisance Estimation

When validating work permit requests, information that allows the administrator to predict the consequences of her validations on nuisances is essential. Thus, to extract this kind of information, simulators for traffic and noise nuisances (among others) are integrated. They are used to make a projection of the new state of nuisance on the territory if new work permits are validated.

Traffic

To estimate the state of traffic, the SUMO simulator is used ([4], [6] and [7]). Traffic is evaluated by taking into account all the planned projects, i.e. projects validated by the administrator. As shown in figure 3, traffic information is displayed in the spatial window (where the map's temporality is designed by the time slider) and in the temporal window (to show the temporal evolution of traffic for the location pointed by the map cursor).

The following colour code is used to distinguish the different states of traffic (depending on the speed of traffic or delays experienced by users): Green to indicate that there is no disturbance; yellow to indicate an average disturbance and Red to indicate a strong disturbance.

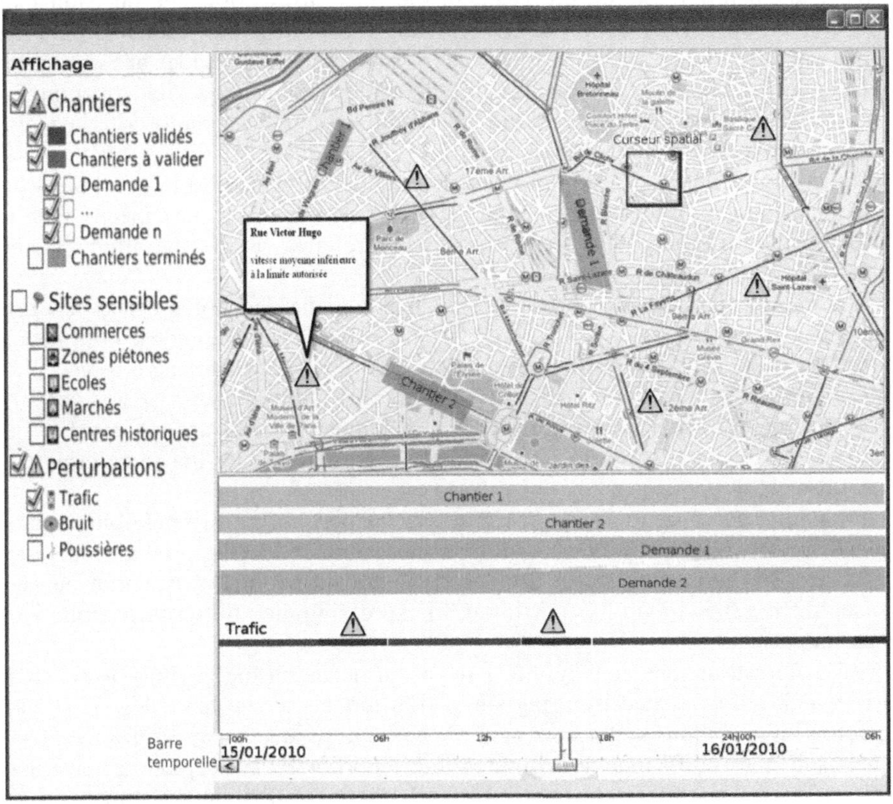

Fig. 3. This figure shows how the system displays traffic information both in spatial and temporal windows. Alerts are represented by graphical indicators. When these indicators are pointed, the corresponding warnings are displayed.

Noise

Like traffic, the estimation and display of noise pollution is achieved through the integration of a simplified model for noise simulation ([3], [5], [8] and [9]) which uses data from the local context and data on the mitigation provisions taken by project owners. Three levels of noise pollution generated by construction sites have been retained: low, medium, high.

3.1.3 Alerts

In the case where the nuisances exceed the levels defined by the administrator in his territorial strategy, the system is programmed to display alerts (figure 3). To compute these alerts, the system uses the simulators (introduced in section 3.1.2) to estimate the nuisance level. Once these estimations are realized, they are compared to the values tolerated.

3.2 Decision Support by Automatically Classifying Actions

In the manual programming presented in the previous section, the decision maker chooses the appropriate action after reviewing the information (about different actions) computed by the system. However, in some cases the decider has no expertise to make such a choice or even the multitude of decision criteria yields a complexity that prevents her from doing so. In these cases, the system uses multi-criteria decision algorithms to automatically classify and prioritize actions.

The object of this paper being to describe the decision support system as a whole, these algorithms will not be detailed here. However, we will expose briefly the modeling of the multi-criteria decision problem while referring to the algorithmic methods for its resolving.

In a multi-criteria approach of decision making, the decisional space is a set of strategies (or actions) S: {S1, S2... Sn}. In our case, this set is composed of the possible actions (as discussed in the previous section) that we reduce here to acceptance (S_1) and refusal (S_2) of a work permit for simplicity.

As every multi-criteria decision problem, different scenarios are evaluated on decision criteria. So each scenario S_i is formally represented with the same criteria set: <C1, C2, ..., Cm>.

Intuitively, deciding to accept or refuse a work permit (the two possible scenarios) depends on the utility of a construction project compared to the produced amount of nuisances. So the criteria set is composed of the estimations of different nuisances (noise, traffic…) and a utility function $u(S_i)$ expressing the utility of accepting a construction project.

Nuisance estimations convey the state of nuisances on the territory for each scenario which are evaluated by using simulation models as we have described above. Therefore, it is obvious that accepting a work permit produces more nuisances. On the other hand, the utility function captures the importance of accepting a construction project. The utility of the refusal scenario $u(S_2)$ is set to 0 while the utility of accepting $u(S_1)$ is comprised between 0 and 1. It is set based on expert information (saved in a table) about the type of the project (for example road network maintenance works have a high utility value).

The conflicting nature of decision criteria can be easily seen: nuisance production criteria favor the refusal scenario (S_2) while the utility criteria favor the acceptance one (S_1). The multi-criteria decision algorithm is about choosing the scenario which is a better compromise considering the whole set of criteria.

In order to chose the best scenario, the ELECTRE [1] family of multi-criteria decision analysis methods are used.

4 Conclusion

In this paper, a decision support system for managing nuisance of urban building sites has been described. First, the human decision-making process for nuisance management has been studied in order to understand the use context of the decision support systems. Two levels where bringing decision support is appropriate have been indentified: at the territorial level for the administrator of the public space and at the building site level the project owner.

The decision support system has been further described at the former level. The interaction interface has been detailed as well as the different functionalities of the system, showing the decision aid brought to the user and the type of processing associated with it.

This system is currently being implemented. The components for decision support by extracting and presenting relevant information such as the interface, the database backend, integration with nuisance estimators, and the basic decision aid functions such alerts are already functional. An evaluation of the system has been started with its main users (the administrators) in order to assess the system's usability.

Currently, multi-criteria decision methods for automatically classifying actions are being integrated.

References

1. Bernard, R.: Classement et choix en présence de points de vue multiples (la méthode ELECTRE). La Revue d'Informatique et de Recherche Opérationelle (RIRO), 57–75 (1968)
2. Chachoua, M.: Toward a qualitative decision theory under uncertainty. International Review on Computers and Software 5(5), 562–568 (2010)
3. CETUR: Bruit et formes urbanises : propagation du bruit routier dans les tissus urbains. Technical report, Centre d'etudes des transports urbains (1981)
4. Krajzewicz, D., Bonert, M., Wagner, P.: The Open Source Traffic Simulation Package SUMO. In: RoboCup 2006 Infrastructure Simulation Competition. RoboCup 2006, Bremen (2006)
5. Felscher-Suhr, U., Guski, R.: The concept of noise annoyance: how international experts see it. Journal of Sound and Vibration, 513–527 (1998)
6. Hoogendoorn, S.: State-of-the-art of Vehicular Traffic Flow Modeling. Special Issue on Road Traffic Modelling and Control of the Journal of Systems and Control Engineering, 283–303 (2001)
7. Kotusevski, G., Hawick, K.A.: A Review of Traffic Simulation Software. Technical Report. Massey University, Auckland (2009)

8. Marquis-Favre, C., Premat, E., Aubrée, D., Vallet, M.: Noise and its effects: a review on qualitative aspects of sound. Part 1: Notions and acoustics ratings. Acta Acustica United with Acustica 91, 613–625 (2005)
9. Miedema, H.: Annoyance cause by two noise sources (letter to the editor). J. Sound Vib. 98, 592–595 (1985)
10. Martin, C.: L'ergonome dans les projets architecturaux. In: Falzon, P. (ed.) Ergonomie, pp. 421–436. PUF, Paris (2004)
11. Six, F.: De la prescription à la préparation du travail: apports de l'ergonomie à la prévention et à l'organisation du travail sur les chantiers du BTP. Université Charles de Gaulle, Lille (1999)

Perfect Product Launching Strategies in the Context of Survival of Small Scale Consumer Products Industries

Ravi Terkar[1], Hari Vasudevan[2], Vivek Sunnapwar[3], and Vilas Kalamkar[4]

[1] Mukesh Patel School of Technology Management & Engineering, Mumbai 400 056 India
[2] Dwarkadas J. Sanghvi College of Engineering Mumbai 400 056, India
[3] Lokmanya Tilak College of Engineering, Navi Mumbai 400 709, India
[4] Sardar Patel College of Engineering Mumbai 400 058, India
hari.vasudevan@djsce.ac.in,
{raviterkar,vivek.sunnapwar,vilas.kalamkar}@gmail.com

Abstract. Consumer products companies are under escalating demands to reduce time-to-market and the cost of introducing new products. As product lifecycles continue to decrease, compressing development cycles and accelerating new product introductions are becoming critical. Product complexity is also increasing substantially, making development and product introduction even more challenging. This paper presents the challenges the Consumer products industry is facing, as increased complexities in the competitive environment are forcing shorter product lifecycles and increasing cost pressures. It evaluates the impact that these complexities have on the product development process and focuses on a few recommendations that Consumer products executives should consider to reduce their time-to-market and increase their return on investment for new product introductions.

Keywords: Consumer Products, Product Lifecycle Management, Product Cannibalization, New Product Development, Return on Investment, Perfect Product Launching.

1 Introduction

Consumer products (CPs) manufacturers are under increased pressure to grow revenues and improve operating efficiency. Challenges in meeting growth targets include changes in consumer's demographics, increased competition in mature markets, increased spending on services, and the rise of private labels and the low success rate of new brands [5]. Market is definitely entering the era of innovation and heavy competition. It is pervasive, it is influencing the way in which companies think about virtually every aspect of research, marketing product development, supplier and materials management, manufacturing, distribution, warranty and defect management, maintenance repair and overhaul, and product end-of-life and disposal[14]. Innovation is global [2]. Innovation knows no boundaries. Its growth is being nurtured by active investments, grants and tax incentive policies of established, industrialized nations and emerging economies. Put in the context of the era of innovation, the "perfect product launch" and lifecycle management is now viewed in a different and expanded way [7].

K. Shah, V.R. Lakshmi Gorty, and A. Phirke (Eds.): ICTSM 2011, CCIS 145, pp. 321–326, 2011.

1.1 Value of Time-to-Market in PLM

The perfect product launch involves managing the development and support of complex products and services throughout the entire lifecycle from product design to product build to post-sale service [15]. It includes the integration of traditional new product with sourcing and procurement, supply chain planning and execution, and service–the entire product lifecycle Management (PLM). The importance of being first to market is discussed extensively in various sources [7]. Besides the instinctive idea that it is best to be first, other measurable benefits are possible for those that get to the market sooner with innovative products and services (see Fig. 1) and they are,

a) Increased sales through longer sales life – The earlier the product reaches the market, relative to the competition, the longer its life can be.
b) Increased margins–The more innovative the product (that is, the longer it remains on the market with little or no competition), the longer consumers will pay a premium purchase price [13].
c) Increased product loyalty – Getting the first opportunity to attract customers, especially early adopters, offers an advantage in terms of customer loyalty; customers will most likely upgrade, customize or purchase companion products [13].
d) More resale opportunities – For components, commodities or products that other companies can Private-label, being first to market can often help ensure.
e) Greater market responsiveness – The faster companies can bring products to market that satisfy new or changing customer needs, the greater the opportunity to capitalize on those products for margin lift and to increase brand recognition [12]. (See Fig. 1.)

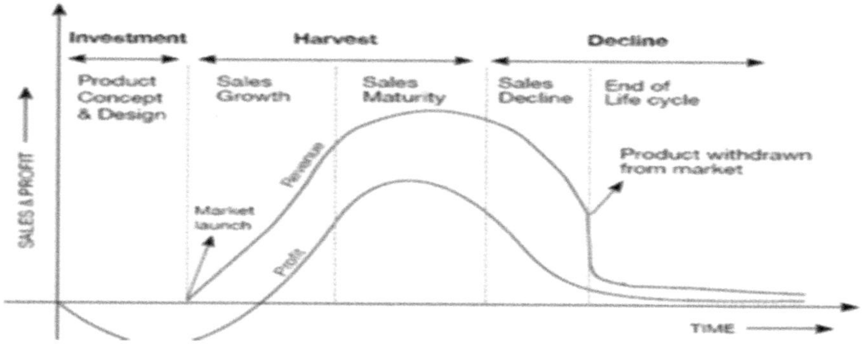

Fig. 1. Market Sooner in PLM

1.2 The Perfect Product Launch

Successful innovation has become a key driver for revenue growth, competitive margins and, in some cases, even survival [1]. The ability to bring this innovation to market quickly, efficiently and ahead of competition is becoming increasingly important. An efficient product launch requires integration and coordination among multiple functional areas, including product design, procurement, planning, manufacturing/process and sales and marketing [11].

In addition, as organizations increasingly leverage core capabilities of other companies, innovation has to be delivered through virtual networks, working with partners in a collaborative environment to bring product and services to market faster, smarter and cheaper [12]. Consequently, organizations now not only need to integrate internally, but also externally with suppliers and customers, creating end-to-end supply chain processes and capabilities which differentiate on product and customer requirements.

2 Strategies for Successful Product Development

For development of new product, four parameters are very important. Product best fit for customer, innovative product, service cost of product and product first to market are deciding factors in development of new product [11]. A study conducted in the case of a seasonal consumer product shows that the results are not so surprising. For launching of new seasonal product, product fit to customer requirement is the crucial factor. Near about 60 % industries are giving the importance to customer satisfaction and need. (See Fig. 2).

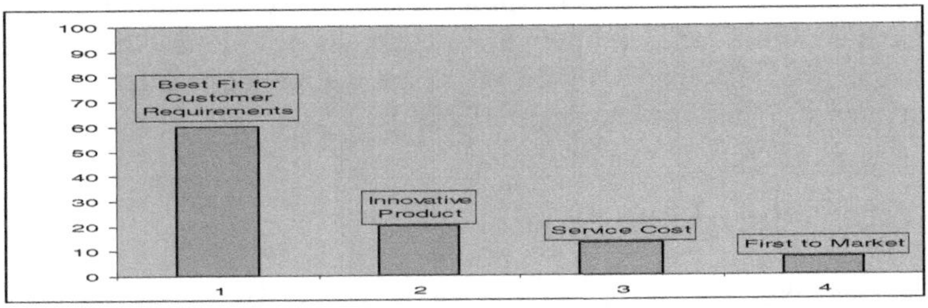

Fig. 2. Primary Strategies for New Product Launching

In view of this, market research is very important area for perfect launch. For production of more and more Innovative Product, company has to concentrate on R & D Sector. If the learning capacity of industry increases, automatically innovative product comes out as per customers requirements. Study of Product Life Cycle in view of customer point of is the key factor for manufacturing the product fit for customer requirements.

3 Product Life Cycle

Fig. 3. Indicates the Product Life Cycle of an air cooler industry product. Total sale of air cooler in 2559 days are 11,021 i.e. on an average more than four air coolers were sold in every day. Atmospheric temperature varies from 26°C to 42°C in twelve months. Rise of temperature starts from February and ends in the June.

February and ends in the June. In hot atmosphere demand of air cooler is more so in the month of March, April and May. As atmospheric temperature goes on decreasing, sale of air coolers keeps coming down. PLC of this seasonal product is shown in Fig. 3 and Fig. 4. It can be seen that in this seasonal product, PLC is continuously increasing or decreasing as per customer demand.

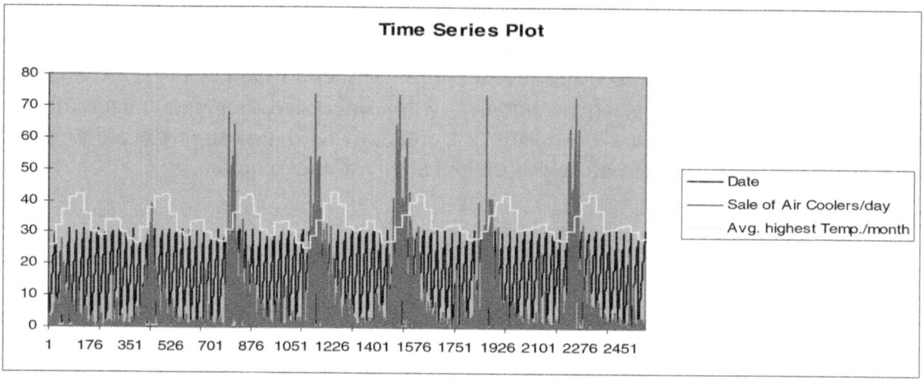

Fig. 3. Effect of atmospheric temperature on sale of air coolers

Customer demand is not only depending upon atmospheric temperature but also quality of product is important. In some initial season, profit on the product increases continuously but after some seasons profit goes on decreasing and decreasing. For getting maximum profit and optimum PLC, company has to introduce new up graded product as per the customer requirements. Hence Product best fit for customer requirement plays a continuous impact on PLC.

4 Perfect Product Launch

Fig. 4, Indicates the overall net profit in every season. Whereas, sale volume of air cooler has not changed drastically but Profit volume reduced drastically. After 901 days profit of air cooler has reduced continuously. Here manufacturer has not thought about new product development. In the fourth season, company has to think about launching of new updated air cooler. But unfortunately company has not thought of developing new version of air cooler. Mean while customer expectation has increased continuously and loyal customer has shifted to buy new version of air cooler from competitors. In PLM, development of new product to satisfy the customer need is the

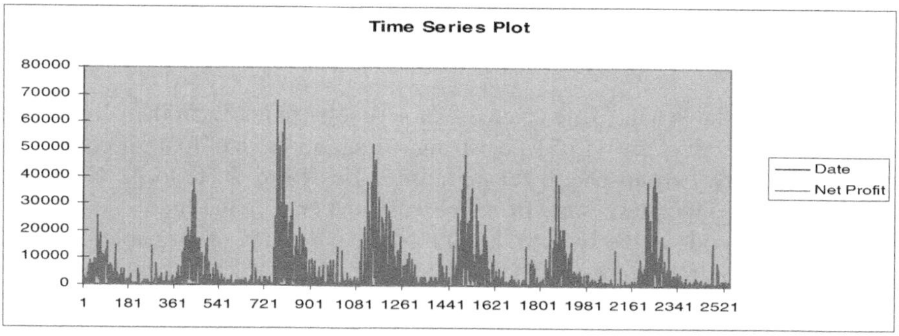

Fig. 4. Profit Trends and Variations

major objective. Here perfect product cannibalization is the trick for getting more profit. Launching time of developed product depends upon number of competitors in the market and customer psychology. Prices of air coolers are more during hot season but in rainy & winter season prices are going down.

After 3 years, prices are constant for one year and it falls sudden for next few years. Point where prices are more constant and variation in prices are little bit is the right time for product cannibalization in product life cycle. As shown in Fig. 4 & 5, overall profit decreases day by day, whereas quantity of sale has not decreased at all.

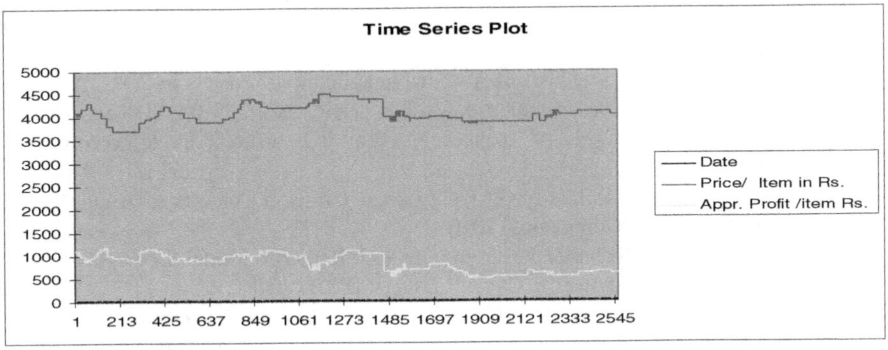

Fig. 5. Profit Trends and Variations

After a three year span, profit margin has decreased continuously; hence the company has to think about product cannibalization. Company is expected to know the customer expectation and hence change in existing product for betterment is essential. Perfect product cannibalization is the key to success in PLM.

5 Conclusion

Before launching new or up grading Consumers Products, industries have to manufacture the products which are best fit to customer's requirement. In Product Life Cycle, up gradation of old product and its replacement at right time are more important for survival of Small Scale Consumer Products Industries. This paper helps to know the importance of launching of up graded product and the need for effective product cannibalization. Small manufacturing industries have to concentrate on launching of new products at right time by improving their learning capacity. Continuous improvement in product development and perfect product cannibalization is the key to success in the present competitive world for better survival in the market.

References

1. Day, G.: The Product Life Cycle Analysis and Application issue. Journal of Marketing 41, 60–67 (1981)
2. Frederic, J.H.: Investigation of Product Life Concept as an Instrument in Marketing Decision Making. University of Pretoria (2001)

3. Feng, K., Lempert, M., Tang, Y., Tian, K., Cooper, K., Franch, X.: Defining Project Scenarios for the Agile Requirements Engineering Product-line Development Questionnaire Technical report, co-published as UTDCS-21-07 (UTD) and LSI-07-14-R, UPC (2007)
4. Gwyneth, D.: Launching your Product: Seven Marketing Musts. Marketing Profs (2006)
5. Ioannis, K.: Product Life Cycle Management. Urban and Regional Innovation Research Unit, Thessaloniki (2002)
6. John, H., Gerard, J., Tellis, A.G.: Research on Innovation: A review and Agenda for Marketing Science. In: Emory-MSI Conference (March 2005)
7. Johns, W.R., Kokossis, A., Thompson, F.: A flow sheeting approach to integrated life cycle analysis. Chemical Engineering and Processing 47, 557–564 (2008)
8. Kotler, P., Wong, G.: Principles of marketing, Power Points, 4th edn. Pearson Education Limited, London (2005)
9. Mulder, J., Brent, A.C.: Selection of sustainable agriculture projects in South Africa: Case studies in the Land Care programmed. J. Sustain. Agric. 28(2), 55–84 (2006)
10. Norman, D.: The Life Cycle of Technology. Why it is difficult for large companies to innovate (1998)
11. Rick, S., Ling, G.: Four Pillars for Product Lunch. Crimson Consulting Group, Best Practices from World-Class Companies (2010)
12. Shah, J., Patil, R.: A Stochastic Programming with Recourse Model for Supply Chain and Marketing Planning of Short Life Cycle Products; IIMB (2008)
13. Terkar, R., Mantha, S., Sen, A.: Effective Strategies of Product Development Phase for Product Cannibalization in Product Life Cycle Management. In: National conference on Design for Product Life Cycle, February 17-18, BITS, Pilani (2006)
14. Krishnan, V., Karl, U.: Product Development Decisions. A Review of Literature Management Science 47, 1–21 (2001)
15. Witty, M., Jay, H., Jason, S.: CPG Manufacturing Industry Update 1Q06. Manufacturing Insights, an IDC company. Document # M 1201026 (2006)

Impact of Information Sharing in Supply Chain Performance

T. Chinna Pamulety[1] and V. Madhusudanan Pillai[2]

[1] Research Scholar
chinna081@gmail.com
[2] Associate Professor, Department of Mechanical Engineering,
National Institute of Technology Calicut, Calicut-673 601, Kerala, India
vmp@nitc.ac.in

Abstract. Supply chain is a network of organizations that are directly or indirectly involved in fulfilling the customer requirements. Bullwhip effect in a supply chain is having negative impact on the performance of the supply chain. One of the reasons for its occurrence in supply chain is lack of customer demand information at all stages. So, the objective of the present study is to know the impact of sharing history of Customer Demand Information (CDI) on bullwhip effect in a four stage serial supply chain and to evaluate its performance by conducting experiments similar to beer distribution game. History of CDI can be shared easily because of advancements in information technology. Various performance measures used for the evaluation are fill rate, variance of orders, total inventory at each stage and Total Holding Cost of the Supply chain (THCS). Results show that sharing history of CDI improves the performance of the supply chain.

Keywords: Supply chain, Bullwhip effect, Customer demand information sharing.

1 Introduction

Supply chain is a network of organizations that are directly or indirectly involved in fulfilling the customer requirements. Various functions performed in it are procurement of raw materials, converting the same into semi finished and finished products, and distributing them to the end customers. Increase in the demand variability as we move from downstream to upstream stage in a supply chain is called bullwhip effect. Forrester was the first person who noticed this phenomenon [1]. Its presence creates excessive inventory investments, poor customer service level, ineffective transportation, unstable production plans and lost revenues. So, it is harmful and deteriorates the performance of the supply chain. The causes of the bullwhip effect are: lack of customer demand information [2], demand forecast updating, order batching, variation in prices, rationing and shortage gaming [3], replenishment rule [4] and lead time [5,6,7]. Procter & Gamble (P&G) observed this phenomenon in one of their best selling products called pampers where as Hewlett – Packard (HP) observed it in their printer product [3]. They tried to reduce it and could see the increase in their profit. Reduction in bullwhip effect has the significant impact on the profitability of the whole supply chain [8]. Elimination of bullwhip effect can increase the profit of a firm [5].

K. Shah, V.R. Lakshmi Gorty, and A. Phirke (Eds.): ICTSM 2011, CCIS 145, pp. 327–332, 2011.

Bullwhip effect can be tamed by various ways such as: avoid forecasting at all stages, reducing order batch size variation, stabilize the prices, supplying goods to the customer based on his past sales record than actual orders during supply shortages [3], reduce the lead time, adapt some strategic partnerships like Vendor Managed Inventory (VMI), etc.

Another way of reducing the bullwhip effect is, by sharing the customer demand information among all stages in the supply chain [2]. In the present study, the performance of a four stage serial supply chain is evaluated under with and without history of CDI sharing by conducting experiments similar to beer distribution game, and is not studied in the existing literature as per the knowledge of authors. Lead time considered is smaller (1 week) than the lead time (4weeks) given in literature [2, 9, and 10] because it is the overriding cause of the bullwhip effect [5]. Various performance measures used for the evaluation are fill rate, variance of orders placed by each stage, total inventory at each stage and total holding cost of the supply chain. Results show that sharing history of CDI improves the performance of the supply chain.

The paper is organized as follows: literature survey and experimentation are described in Section 2 and 3 respectively. Statistical test conducted is explained in Section 4. Discussion and conclusions are given in Section 5.

2 Literature Survey

Since the presence of bullwhip effect reduces the performance of a supply chain, many researchers tried to find the ways by which the bullwhip effect can be reduced or controlled. They used different tools like analytical methods, simulation, and experimentation for the same. Experiments are conducted by using beer distribution game which was developed by MIT, USA. The Beer Distribution Game is a simulation of a supply chain with four co-makers (retailer, wholesaler, distributor and factory). The details of the game can be seen in [2, 11]. Sterman [2] is the first person who used the beer distribution game for studying the managerial behavior in a supply chain experimentally. This experiment involves a supply chain with four players namely retailer, wholesaler, distributor and factory. Each player receives the orders from their immediate downstream member and takes decisions about the order quantity and shipment quantity independently without consulting the other players. The players reported that the reason for larger variability in their orders is their inability to predict the pattern of customer demand. Impact of customer demand distribution, Point of Sale (POS) data and inventory information sharing are studied by Croson and Donohue [9, 10]. Lead time considered in these studies is 4 weeks which is the source for bullwhip effect. In the present study, history of CDI with small lead time (1 week) is shared and its impact is studied. Various performance measures used in the present work are also different and were not measured in the above experimental studies. Steckel, et al. [12] also used beer distribution game to know the impact of POS and lead time in the performance of supply chain.

3 Experimentation

A supply chain role play game is developed and used for experimentation. Its features and design details can be seen in Pamulety and Pillai [13]. It works similar to the beer distribution game. It simulates the operational decisions taken at each stage in a four stage serial supply chain and evaluates its performance in terms of variance of orders placed by each stage, customer delivery level (fill rate), total inventory at each stage and total holding cost of the supply chain. The decisions taken at each stage are shipment quantity and size of the order to be placed. These decisions are taken with the objective of maximizing the fill rate and minimizing the inventory.

There are 56 post graduate and under graduate students who have participated in the experiments and most of them are from Industrial Engineering and Management background. Each student acts as a stage manager in a 4-stage serial supply chain and formed 14 identical supply chains. Among them, 7 supply chains are tested under without any information sharing and remaining are tested under history of CDI sharing. A trial game for a period of 10-week is played before the actual experiment. The duration of the actual experiment was not revealed to the participants and was continued for 55 weeks. In those 55 weeks, first 6 periods are considered as trial period and the periods from 7 to 48 are considered for performance analysis purposes. This is considered to reduce the end game effect [12]. The performance of each supply chain under each setting is evaluated. Variance of orders placed by each stage under no information sharing and with history of CDI sharing is shown in Figures 1 and 2

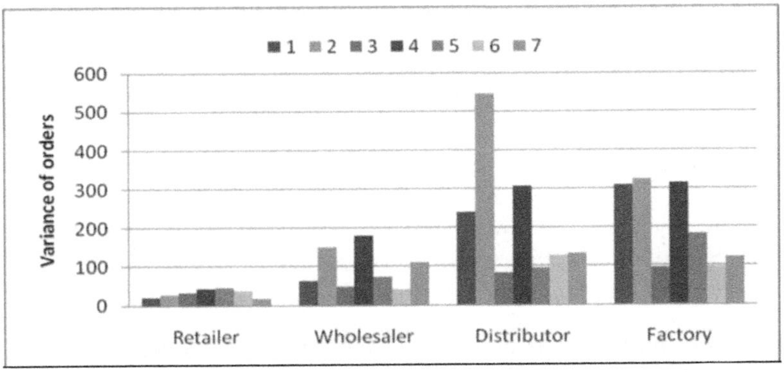

Fig. 1. Variance of orders placed by each stage under no information sharing

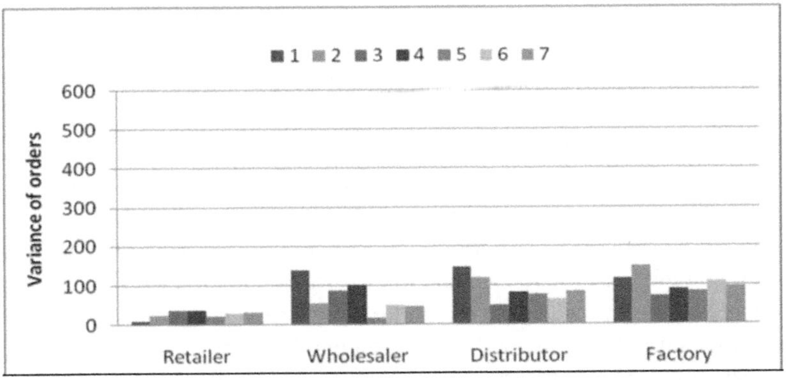

Fig. 2. Variance of orders placed by each stage under history of CDI sharing

Table 1. Average values of performance measures of supply chains

Performance measures	Stage Name			
	Retailer	Wholesaler	Distributor	Factory
Supply chain under no information sharing				
Variance of orders placed	31.74	94.85	217.97	205.62
Fill rate	0.931			
Total end period inventory	388.14	581.85	786.85	1034.71
THCS in $	1395.82			
Supply chain under history of CDI sharing				
Variance of orders placed	27.24	69.52	87.94	101.76
Fill rate	0.973			
Total end period inventory	587.42	693.57	863.85	433.28
THCS in $	1289.07			

respectively. The average value of each performance measure over seven supply chains under each setting is tabulated in Table 1. The details for the calculation of the performance measures can be seen in Pamulety and Pillai [13].

4 Statistical Test for Evidence of Bullwhip Effect

Since the variable under consideration (variance) has continuous distribution, sign test can be used to test the presence of bullwhip effect [9, 14]. This non-parametric test is used to know whether the bullwhip effect is present or not in a supply chain under sharing the history of CDI with small lead time. The experimental results of variance of orders placed by each stage under history of CDI are shown in Figure 2.

Hypothesis: The Bull whip effect will not occur under sharing the history of CDI with small lead time and no backorders.

The procedure of the sign test is as follows. For each supply chain, we code an increase in the variance of orders placed between stages as a success and a decrease as a failure. If the variance of orders between stages is same, we code it as zero. The sum of the possible successes and failures forms sample size (N). The probability of success or failure is 0.5. Success is represented by a plus (+) sign and a failure with minus (-) sign. If X represents the number of plus signs, then the probability of getting X or more plus signs is calculated from the Binomial distribution. If this probability is less than the significance level ($\alpha = 0.05$) fixed, the null hypothesis must be rejected. The details of the sign test can be found in [14, 15]. For the problem described here, N = 21 and we got X as 17 and hence the Probability (X >= 17) is 0.0036 which is less than the significance level set for the present problem. So the above hypothesis must be rejected and we conclude that bullwhip effect is present under history of CDI sharing with a small lead time and no backorders. Similarly, it is possible to establish the presence of bullwhip effect with no information case also.

5 Discussion and Conclusions

The performance of a four-stage serial supply chain is evaluated under with and without history of CDI sharing. In a traditional supply chain, only order information is shared between the stages and the members at upstream stages decide the size of orders to be placed by using this information. This way of ordering leads to bullwhip effect. The CDI sharing with upstream stages helped the members to take better decisions and hence the magnitude of bullwhip effect is less in the CDI sharing. Information can be shared easily at less cost because of the advancements in information technology and it reduces the lead time also. From Table 1 and sign test, we conclude that sharing history of CDI reduces the bullwhip effect but does not eliminate completely. It may be noted that the experiments are conducted with small lead time and hence the effect of lead time is less in the bullwhip effect. The prominent aspect of bullwhip effect in this study is the behavior aspect of the decision maker of each stage.

Acknowledgement. The authors are thankful to the Post graduate students (2010-12 batch) and Production Engineering under graduate students (2007-2011 batch) of NIT Calicut for participating in this experiment.

References

1. Wu, D.Y., Katok, E.: Learning, Communication and Bullwhip Effect. Journal of Operations Management 24, 839–850 (2006)
2. Sterman, J.D.: Modeling Managerial Behaviour: Misperceptions of Feedback in a Dynamic Decision Making Experiment. Management Science 35, 321–339 (1989)
3. Lee, H.L., Padmanabhan, V., Whang, S.: The Bullwhip Effect in Supply Chains. Sloan Management Review 38, 93–102 (1997)
4. Dejonckheere, J., Disney, S.M., Lambrecht, M.R., Towill, D.R.: Measuring and Avoiding the Bullwhip Effect: A Control Theoretic Approach. European Journal of Operational Research 147, 567–590 (2003)
5. Metters, R.: Quantifying the Bullwhip Effect in Supply Chains. Journal of Operations Management 15, 89–100 (1997)
6. Simchi-Levi, D., Kaminsky, P., Simchi-Levi, E., Shankar, R.: Designing and Managing the Supply Chain: Concepts, Strategies and Case Studies. Tata McGraw-Hill, New Delhi (2008)
7. Wang, X., Liu, Z., Zheng, C., Quan, C.: The Impact of Lead Time on Bullwhip Effect in Supply Chain. In: 2008 ISECS International Colloquium on Computing, Comunication, Control, and Management, pp. 93–97. IEEE Transactions, China (2008)
8. Bottani, E., Montanari, R., Volpi, A.: The Impact of RFID and EPC Network on the Bullwhip Effect in the Italian FMCG Supply Chain. International Journal of Production Economics 124, 426–432 (2010)
9. Croson, R., Donohue, K.: Impact of POS Data Sharing on Supply Chain Management: an Experiment. Production and Operations Management 12, 1–12 (2003)
10. Croson, R., Donohue, K.: Behavioural Causes of the Bullwhip Effect and the Observed Value of Inventory Information. Management Science 52, 323–336 (2006)
11. Massachusetts Institute of Technology, `http://web.mit.edu/jsterman/www/SDG/beergame.html`

12. Steckel, J.H., Gupta, S., Banerji, A.: Supply Chain Decision Making: will Shorter Cycle Times and Shared Point-of-Sale Information Necessarily Help? Management Science 50, 458–464 (2004)
13. Pamulety, T.C., Pillai, V.M.: Evaluating the Performance of a Multi-echelon Supply Chain. In: 4th International Conference on Advances in Mechanical Engineering, Surat, India, pp. 271–275 (2010)
14. Siegel, S., Castellan, N.J.: Nonparametric Statistics for the Behavioural Sciences. McGraw-Hill, New York (1988)
15. Johnson, R.A.: Probability and Statistics for Engineers. Prentice-Hall, India (2004)

Author Index

GPSR Compliance

The European Union's (EU) General Product Safety Regulation (GPSR) is a set of rules that requires consumer products to be safe and our obligations to ensure this.

If you have any concerns about our products, you can contact us on ProductSafety@springernature.com

In case Publisher is established outside the EU, the EU authorized representative is:

Springer Nature Customer Service Center GmbH
Europaplatz 3
69115 Heidelberg, Germany

Batch number: 10844619

Printed by Printforce, the Netherlands